Virtual Bio-Instrumentation

D1279048

ISBN 0-13-065216-4

9 790130 652163

NATIONAL INSTRUMENTS | **VIRTUAL INSTRUMENTATION SERIES**

Virtual
Bio-Instrumentation

Biomedical, Clinical, and Healthcare Applications in LabVIEW

▲ Jon B. Olansen, Ph.D.
▲ Eric Rosow, M.S.

Prentice Hall PTR, Upper Saddle River, NJ 07458
www.phptr.com

Library of Congress Cataloging-in-Publication Data

Olansen, Jon B.
  Virtual bio-instrumentation: biomedical, clinical, and healthcare applications
in LabVIEW / Jon B. Olansen, Eric Rosow.
    p. cm.
  Includes bibliographical references and index.
  ISBN 0-13-065216-4 (pbk.)
  1. Scientific apparatus and instruments—Computer simulation. 2. Computer graphics.
3. Engineering instruments—Data processing. 4. Biomedical engineering. I. Rosow, Eric.
II. Title.
R856 .O434 2002
610′.28′0113—dc21

                                                   2001052399

Editorial Production/Composition: *G & S Typesetters, Inc.*
Acquisitions Editor: *Bernard Goodwin*
Editorial Assistant: *Michelle Vincenti*
Cover Design: *Anthony Gemmellaro*
Cover Director: *Jerry Votta*
Marketing Manager: *Dan DePasquale*
Marketing Manager: *Alexis R. Heydt-Long*
Project Coordinator: *Anne R. Garcia*

© 2002 by Prentice Hall PTR
Prentice-Hall, Inc.
Upper Saddle River, New Jersey 07458

Prentice Hall books are widely used by corporations and government agencies for training, marketing, and resale.
The publisher offers discounts on this book when ordered in bulk quantities. For more information, contact:

  Corporate Sales Department
  Prentice Hall PTR
  One Lake Street
  Upper Saddle River, NJ 07458
  Phone: 800-382-3419; FAX: 201-236-7141
  E-mail (Internet): corpsales@prenhall.com

Printed in the United States of America

10 9 8 7 6 5 4 3 2 1

ISBN 0-13-065216-4

Pearson Education Ltd.
Pearson Education Australia PTY, Ltd.
Pearson Education Singapore, Pte. Ltd.
Pearson Education North Asia Ltd.
Pearson Education Canada, Ltd.
Pearson Educación de Mexico, S.A. de C.V.
Pearson Education—Japan
Pearson Education Malaysia, Pte. Ltd.

For our families:
Nancy, Kathy, Jon, Kristin
Pamela, Jonathon, Rachael, Julia

Contents

▼ **2**

Research Applications   27

▼ 3

Clinical Applications   137

▼ 4

Medical Device Development Applications   257

▼ **5**

**Healthcare Information
Management Systems   343**

▼ **6**

Advanced Topics   437

11 Mathematical Modeling/
   Simulation of Physiologic Systems 439

Disclaimer

Warning Regarding Medical and Clinical Use of National Instruments Products

National Instruments products are not designed with components and testing for a level of reliability suitable for use in or in connection with surgical implants or as critical components in any life support systems whose failure to perform can reasonably be expected to cause significant injury to a human. Applications of National Instruments products involving medical or clinical treatment can create a potential for death or bodily injury caused by product failure or by errors on the part of the user or application designer. Because each end-user system is customized and differs from National Instruments testing platforms and because a user or application designer may use National Instruments products in combination with other products in a manner not evaluated or contemplated by National Instruments, the user or application designer is ultimately responsible for verifying and validating the suitability of National Instruments products whenever National Instruments products are incorporated in a system or application, including, without limitation, the appropriate design, process, and safety level of such system or application.

Preface

Graphical Programming and Virtual Instrumentation: Applying Revolutionary Techniques to Advance the Healthcare Industry

Over the last decade, the graphical programming revolution has empowered engineers to develop customized systems the same way the spreadsheet has empowered business managers to analyze financial data. This software technology has resulted in another type of revolution—the virtual instrumentation revolution, which is rapidly changing the instrumentation industry by driving down costs without sacrificing quality.

Virtual instrumentation can be defined as

A layer of software and/or hardware added to a general-purpose computer in such a fashion that users can interact with the computer as though it were their own custom-designed traditional electronic instrument.

The major benefits of virtual instrumentation include increased performance and reduced costs. Because the user controls the technology through software, the flexibility of virtual instrumentation is unmatched by traditional instrumentation. The modular, hierarchical programming environment of virtual instrumentation is inherently reusable and reconfigurable.

Virtual instrumentation applications have encompassed nearly every industry, including the telecommunications, automotive, semiconductor, and biomedical industries. In the fields of healthcare and biomedical engineering, virtual instrumentation has empowered developers and end-users to conceive of, develop, and implement a wide variety of research-based biomedical applications and executive information tools. These applications fall into several categories, including clinical research, equipment testing and quality assurance, data management, and performance improvement.

This book opens the boundless potential of virtual instrumentation (VI) into the wide variety of disciplines that exist within the biomedical domain. The power of virtual bio-instrumentation (VBI) is demonstrated not only through the interfacing of VI with traditional medical instruments and devices but also by effectively leveraging other technologies, including the Internet, machine vision, ActiveX components, and integrated database applications. We use specific examples within this book to highlight VBI applications in the laboratory and clinical environment, connectivity to patient information systems, computerized maintenance and management systems (CMMS), and business intelligence and decision support applications. Each VBI application consists of detailed descriptions and, in many cases, interactive demonstrations of how virtual instrument solutions have been conceived and developed to meet specific end-user requirements within the biomedical and healthcare arena. Collectively, these applications support better, faster, and data-driven decisions, thereby enhancing clinical outcomes and reducing costs to the participating healthcare institutions.

As practicing biomedical engineers and virtual instrumentation "evangelists," we wrote this book to inform and, hopefully, inspire you about the ever-expanding capabilities of virtual instrumentation systems within the biomedical and healthcare fields. Many traditional books on bio-instrumentation concentrate on theoretical principles—this book focuses entirely on real-world applications. We refer to these applications as virtual bio-instrumentation, or VBI. Throughout each section and chapter, you'll discover many practical biomedical applications that have been created with LabVIEW™. Each example will provide detailed explanations of its design, implementation processes, and utility. We particularly emphasize methods for measurement, analysis, presentation, and distribution of biomedical and health system information. Throughout this book, we have striven to identify common challenges associated with the measurement, analysis, and presentation of information;

and we provide you with practical solutions and proven problem-solving techniques from experienced scientists, engineers, clinicians, and healthcare administrators.

Regardless of your application or your experience with LabVIEW, it is our sincere wish that, through this book and the virtual instrument (VI) examples contained on the accompanying CD-ROMs, you will gain insight and appreciation for the many ways in which virtual instrumentation can be applied to the biomedical and healthcare industry.

Acknowledgments

This book has been made possible by the contributions and support of many people. We acknowledge them here as a mere token of the sincere respect, admiration, and appreciation we have for their efforts.

Thanks to Bernard Goodwin at Prentice Hall for believing, and helping us to believe, that we had something worthwhile to offer the biomedical and virtual instrumentation communities. Special thanks also go to Anne R. Garcia, Alison Rainey, and Joshua Goodman for all of their efforts in producing this book.

Several people at National Instruments helped make this book a reality. The efforts and support of Hall T. Martin and Ravi Marawar from the initial inception of the book through its completion was indispensable. Morten Jensen's and Newton DeFaria's efforts at reviewing and contributing to the book helped ensure its currency and were greatly appreciated.

There are numerous components of this book contributed by other authors that significantly broadened the scope and helped us realize our goal of demonstrating the wide-ranging opportunities available for virtual bioinstrumentation. Jeffrey Travis, author of *Internet Applications in LabVIEW*, contributed Chapter 12, demonstrating concepts for pushing VBI across the Internet and into the future. Greg Swanson, president of TimeSlice, Inc., supplied the content for Chapter 8, "LabVIEW in a Regulated Environment," a critical area of concern for those developing or testing medical devices with LabVIEW. Several references to work accomplished by Premise Development Corporation appear in the book. The efforts of Joseph Adam and Chris Roth in contributing those works are greatly appreciated. Some of the research applications were adapted from courses developed and taught by Drs. John Clark and David Caprette of Rice University and Dr. Akhil Bidani of the University of Texas Medical Branch at Galveston. All were mentors to Jon Olansen during his doctoral studies and gave him the background and confidence in performing the type of research that led to the writing of this book.

We have tried to include in each chapter relevant samples of works done by other people in the various biomedical fields. Special thanks are extended to Roman Pichardo and Orlando Torres for their contributions to Chapter 5 on the Cardiopulmonary Measurement System and the Vibrotactile Stimulation System to Treat Apnea of Prematurity, and Keena Patel and Jennifer Jackson for their contributions on the FluidSense IV Pump Testing in Chapter 7. Others who deserve special recognition include Marc Palter, MD, Thomas Murt, MD, and Witold Waberski, MD, for their contributions in Chapter 9 on medical informatics, and Mark Leggitt for his contributions in Chapter 10 on performance indicators in healthcare.

Lastly, we are indebted to the many reviewers who devoted substantial time and effort to provide valuable feedback and recommendations for this book. Special thanks are extended to Joseph D. Bronzino, P.E., Ph.D., professor and mentor to Eric Rosow, and a true icon in the field of biomedical engineering.

From Eric:
For Pamela, Jonathon, Rachael, and Julia—without whom I could have finished this book in one fourth the time, but it would have only been one fourth as good—thanks for all your support, understanding, and love. To Mom and Dad who gave me the tools, confidence, and love to take on challenges, and to the extraordinary staff at Hartford Hospital who provided me with the support and latitude to "evangelize" the power of virtual instrumentation within the healthcare environment. Finally, this book is also dedicated to all the healthcare professionals, biomedical and software engineers, professors, and medical device manufacturers who combine engineering principles and technology with compassion and creative energy to make truly meaningful differences in people's lives.

From Jon:
The encouragement and support of family and friends is crucial in completing a task like this. To Mom and Dad, I am who I am now because you took the time to teach me. Thanks to my brother, Brian, and sisters, Debbie, Julie, and Dawn and their families for their continuous support. And to my never-ending inspiration, Jennifer Kreykes Pohl, a great friend who supported and encouraged me throughout my life even as she fought and eventually lost her battle with lymphoma at age 32. Her spirit lives on through many who knew her; I see it in my children's eyes every time they ask to bring flowers to her grave.

Most importantly, my deepest love and appreciation belongs to my wonderful wife, Nancy, who endured with a smile the long hours that I would work. And to my three children, Kathy, Jon Jr., and Kristin—their love and understanding, not only through this effort but also through my schooling, traveling, or any other responsibility that has taken me away from them, makes me realize how important it is to make the most of my time with them. Projects such as this will come and go, but it is the love shared with family and friends that gives true meaning to life.

Part I
Preliminaries

Introduction

1

General Goals of VBI Applications
Educational Objectives
Professional Objectives
Previous Knowledge Requirements

Organization of the Book
Book Conventions
What This Book Is
What This Book Is Not
Research and Clinical Applications
Medical Device Development and Test Applications
Healthcare and Informatics Applications
Advanced Applications

Contents of the CD-ROM

Virtual bio-instrumentation (VBI) is a phrase we coined to encompass the nearly unbounded potential for innovative utilization of virtual instrumentation in biomedicine, healthcare, and related industries. Thus, this book is designed as a reference tool for a wide variety of biomedical and healthcare professionals, from practitioners to clinical engineers, to administrators, to designers and manufacturers of medical devices, to students studying within the biomedical spectrum. The real-world examples described herein serve merely as representative samples of the types of applications possible with virtual instrumentation. Concepts include

- Reenactment of classical biomedical experiments for educational purposes,
- Applications in clinical research,
- Applications in clinical engineering and hospital management, and
- Advanced applications such as mathematical modeling and Internet solutions.

The common thread through all of these applications is their computer-based nature, all of them being developed and conducted in a virtual instrumentation (VI) environment. This is accomplished using LabVIEW™ or BioBench™ software from National Instruments (Austin, TX).

Representative applications are discussed in detail to assist professionals and students alike in their project development. These applications are intended to introduce readers to the benefits of virtual instrumentation in the various fields of biomedical experimentation, from measurement and data acquisition requirements to subsequent data analysis techniques. VBI projects included in the book fit one of the following categories. They

- Demonstrate fundamental physiological properties,
- Demonstrate advanced analysis capabilities that explore potential research topics,
- Demonstrate clinical utilization of virtual instrumentation,
- Demonstrate functions related to medical device development and tests, or
- Demonstrate hospital management or clinical engineering concepts.

General Goals of VBI Applications

Because this book has been written for a broad audience, we have incorporated a variety of VBI applications. Let's review the diverse objectives of these applications in the following sections.

Educational Objectives

It is difficult to design laboratory exercises using complex scientific equipment and still have the student learn the concepts being demonstrated. Using LabVIEW to facilitate the data generation and collection changes the focus from learning how to use the equipment to learning the physiological concepts being presented in the lab. Because LabVIEW allows designing simple, easy-to-use interfaces, however, care must be taken not to make the virtual instruments (VIs) so simple to use that students may not see the details of the measurement process. Instructors must ensure that students do not leave the lab thinking that measuring physiologic phenomena means you click once on a button and the data magically appears. Many of the VIs included with this book are, in fact, somewhat advanced in that they automate many of the routine portions of data collection. They are intended not only for teaching the appropriate lab but also as a robust basis for building more advanced research projects. It is therefore incumbent on the reader to use the VIs, or portions thereof, appropriately and to ensure that they understand the efforts and issues involved in collecting such data.

Although experimental projects need not focus on basic science, it is an important approach to demonstrating these classical topics. The biological theories involved are interesting and informative. Combining that with experience in experiment development, live tissue dissection and preparation, and experimental control and data acquisition would only enhance the learning environment and the quality of education that the students obtain. The VBI exercises described herein thus allow students to learn the concepts of the phenomena first, and to learn the measurement system second.

Professional Objectives

LabVIEW is a powerful tool for students during their education. It also is a tool for use after graduation in the workplace, where LabVIEW can be applied to procedures specific to the graduate's job duties. The applications in this book introduce students and professionals alike to LabVIEW-based systems but do not teach them everything there is to know about LabVIEW programming. The LabVIEW programs, called virtual instruments, or VIs, that accompany this book demonstrate the potential uses of VBI throughout the healthcare industry. Our goal is to provide you with a general background in applying LabVIEW to address a broad spectrum of biomedical needs. This book can thus serve as a reference to guide the medical professional in development of VBI applications or as a resource detailing the potential enhancements VBI can convey to his or her workflow model through improved data collection, analysis, or management capabilities.

Previous Knowledge Requirements

The tutorials in this book were developed with a target education level equivalent to that required for a junior-level college course in biomedical engineering. Before utilizing or building upon the applications contained in this book, readers should understand the scientific and mathematical principles relevant to their field of study. In particular, the following list details some commonly considered prerequisites for this type of development work:

- Differential and integral calculus
- The concept of temperature
- Spreadsheet analysis of data, charts, linear regression, and semi-log graphs
- Newton's laws and vector forces
- The ideal gas law
- Concepts of current, voltage, and resistance
- Measurement of current, voltage, and resistance with a digital multimeter

Organization of the Book

The following pages present information regarding general educational goals of the book, any unique laboratory or clinical design information, and layout and setup descriptions of a computer-based instrumentation workstation. The main body of the book consists of descriptions of established applications, including notes specifying how the application or experiment may be used and any lab setup, objectives, and background information that may be required.

Book Conventions

The following text conventions are used in this book:

Bold	Words in bold refer to LabVIEW menus, menu items, palettes, subpalettes, functions, and VI's. For example, **File.**
Italics	Words in italics are for emphasis.
`Courier`	Words in Courier indicate drive names, libraries, directories, pathnames, filenames, and sections of programming code.
<Shift>	Angle brackets enclose names of keys.

What This Book Is

This book is a tool for engineers, scientists, practitioners, investigators, or instructors conducting biomedical or clinical research, developing or testing medical devices, or practicing clinical engineering or management functions. It is intended as a resource guide for developing computer-based applications using LabVIEW and other National Instruments products.

The content of this book is intended to be useful at a variety of educational levels including undergraduate engineering/preprofessional or graduate/medical school students through practicing biomedical professionals. Numerous projects are described herein, with concepts ranging from basic neural electrophysiology to cardiovascular hemodynamics to Internet-based biomedical applications. Substantial effort has gone into ensuring that all of the projects are well documented with many applications included on the accompanying CD-ROM. These programs should serve as a learning tool upon which you can build your own virtual applications. Background material on the pertinent physiology and notes and tips on laboratory setup and programming techniques are also included in each description. Due to the modular nature of the LabVIEW software being used, many of the programs included with this book can be readily adapted to similar studies that other instructors or researchers may be more interested in pursuing.

What This Book Is Not

The applications described in this book are not intended to be out-of-the-box or plug-and-play type exercises. Some modifications will be necessary depending on your specific setup, curriculum, and pedagogical style. This book is not intended to dictate a specific pedagogy. Some pedagogical elements are inherent in the applications. But these elements are minimized because each instructor, each institution, and each program has specific needs and systems. For example, some of the units discussed herein were designed as a subset of the laboratory portion of a three-credit undergraduate biomedical engineering course. Students were instructed to study material in lecture and outside of the lab and classroom. If you are designing a standalone lab course, you will need to change the exercises presented.

Also, the VIs included with this book carry no warranty expressed or implied. As with the book itself, the software is meant to be a starting point for the user developing her or his own applications.

Research and Clinical Applications

Part II discusses specific biomedical laboratory and research applications using virtual instrumentation. A particular emphasis is placed on biopotentials and cardiopulmonary dynamics as representative samples of biomedical research work that can be aided and accomplished through the use of VBI. Part III provides real-world clinical applications ranging from virtual instrumentation systems that measure lung diffusing capacity and oxygen consumption to cardiac output measurement systems to machine-vision systems that can measure wounds and quantify the rates at which they heal.

Medical Device Development and Test Applications

Part IV discusses how virtual instrumentation is used to test and validate medical devices. Examples provided in Chapter 7 range from test and measurement systems used in the manufacturing environment to those used in the clinical environment. Chapter 8 has been devoted specifically to discussing the testing and validation processes required by the U.S. Food and Drug Administration (FDA). Particular attention will be placed on general validation principles that the FDA considers to be applicable to the validation of medical device software or the validation of software used to design, develop, or manufacture medical devices. Practical examples of LabVIEW-based automation tools that assist developers with the challenge of creating large, complex, and mission-critical applications will also be demonstrated. For this chapter, we have drawn upon the expertise of established LabVIEW professionals engaged in the fields of regulatory process development and quality management.

Healthcare and Informatics Applications

Part V is dedicated to the growing field of medical informatics and decision support applications. The healthcare industry has evolved into a multibillion-dollar industry with a multitude of stakeholders including service providers and vendors, payers, the government, employers, and regulatory agencies. Within the healthcare information environment, technology is one of the fastest changing components. New hardware, advanced communications technologies, and new methodologies are enabling process changes that will dramatically impact the way healthcare is delivered in our hospitals and communities. Today, healthcare organizations can harness the power of information to provide comprehensive, *actionable information* at the point of decision making. The result of leveraging information technology will be to improve mission-critical clinical, financial, and customer-satisfaction outcomes.

Chapters 9 and 10 explore recent technological innovations as well as established technologies, such as virtual instrumentation, now finding new applications in healthcare. It is important to note the critical role of the physician when considering the many processes to acquire, represent, process, and manage knowledge and data within healthcare and the biomedical sciences. Providing physicians with the right information, at the right time, in the right way will help them provide the highest quality care possible. In preparing this section of the book, we have drawn upon the insights and experiences of a wide variety of healthcare professionals, including physicians, nurses, computer and information scientists, biomedical engineers, medical librarians, and academic researchers, educators, and administrators. Collectively, we provide examples of how virtual instrumentation can play an important role in enterprise-wide clinical management, access management, patient financial management, health information management, strategic decision support, resource planning management, and enterprise application integration solutions to healthcare organizations.

Advanced Applications

Part VI looks at some more advanced tools and their application in the biomedical field. Chapter 11 presents an in-depth review of mathematical modeling and simulation techniques within LabVIEW. The efforts described herein are applied specifically to the cardiopulmonary systems, but their design is readily adaptable to other physiologic subsystems. Chapter 12 provides a good overview of techniques available for bringing VBI onto the Internet. The book is completed with Chapter 13, which essentially challenges the reader to advance VBI with some possible future applications.

Contents of the CD-ROM

The CD-ROM that accompanies this book includes the following files:

- The LabVIEW VIs for several of the laboratory exercises described in detail within the text
- Demonstration software for several commercially available products described within the text
- Demonstration versions of LabVIEW 6.0i and BioBench 1.2

Basic Concepts

2

Data Acquisition (DAQ) Basics

LabVIEW™ Basics

BioBench™ Basics

This chapter provides a cursory overview of basic concepts fundamental to the establishment of a modern computer-based instrumentation laboratory or clinical workstation. Many issues associated with analog data collection are mentioned. Data collection via computer serial ports from existing medical devices is also discussed. Additionally, brief introductions to LabVIEW™ and BioBench™ are provided as a starting point for those unfamiliar with the software packages.

It is not the intent of this book to necessarily teach these fundamentals, however. Numerous other sources are available to do that, and the writings herein assume some level of understanding by the reader. Rather, this book focuses on applying medical instrumentation concepts to various biomedical, clinical, and healthcare issues, using the tools and techniques afforded by virtual instrumentation. *Medical Instrumentation: Application and Design*, edited by John Webster, is one excellent reference for the physical development of

11

medical instrumentation. Further information on the basics of LabVIEW development can be found in a number of texts, such as *LabVIEW For Everyone*, written by Lisa Wells and Jeff Travis.

Data Acquisition (DAQ) Basics

There are two basic aspects of data acquisition (DAQ) that need to be addressed by anyone attempting to gather data, whether for a basic laboratory experiment or a clinical research trial. The first aspect is to understand what data collection is and all of the components that together define a data acquisition system. Second, it is essential to understand the data that is being collected.

The primary component of any data collection system is the actual device (sensor) used to detect the phenomena in question. These devices may include pressure transducers, infrared CO_2 sensors, or phased array ultrasonic imaging systems used in echocardiography. The wide variety of sensor types and models available necessitate that an investigator have at least a rudimentary understanding of the relevant sensors involved and the principles upon which they operate. Typically, a sensor will convert changes in its environment (pressure, CO_2, etc.) to variations in voltage, current, or both. Reading these electrical analog signals constitutes the basis for a standard DAQ system. Therefore, an understanding of the principles underlying the operation of such a unit will greatly facilitate the interpretation of the resulting data.

All computer-based data acquisition systems, such as that depicted in Figure 2-1, digitize real-world analog signals and then process them in digital form for data logging, analysis, and display. As a result, analog/digital (A/D) conversion can be considered the backbone of computer-based DAQ. In order for a researcher to understand the data that is being obtained, basic A/D concepts such as resolution, the least significant bit, and sampling frequency and its relation to the Nyquist frequency theory* must be understood. (Detailed discussion of these concepts is beyond the scope of this book, and you are referred to your favorite digital signal processing text or National Instruments DAQ course material for more information.)

---

*Nyquist frequency theory states that any signal should be sampled at a frequency > two times the highest component frequency within the signal to sufficiently replicate the original signal.

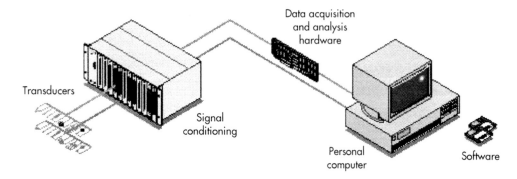

Figure 2-1.
A typical computer-based data acquisition system.

These digital signal processing (DSP) issues are critical in the development of laboratory experiments. Most of the research applications discussed in this book are conducted without the use of intermediate commercial components. In other words, the signals being studied are acquired directly using National Instruments (NI) DAQ hardware. It is therefore up to the VI developer to establish the proper amplification and sampling frequency without relying on an existing black box.

Aside from the hardware components of DAQ, it is important to know how to ensure that the signal being studied is a true representation of the phenomenon it represents. To accomplish this, it is necessary to calibrate the transducers used. Many commercial components, including the NI DAQ boards, have self-calibration routines that may be sufficient if the investigator understands the methods and potential error sources involved. Standalone DAQ systems, however, will require separate calibration procedures. Additionally, the signal must be properly conditioned through various means, including isolation, amplification, and filtering.

Armed with an understanding of the principles of data acquisition, the researcher must now integrate this knowledge with a proper comprehension of the data that is being pursued. For example, if you do not know what a typical waveform looks like, you will not be able to accurately determine amplification and resolution requirements or sampling frequency. It is necessary to estimate peak inputs to the DAQ system in order to determine the proper amplification to maximize resolution yet prevent the unwanted saturation of a signal. Understanding and minimizing interference or measurement artifacts will also enhance the quality of the relevant signal. This could be as simple as modifying the location of the sensor. For example, an esophageal balloon being used to estimate pleural pressure can produce a very noisy signal if placed too near the heart. Cardiogenic interference can be nearly eliminated in many subjects, however, with slight repositioning of the balloon.

LabVIEW™ Basics

A multidimensional data acquisition system requires a flexible programming environment to access and configure various hardware components via software. Traditionally, text-based languages such as C/C++ or BASIC have been used to support these complex systems, with additional overhead required to implement a graphical user interface. LabVIEW is a different program development application that combines the required flexibility with a built-in user-friendly graphical interface. LabVIEW was originally released in 1986 as a Macintosh-based DAQ system and was the first software application to apply the 'G' programming language, which utilizes graphical and iconic scripting techniques. Since then, the National Instruments suite of 'G'-based DAQ, Test, and Measurement applications, of which LabVIEW is one, has evolved into a standard in automated DAQ and test development. Most of these programs are available for Windows 95/NT, Macintosh, or Sun workstations.

The flexibility of LabVIEW begins with underlying machine-level code that enables the software to communicate with several sources. A LabVIEW-equipped PC/Mac can perform analog or digital data acquisition, processing, and instrument control over a GPIB (IEEE-488.2) interface. Built-in (or subsequently added) drivers enhance communication with an extensive range of standalone instruments. Digital DAQ can be accomplished via an RS-232 serial port or GPIB interface. Analog data (along with single TTL digital channels) can be obtained using any number of quality analog-to-digital (A/D) conversion boards developed by National Instruments (or other vendors). Recent advancements include the addition of an image acquisition board with a digital image processing toolkit extension for LabVIEW as well as the addition of specialized DSP boards for system control applications (originally developed by nuLogic of Boston, MA).

Although LabVIEW is primarily considered a programming environment for acquiring signals, its capabilities extend far beyond that. LabVIEW 6.0 ships with numerous function libraries that enhance data processing and display. These include various digital signal processing (DSP) algorithms such as Fast Fourier Transform routines for frequency-based analysis. Extensive statistical functions are also available from a moving average to peak detection to linear or polynomial curve fitting. A wide variety of digital filters are included as well. To meet specific needs, most of these features can be modified directly by the user.

LabVIEW applications are denoted as *virtual instruments* (VIs), due to their visual imitation of actual hardware instruments. Because of the graphical basis of LabVIEW development, all applications are developed with a ready-made graphical user interface (GUI). Additionally, the source code is developed graphically in a block diagram format. Components of the block diagram can include any of the features of text-based programs such as FOR . . . NEXT or WHILE loops or conditional branching (IF . . . THEN) statements. Each of these can be represented by a graphical symbol, with the contents of each feature literally located within the appropriate symbol. Subroutines within the code appear as icons on the block diagram, with terminals for the passing of data/variables between the main code and the subroutine.

VIs are designed to be hierarchical and modular. LabVIEW is essentially a modular programming environment within which VIs are created. Complex programs are broken down into simpler tasks, each of which is coded in a VI. Icons representing these subroutines are incorporated into the block diagram of the higher level VIs, and so on, until the top-level code is completed. This hierarchical structure permits easy modification of program elements, as well as the ability to use component VIs to develop a completely unrelated program.

Another benefit of the 'G' programming environment is the execution order. LabVIEW applications do not run line by line as standard text-based C/C++ or BASIC applications do. Rather, LabVIEW applications are data flow driven; that is, steps within the code do not execute until all of the required inputs are present. This semi-parallel execution scheme ensures that, for example, data processing does not occur until the desired data is read from the DAQ buffer.

It should be noted that LabVIEW, being graphics based, does require significant memory and computer power to operate efficiently. Newer Pentium or PowerPC computers, especially those with Advanced Graphics Port (AGP) capability, will have no problem running the latest versions of LabVIEW. Another issue to consider is the amount of data to be collected. Collecting large amounts of data will require streaming the data to a hard disk or Zip drive, whereas lesser amounts of data can be maintained in memory as the program is run.

BioBench™ Basics

The biomedical industry relies heavily on the ability to acquire, analyze, and display large quantities of data. Whether researching disease mechanisms and treatments by monitoring and storing physiological signals, researching the effects of various drug interactions, or teaching in labs where students study physiological signs and symptoms, it was clear that there existed a strong demand for a flexible, easy-to-use, and cost-effective tool. In a collaborative effort, biomedical engineers, software engineers, clinicians, and researchers created a suite of virtual instruments called BioBench™.

BioBench™ is a software application designed specifically for physiological data acquisition and analysis. It was built with LabVIEW, the world's leading software development environment for data acquisition, analysis, and presentation.* Coupled with National Instruments data acquisition (DAQ) boards, BioBench integrates the computer with data acquisition for the life sciences market.

Many biologists and physiologists have made major investments over time in data acquisition hardware built before the advent of modern personal computers. While these scientists cannot afford to throw out their investment in this equipment, they recognize that computers and the concept of virtual instrumentation yield tremendous benefits in terms of data analysis, storage, and presentation. In many cases, traditional medical instrumentation may be too expensive to acquire or maintain. As a result, researchers and scientists are opting to create their own computer-based data monitoring systems in the form of virtual instruments.

Other life scientists, who are just beginning to assemble laboratory equipment, face the daunting task of selecting hardware and software needed for their application. Many manufacturers for the life sciences field focus their efforts on the acquisition of raw signals and converting these signals into measurable linear voltages. They do not concentrate on digitizing signals or the analysis and display of data on the PC. BioBench is compatible with any isolation amplifier or monitoring instrument that provides an analog output signal. The user can acquire and analyze data immediately because BioBench automatically recognizes and controls NI DAQ hardware, minimizing configuration headaches.

---

*BioBench was developed for National Instruments (Austin, TX) by Premise Development Corporation (Avon, CT).

Figure 2-2 illustrates a typical setup of a data acquisition experiment using BioBench. Some of the advantages of computer-based data monitoring include

- Easy-to-use-software applications
- Large memory and the PCI bus
- Powerful processing capabilities
- Simplified customization and development
- More data storage and faster data transfer
- More efficient data analysis

BioBench also features pull-down menus through which the user can configure devices, therefore those who have made large capital investments can easily migrate their existing equipment into the computer age. Integrating a combination of old and new physiological instruments from a variety of manufacturers is an important and straightforward procedure. In fact, within the clinical and research setting, it is a common requirement to be able to acquire multiple physiological signals from a variety of medical devices and

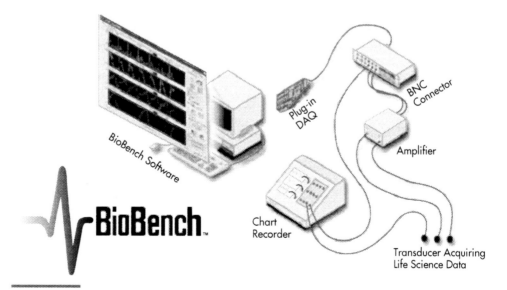

Figure 2-2.
A typical biomedical application using BioBench (courtesy of National Instruments).

instruments that do not necessarily communicate with each other. Often, this situation is compounded by the fact that end-users would like to be able to view and analyze an entire waveform and not just an average value. In order to accomplish this, the end-user must acquire multiple channels of data at a relatively high sampling rate and have the ability to manage many large data files. BioBench can collect up to 16 channels simultaneously at a sampling rate of up to 1,000 Hz per channel. Files are stored in an efficient binary format, which significantly reduces the amount of hard disk and memory requirements of the PC. During data acquisition, a number of features are available to the end-user. These features include data logging, event logging, and alarming.

Data Logging: Logging can be enabled prior to or during an acquisition. The application will either prompt the user for a (descriptive) filename or it can be configured to automatically assign a filename for each acquisition. Turning the data logging option on and off creates a log data event record that can be inspected in any of the analysis views of BioBench.

Event Logging: The capacity to associate and recognize user commands associated with a data file may be of significant value. BioBench has been designed to provide this capability by automatically logging user-defined events, stimulus events, and file logging events. With user-defined events, the user can easily enter and associate date- and time-stamped notes with user actions or specific subsets of data. Stimulus events are also date- and time-stamped and provide the user information about whether a stimulus has been turned on or off. File logging events note when data has been logged to disk. All of these types of events are stored with the raw data when logging data to file, and they can be searched for when analyzing data.

Alarming: To alert the user about specific data values and thresholds, BioBench incorporates user-defined alarms for each signal that is displayed. Alarms appear on the user interface during data acquisition and notify the user if an alarm condition has occurred. Figure 2-3 is an example of the Data Acquisition mode of BioBench.

Once data has been acquired, BioBench can employ a wide array of easy-to-use analysis features (see Figure 2-4). The user has the choice of importing recently acquired data or opening a data file that had been previously acquired for comparison or teaching purposes. Once a data set has been selected and opened, BioBench allows the user to simply select and highlight a region of interest and choose the analysis options to perform a specific routine.

BioBench implements a wide array of scalar and array analyses. For example, scalar analysis tools will determine the minimum, maximum, mean,

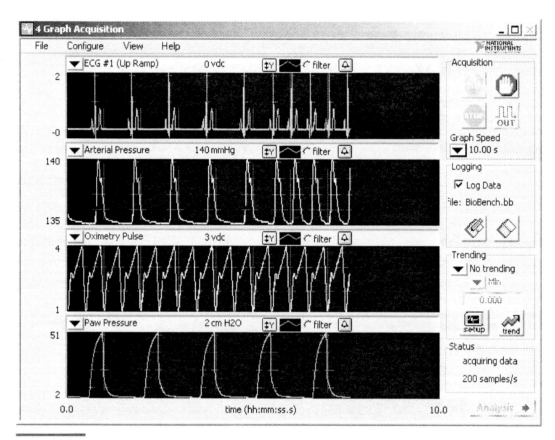

Figure 2-3.
BioBench Acquisition mode with alarms enabled.

integral, and slope of a selected data set, while the array analysis tools can employ Fast Fourier Transforms (FFTs), peak detection, histograms, and X versus Y plots. Additionally, BioBench version 1.2 added the ability to use custom-designed analysis VIs, built with LabVIEW, as plug-in routines. This allows users to take advantage of the simple turnkey DAQ aspects of BioBench, while applying their own analysis methodologies.

The ability to compare multiple data files is very important in analysis, and BioBench allows the user to open an unlimited number of data files for simultaneous comparison and analysis. All data files can be scanned using BioBench's search tools in which the user can search for particular events that

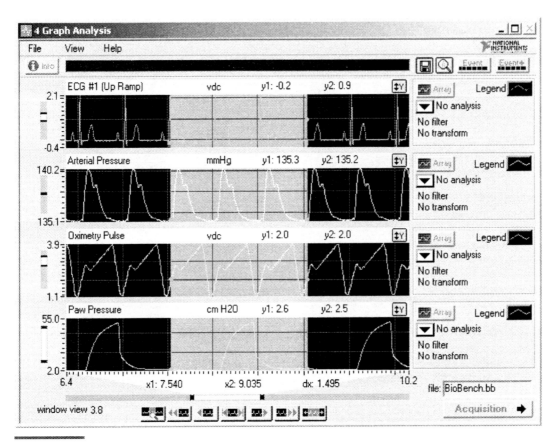

Figure 2-4.
BioBench Analysis mode.

are associated with areas of interest. In addition, BioBench allows users to employ filters and transformations to their data sets, and all logged data can be easily exported to a spreadsheet or database for further analysis. Finally, any signal acquired with BioBench can be played back, thus taking lab experience into the classroom.

Independent Solution Articles

Ensemble Averaging of Physiologic Signals: A LabVIEW-based Software Package Assisting the Analysis of Cyclic Data

By Morten Jensen, Georgia Institute of Technology

The Challenge: Developing a system for triggering and ensemble averaging of periodic flow and pressure signal data acquired during *in vivo* drug discovery animal experiments at Mercer University Hospital, Georgia, USA.

The Solution: LabVIEW was used to develop a program for triggering *in vivo* signals. This allowed for analysis of flow and pressure curves based on an average heart cycle.

Introduction

The LabVIEW-based system was developed for analyzing the effects from certain drugs administered to rats that were known to affect the heart rate and blood velocity in the abdominal aortic artery. The rat's heart rate is about 300 to 400 beats per minute, thus requiring a sample rate of 1 kHz or more. In order to assess the effects of the drugs, electrocardiogram (ECG), blood pressure, and blood velocity data had to be collected, collated, and averaged. Ten-second windows of data collection were typically used, which would yield 50–70 heart cycles (HC) for analysis.

To perform an ensemble average, a point at the beginning of every HC and the length of the HC needs to be determined. The program separates all HCs by triggering on the ECG signal or any other appropriate signal, then enabling the user to deselect the bad HCs, and at last performing the ensemble average. HCs are defined by setting criteria on one of the signals recorded. For example, if the ECG was recorded, the program would enable definition of a heart cycle as the time interval between spikes (Figure 2-5).

The trigger channel and the data channel (Figure 2-5) controls enable the user to set any acquired channel as data or trigger. The example shown in Figure 2-5 has the ECG as trigger signal, while the blood velocity is the data channel. The controls in the upper right and the lower left corners enable the user to zoom into areas of interest on the two graphs. Figure 2-6 shows an example of this zoom function.

The user sets the trigger level on the upper graph. The horizontal line intersecting any positive slopes defines the trigger points. The zoom function together with the vertical cursor enables the user to estimate where the trigger points on the upper graph will take action on the lower graph.

The max/mean control on top of the trigger channel graph enables the user to choose between two methods to determine the length of the trigger interval. When the trigger

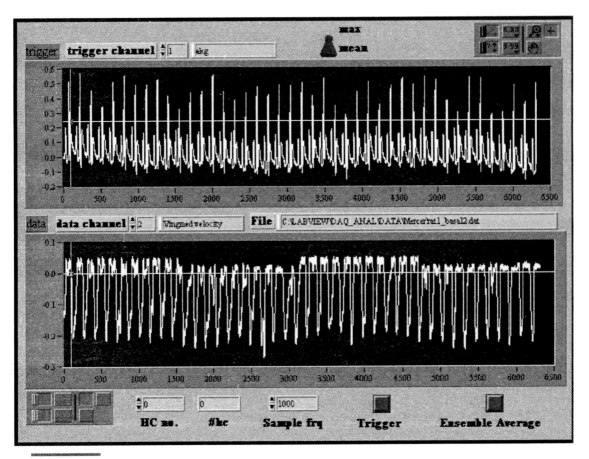

Figure 2-5.
The New Data File screen. Here the trigger level is set on the upper graph.

button is pressed, the length of the ensemble averaged HC is calculated. Even in the case of the trigger channel containing noise or other artifacts, it is possible to obtain a reasonable result using the mean setting. If the trigger channel data is clearly defined without noise, the max value is preferable. Then, using the max distance between the trigger points in an ECG, the user should be careful to include all the R-waves. If one R-wave is below the trigger level, the index interval in the triggered signal is two times the actual HC length.

When the trigger button is pressed, the trigger function is performed, and all the triggered data channel HCs are now showed on the lower graph (Figure 2-7). The HC no. control (below the lower graph) enables the user to scroll through all the triggered HCs.

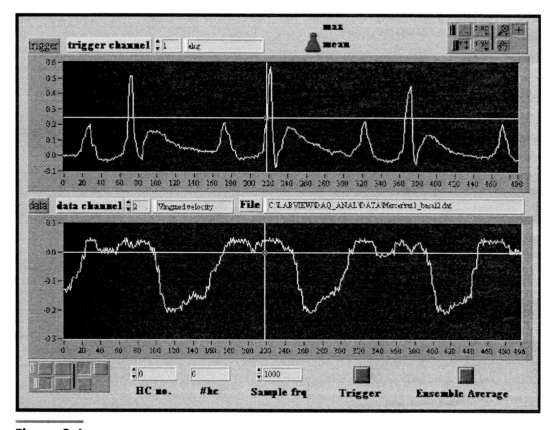

Figure 2-6.
Example of the zoom function. The user can determine if the trigger signal is feasible to be used as a trigger on the data channel by moving the x value cursor along the graphs.

Deselect Bad Heart Cycles

Breathing artifacts, movement of instruments, and so on, often contaminate *in vivo* experiment data. Therefore, the researcher might want to deselect obvious bad heart cycles for further analysis. Figure 2-8 shows a part of the triggered blood velocity measurements, where HC number 18 is deselected.

Result Panel

When the bad HCs are deselected, the ensemble average result panel shows the result of the analysis (Figure 2-9). The filename, heart rate, number of heart cycles used in the ensemble

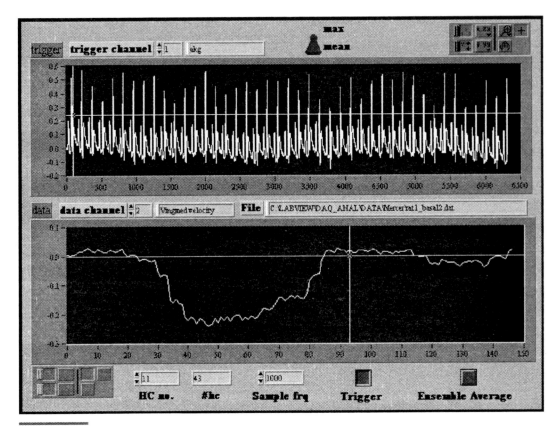

Figure 2-7.
The lower graph holds the triggered HCs after the trigger operation is performed.

average calculation, the used sample frequency, and the max and mean values of the depicted graph together with the $+/-$ standard deviation at their indexes are shown at the result screen. The user can choose to print the displayed graph with the additional information on the result front panel. The data can also be exported to an Excel spreadsheet.

Conclusion

The LabVIEW programming language was used to generate a generic tool in a very short amount of time that can be used in applications for triggering signals acquired *in vivo*. Researchers and scientists at Mercer University Hospital currently use the system for analyzing data from various *in vivo* experiments and have published these data in peer-reviewed

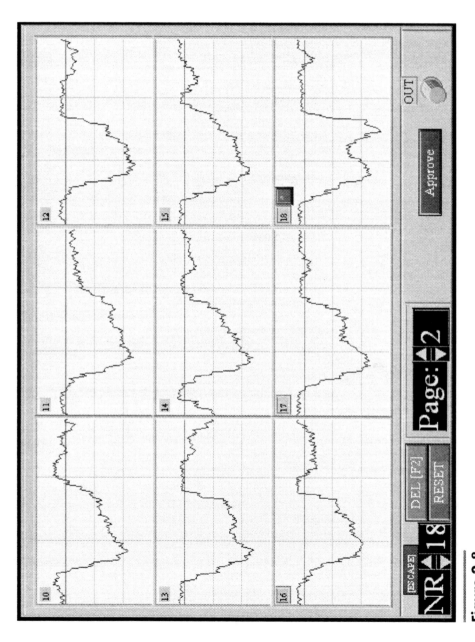

Figure 2-8.
Deselecting bad HCs. Here HC number 18 is deselected.

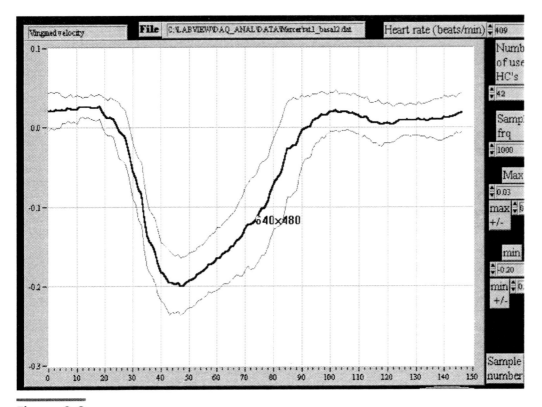

Figure 2-9.
The result panel shows parameters of interest together with the ensemble averaged curve
and the standard deviation curves. In the example shown here, none of the bad HCs was
deselected. This explains the relatively large standard deviation.

journals. The program will in the future be tied together with the data-acquisition program
to create one solution system for acquiring, analyzing, and presenting *in vivo* data to assist
the development of better and safer drugs.

Part II
Research
Applications

Biopotentials

3

This chapter discusses representative applications relevant to the measurement and analysis of bioelectric events emanating from within living creatures. *Excitable cells* serve as the source of bioelectric potentials due to their electrochemical activity. These cells are fundamental components of neural, muscular, or glandular tissue. Measurement of these bioelectric phenomena can be readily accomplished with the appropriate monitoring equipment. Typical bioelectric recordings used in the clinical environment include the electrocardiogram (ECG), electroencephalogram (EEG), electroneurogram (ENG), electroretinogram (ERG), electrooculogram (EOG), and the electromyogram (EMG).

This chapter begins with some generic discussions relative to laboratory setup, experiment preparation, and VI development. It then proceeds to document representative examples in neural, neuromuscular, and cardiac electrophysiology. All of the data acquisition and analyses described in these labs are automated through the VIs described, which are also included on the accompanying CD-ROM. While that was an important part of the development of these labs, it may not be the proper pedagogical style for everyone. Due to the modular design of the VIs, it should be relatively straightforward to strip away some of the automated analysis and enable students to develop their own routines.

Typical Laboratory Workstation

A standard workstation used for the laboratory projects described herein is illustrated in Figure 3-1 (Olansen, Ghorbel, Clark, & Bidani, 2000). A 12-bit PCI-MIO-16E-4 board from National Instruments (NI) integrated into a personal computer has worked well for all of the experiments discussed in this section. An NI Signal Conditioning eXtensions for Instrumentation (SCXI) system is also included to perform much of the signal conditioning required, particularly when dealing with very small voltage recordings. The SCXI-1000 chassis houses an SCXI-1100 32-channel differential input multiplexer mod-

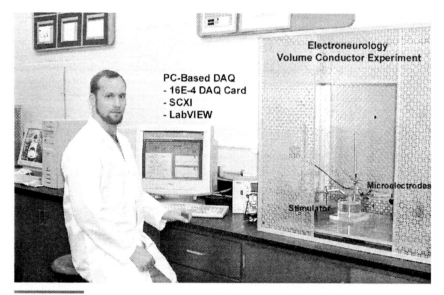

Figure 3-1.
Typical biomedical systems and instrumentation laboratory workstation at Rice University. It is shown here configured for an experiment demonstrating the effects of a volume conductor on electroneurology recordings. (Reprinted with permission by Olansen, Ghorbel, Clark, & Bidani, 2000.)

ule, an SCXI-1120 8-channel isolated input module, and an SCXI-1180 feed-through panel. The SCXI-1100 and SCXI-1120 are used to condition input signals, while the SCXI-1180 allows for simultaneous output from the DAQ board (used as a stimulator when required for a specific preparation). The SCXI-1100 and SCXI-1120 modules do require some specific setup to be compatible with the included tests. The only physical setup required for the SCXI-1100 module is to set the filters as desired for the experiment at hand. The filter/gain settings for the SCXI-1120 module will vary depending on the trial being conducted. The SCXI-1120 module has an SCXI-1305 BNC interface to allow for easy reconfiguration and to provide AC coupling as required.

Secondary to the lab workstation is a mobile workstation consisting of a laptop with a DAQCard-AI-16E-4 PCMCIA card. Combined with the same SCXI chassis described earlier, the DAQ capabilities become portable, allowing experiments to be conducted at a variety of hospitals, clinics, or laboratories. (Note: Experiments using stimulus output also require an external stimulator in the mobile configuration since the DAQCard does not have analog output capability. Stimulus output can still be triggered digitally.)

Lab Layout and Design

You should consider the following items when designing the layout of a laboratory. We have also listed a few items you may want to have around the lab.

- Safety first! You must become familiar with the OSHA requirements for a physical science lab. Ensure that the appropriate safety equipment, such as safety glasses, is available and its proper use is explained.

- The layout of lab benches depends on your instruction style. Two important questions to deal with are, "Do I want to see everyone in the lab?" and "Is space at a premium, and do I need as many stations as I can get into this room?" Whatever your answers, you will always want room to move about the lab and teach!

- There must be plenty of storage space for equipment not in use. Also, students usually carry backpacks and coats and such to class, so there should be an out-of-the-way place for them to put their personal items.

- Use wiring raceways to keep network cable and power cords out of the way.

- Consider sharing more expensive equipment between two or more stations. This means the stations must be positioned such that the equipment can be reached from each station.

- How many different courses will use the lab in a given semester? If you share lab space, make sure you organize the lab to switch setups easily.

- Do you want the students to take care of setup and tear down? If you have large numbers of sections and students, you will probably want to limit student handling of equipment to save on the wear and tear of the equipment.

- For equipment protection, provide robust connections that are easily connected and removed. The next section describes the design of a breakout box.

- Make each station *exactly* the same. This helps tremendously when debugging lab exercise setups; there are no "one-off" lab stations that need separate setup. Also, you don't get problems such as, "Oh! That's the station with the XYZ module—it has to be booted while tilting the monitor to the left."

- The caveat to the previous item is that to save capital equipment costs, you can split your lab into two sections such that two labs are run at one time. This means you will need half as many setups of each lab. Half of the students can perform one lab one week and then trade and perform the other lab the next week. This requires a major adjustment in pedagogy, and the decision should not be made lightly. The major problem is coordinating the lecture material with the lab when you are doing two experiments over a two-week period.

- We recommend you have this standard equipment available in the lab:

  - Various connectors and wiring: banana-to-banana and banana-to-alligator jacks, BNC-to-banana converters, BNC male-to-male and male-to-female, BNC T-connectors, and so on.

  - Last, but certainly not least, is the appropriate safety equipment such as safety glasses. Again, consult OSHA regulations for instructional laboratories.

Generic Instrumentation/Data Acquisition Issues

With the setup just described, a number of issues had to be addressed. As in any experimental research, the calibration of the equipment being used is of utmost importance. These experiments are no different. In conducting electroneurology trials, significant amplification may be required depending on the particular experiment. It is therefore necessary to properly calibrate the electrodes and data acquisition system. If possible, the system should be calibrated with the calibrator in series with the recordings, so as to account for electrode and junction resistances, and other interference phenomena. The system should be calibrated prior to each trial series, and the calibration coefficients should always be saved with the data. Additionally, the entire preparation should be enclosed in a Faraday cage (as portrayed in Figure 3-1) to minimize the effects of electrical interference, particularly 50–60 Hz noise, on the recordings.

The electrophysiological studies discussed herein involved several technical issues that needed resolution. One was the frequency response of the input amplifiers. A frequency response VI was developed using simultaneous

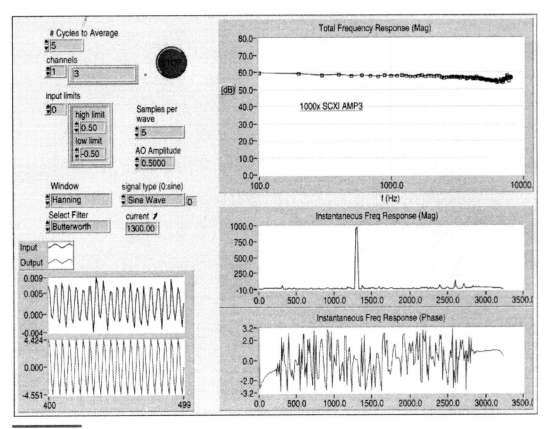

Figure 3-2.
Front panel of **FreqResp** VI depicting typical results used in determining the frequency response of recording amplifiers.

I/O to test the SCXI/DAQ amplifiers. The **FreqResp** VI front panel is shown in Figure 3-2. [Note: For assessment of frequency response at higher gains, the computer-generated output signal first had to be attenuated (by a voltage divider) prior to passing it to the SCXI/DAQ amplifiers in order to prevent the amplifiers from being saturated. If high-value resistors were chosen in implementing the voltage divider, the frequency response of the amplifier would be altered due to altered input impedance. Using 10 k and 1 M resistors for a 10^2 division worked well for this case.] The frequency response so examined was found to be flat to 10 kHz (maximum tested) for all gain settings.

An independent Fourier analysis (via FFT) was conducted on an oversampled (50 kHz) nerve compound action potential (CAP), which indicated that

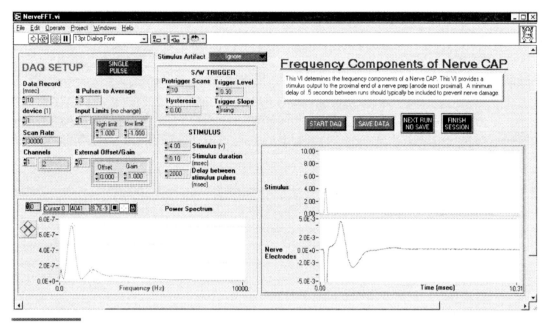

Figure 3-3.
Front panel of NerveFFT VI depicting typical results used in determining the frequency components of a nerve compound action potential (CAP).

there were no significant frequency components above 6 kHz. Therefore, the frequency response of the amplifiers was considered adequate to provide appropriate gain even for the higher frequency components of a nerve compound action potential (see Figure 3-3).

Another issue with the amplification is voltage offset. The common mode voltage rejection of the SCXI-1120 is excellent, but problems can be encountered if the steady potential difference between the recording electrodes and ground is relatively large (e.g., 300 mV). In such a case, amplification by a factor of 20 would saturate the SCXI-1120 output (± 5 V). Usually, CAPs are recorded using AC coupling in order to filter out any DC offset. The SCXI-1304 and SCXI-1305 terminal boards have AC coupling capability onboard with a fixed lower frequency cutoff level of 0.167 Hz. This is sufficient for nerve electrophysiology studies but should be reviewed for different biomedical research applications.

Recording electrical signals in the microvolt (µV) to millivolt (mV) range may also require additional measures to obtain data of sufficient quality. All physical instruments related to the data collection should be referenced to a single well-established ground reference (this is called a star type connection). This precaution, one ground and one connection to that ground from each piece of equipment, eliminates the presence of ground loops, which can induce oscillations and corrupt the signals being acquired. When dealing with very small signals in the μV range, it is often necessary to use cumulative averaging (N trials) where the signal-to-noise ratio is improved by $\sqrt{N}$. This type of automated averaging routine is included in some of the VIs that have been developed (reference the **Neuro.llb** library on the CD for coding examples).

Another problem often encountered in electrophysiological studies is the presence of stimulus artifact. Several methods have been developed for estimating the artifact and removing it from the corrupted signal (McGill et al., 1982). A stimulus artifact rejection routine can be used in conjunction with the VIs developed for these labs. Care must be taken to use the correct estimated artifact waveform for cancellation, which presumably has been developed using experimental settings that are identical to those used in the experiment.

The effect of the volume conductor within which the bioelectric signals are generated is also an issue commonly encountered. The volume conductor has a low-pass filtering effect on the CAP recorded at the surface of the nerve. The filtering characteristics of the volume conductor can be determined based on knowledge of the material properties of the volume conductor medium (bulk-specific conductivity) and the geometry of the nerve trunk and surrounding medium (Greco & Clark, 1977). The average CAP conduction velocity is also required, in order to predict the field potential everywhere in the volume conductor.

Having addressed some of the generic issues fundamental to this type of laboratory setup, the focus is now directed toward the design and implementation of individual experiments.

Electroneurology

Vertebrate neurons consist of cell bodies, long fiber-like processes called axons, and specialized nerve terminal endings called synapses. Individual nerve fibers are gathered together in nerve trunks or simply nerves. Neurons are responsible for a remarkable variety of communication functions, including delivery of sensory information to the brain, mediating local reflexes, and autonomic functions such as control of heart rate, breathing, and visceral function. Neurons also control muscle function, which is the focus of the neuromuscular electrophysiology labs. The frog sciatic nerve bundle is often used as the prototypical nerve preparation, and it has historically served as one of the most useful models for learning the properties of nerve conduction.

The objectives of these laboratories incorporate fundamental animal preparation and data acquisition techniques in addition to the study of specific physiological phenomena. The properties of vertebrate (frog) nerve axon bundles that are studied include recruitment, excitability (strength-duration relationship), conduction velocity, and refractory period. These excitability properties of nerves are addressed in greater detail next.

Physiological Basis

A steady potential difference has been demonstrated to exist between the inside and outside of a wide variety of individual excitable cells, such as neurons, cardiac, skeletal, and smooth muscle cells. This transmembrane potential difference is called the *resting potential* (e.g., -70 mV, inside negative relative to outside of cell) and is the result of an ionic imbalance in sodium (Na^+), calcium (Ca^{2+}), potassium (K^+), and other ions that are maintained in steady state between the intracellular and extracellular fluid media. The cell is said to be *polarized* when in the resting state. External stimuli applied to the membrane can alter this resting potential in the form of *depolarization* (i.e., potential becomes less negative than resting state) or *hyperpolarization* (i.e., potential becomes more negative). The term *repolarization* describes the return of the membrane potential from a stimulated state to its original resting state.

The nerve fiber membrane contains voltage and time-dependent ion channels (ion-specific passageways through the membrane), notably Na^+ and K^+ channels (see Figure 3-4). When a region of membrane is partially depolarized so that voltage-gated sodium channels open, Na^+ ions enter and rapidly depolarize the membrane fully. This change in potential causes voltage-dependent ion channels to open allowing ions to flow across the membrane into the intracellular medium of the cell, resulting in the generation of the action potential. This effect is short-lived, however, since (a) the change in membrane permeability to Na^+ is brief and transient, and (b) there is a delayed increase in membrane permeability to K^+. Both of these latter effects bring about the repolarization of the membrane to its resting state. The peak change in membrane potential is on the order of 100 mV and the duration of

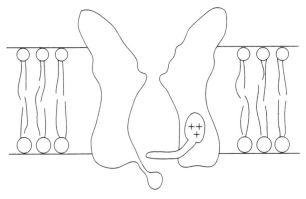

Figure 3-4.
Depiction of typical neural gated transmembrane ion channel.

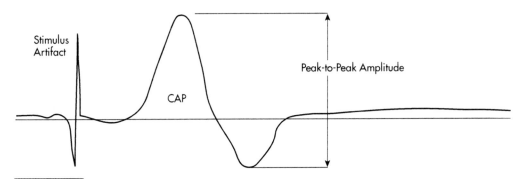

Stimulus Artifact

Peak-to-Peak Amplitude

CAP

Figure 3-5.
Depiction of a typical triphasic nerve compound action potential (CAP).

the response (the action potential) depends upon the particular cell type. For nerves, it is on the order of 1 msec.

The summation of activity from the many component nerve fibers of a nerve trunk gives rise to a triphasic potential waveform known as a *compound action potential* (CAP). A typical CAP is shown in Figure 3-5. The extent to which the membrane must be depolarized in order to trigger an action potential is called its *excitability threshold,* and this threshold is inversely proportional to the diameter of the nerve fiber. Thus, larger diameter nerve fibers are stimulated more easily than smaller fibers. An action potential is conducted along an axon when a stimulated active region sources current to an adjacent segment of membrane, thus depolarizing it. With depolarization of adequate strength, that region of the axon fiber exceeds the voltage threshold for excitability and generates an action potential. That CAP, in turn, activates adjacent segments.

The stimulus itself, due to the volume conductor, contributes a field effect, resulting in a stimulus artifact that is recorded by the recording amplifier. This artifact occurs simultaneously with the time of the stimulus, whereas the conducted nerve impulse follows some time later (i.e., has a delay). One can verify that the conducted waveform is not a second stimulus artifact by reversing the polarity of the stimulus. The artifact changes polarity, whereas the CAP waveform does not. Stimulus artifact is a persistent problem in recording CAPs from nerves. This is particularly true when the stimulus duration is long. As a practical matter, the problem can be observed by first evoking a CAP at a specified stimulus amplitude and duration (e.g., 2 V and 0.05 msec, respectively), and then gradually increasing the duration (with constant amplitude) to about 1 msec. Eventually, the stimulus artifact "walks up" the rising portion of the CAP, interfering with the waveform. To minimize artifact, stimulating and recording compartments should be formed and separated to the greatest extent possible. Often placing a ground electrode between the compartments is helpful in minimizing artifact. In selecting stimuli, it is better to select a brief intense stimulus, rather than a long weak pulse.

Experiment Setup

Typically, nerve studies are performed on an isolated frog nerve preparation established by mounting a nerve segment across a parallel array of electrodes in a moistened Plexiglas chamber. Compound action potentials (CAPs) thus recorded are expected to be in the 1–10 mV range and therefore require some amplification to provide an input signal that covers the input dynamic range of the analog-to-digital (A/D) converter. Performing these studies *in vivo* (i.e., nerve situated within the leg but recording made from the surface of the leg), or with the nerve encased in another type of conducting medium results in field potential recordings on the order of 10 µV. In this case, a gain of 10^5 is typically necessary to boost the signal into the ± V dynamic range of the A/D converter. This can be accomplished using an amplifier with high input impedance (> 10 Mohms) and excellent common mode rejection properties (CMRR = 100 dB). As discussed previously, amplifier bandwidth should be flat to 10 kHz. Cumulative averaging is usually required to pull the small signal out of its noisy environment.

The DAQ setup described earlier can be used to prototype these experiments. The data acquisition board is used to trigger the stimulation pulses and record the data from each electrode with the SCXI interface performing the required signal preconditioning. The SCXI-1120 provides isolation amplifiers with individual gains of up to 2,000 and 10 kHz filters for eight differential input channels. The DAQ board adds a maximum gain of 100, for a total possible gain of 2×10^5. The SCXI-1180 feed-through panel enables the analog output of the DAQ board to be used as the stimulus.

Dissection

For the following series of experiments, two different preparations can be utilized. An *in vivo* preparation that is also suitable for these experiments will be discussed in the neuromuscular write-up. In this section, we document a preparation in which a sciatic nerve (actually a bundle of individual nerve axons) is isolated from a large, freshly anesthetized (pithed) *Rana Catesbeianas*. The nerve is to be placed across a series of stimulating and recording electrodes in a specially constructed nerve chamber, which will be discussed later. The proximal end of the nerve is stimulated and the CAPs are recorded from electrodes further down the nerve. It is desirable to recover as long a piece of the sciatic nerve as possible, since the longer the section of nerve, the more experiments that can be performed. With large frogs, it may be possible to salvage the experiment if the upper sciatic or the extreme lower parts of the nerve are lost. This setup is adapted from a course developed and taught by Dr. David Caprette of Rice University (1998). *Refer to Figure 3-6 during the dissection.*

Step 1. Make an incision through the skin of the back of a frog that has been pithed by the instructor. Push your closed large scissors through the opening and spread the scissors so as to separate the skin from the underlying muscle. Continue this blunt dissection around the waist, occasionally cutting the skin to provide more room, until you have cut all the way around the waist. Now remove the skin by grasping the frog's upper body and grabbing the skin with forceps or fingers. If you have not cut the muscles, a strong pull on the skin should remove it without damaging any of the structures underneath. Pull the skin all the way off the ends of the flippers. Place the carcass on a moistened paper towel and wash hands and instruments.

Step 2. The sciatic nerve in the thigh lies along the femoral artery, either of which can be reached from the lateral surface. Start with blunt dissection as above to begin to separate the *semimembranosus* muscle from the *iliofibrularis*. Cut through the fascia up and down the muscles and use blunt dissection (pushing and spreading the scissors) to reveal the artery and nerve. You may pin the muscles as in Figure 3-6 to expose the nerve.

CAUTIONS:

(1) Avoid cutting any blood vessel, especially the femoral artery, as bleeding will obscure your view.

(2) *Always* see where you cut (conversely, don't cut where you can't see). It is a near certainty that if you violate this rule, you will destroy the very structures that you are trying to preserve.

Step 3. Wherever you expose the nerve, keep it moist with frog-Ringer's solution. *Frog-Ringer's solution* is prepared using the following recipe (Whitfield, 1964):

> 10 ml of 1% Sodium bicarbonate
>
> 10 ml of 1% Calcium chloride
>
> 7.5 ml of 1% Potassium chloride

Diluting to:

> liter with 0.6% Sodium chloride

Once you have located the position of the nerve in the upper thigh, trace it to the knee area. Here it branches and is obscured by connective tissue. Expose the branches by cutting the tendon at the insertion of the *gastrocnemius* and reflecting the muscle. Clear the connective tissue (carefully) to expose the branches through the knee area.

CAUTION: Do not stretch the nerve, crush it with forceps, or let it dry out at any time.

Step 4. Cut underneath the urostyle (the loose bone that projects from the end of the backbone) and cut the muscles on both sides while lifting the end of the urostyle. You can now trace a sciatic nerve to its exit point from the spinal column. Trace the nerve and artery through the hip region. Tie and cut the proximal end of the nerve at as high a point as possible. Now carefully free the nerve past the hip and down past the knee as far as possible, cutting small branches as you go. You may want to tie the largest distal branch before cutting it off to facilitate placing it in the nerve chamber.

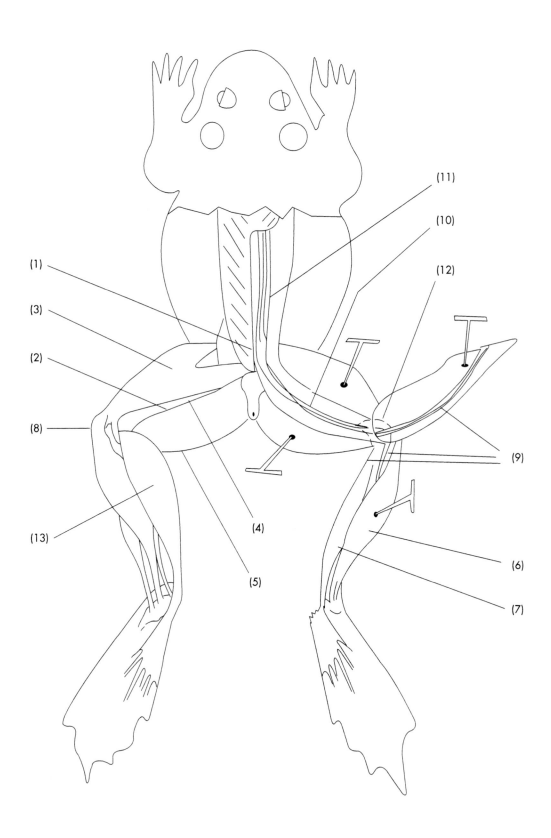

Figure 3-6. (left)
Frog sciatic nerve dissection. (Adapted with permission of Caprette, 1998.)

(1) Lift here with forceps and cut muscles underneath urostyle and on both sides toward head. Reflect the cut muscles to expose the upper sciatic. In the hip region, only muscle need be removed in order to expose the nerve.
(2) Separate the *iliofibrularis* and *semimembranosus* to reveal sciatic nerve and companion artery.
(3) *Triceps femoris*
(4) *Iliofibrularis*
(5) *Semimembranosus*
(6) *Tibialis anticus longus*
(7) *Tibialis posticus*. Between this muscle and bone is one of the distal branches of the sciatic (with companion artery).
(8) Sciatic divides into three major branches in this region. The dissection is tricky—tough connective tissue must be removed to expose the nerves without cutting them. The best branch is probably the one that runs parallel to the *tibialis anticus longus*. The *gastrocnemius* branch is the most superficial and is also relatively easy to recover.
(9) Sciatic branches. Dotted lines show path beneath connective tissue.
(10) Sciatic nerve and companion artery
(11) Upper sciatic
(12) Path of one distal branch runs beneath connective tissue on the kneecap, behind reflected *gastrocnemius*.
(13) *Plantaris longus* (*Gastrocnemius*)

Step 5. The nerve should be free now and tied at least at one end. Remove it to a Petri dish with frog-Ringer's solution (without stretching). Save the carcass on ice—the other sciatic will remain usable for some time.

Nerve Chamber Preparation

A nerve chamber is simply an encased array of electrodes upon which a nerve bundle may be placed for study. The nerve chamber used in these experiments was made of milled Plexiglas with silver wire electrodes inserted at 0.5-inch intervals across the opening (see Figure 3-7). The electrodes were attached to banana plug sockets for easy connection to the DAQ setup.

To prepare the chamber, insert a precut piece of filter paper into the bottom of the chamber, lying flat, and pipette a liberal amount of distilled water onto the paper (without letting the surface contact the electrodes). This helps maintain a moist environment, extending the viability of the nerve. Pick up the

Figure 3-7.
Depiction of typical nerve chamber with excised nerve laid across equidistant electrodes.

nerve by the thread and place one end across an electrode near one end of the chamber. Allow the nerve to lay across the electrodes, covering as many as possible. The nerve can be stretched *slightly* to fit across the electrodes. Now moisten the surface of a glass cover (e.g., standard glass slides) with water, and place it over the opening so the water forms a seal. It should not be necessary to disturb the nerve again for the remainder of the experiment. The entire preparation is enclosed in a Faraday cage (as shown in Figure 3-1) to minimize interference effects on the recordings.

The stimulating electrodes are applied to the proximal portion of the nerve with the anode, or positive, terminal most proximal. (Having the cathode most proximal artificially delays the compound action potential. If necessary, this tactic can be used to separate stimulus artifact from the actual response.) It is typically preferable to use an isolation amplifier to apply the stimulus, so that the stimulus itself is actually floating. This is because the stimulus is orders of magnitude larger than the signal being recorded and can thus create significant interference. (With this setup isolation was not required. However, isolation should be strongly considered for quality research projects.) Applying a ground between the isolated stimulating electrodes and the recording electrodes yields the optimum solution. The stimulus artifact can be identified by comparison to the input pulse. It results from conduction of the stimulus through the annular medium surrounding the nerve and varies directly with the duration of the input.

In these experiments, the nerve is laid across ten electrodes in the trough, each separated by 12 mm. The distal end of the nerve could be damaged to prevent the propagation of the action potential to that region, and that region would serve as the reference potential for the active recording electrodes. Alternatively, the recording electrodes could be referenced to each other. This would result in a different waveform, though, since both electrodes would see CAP activity.

Numerous tests can be conducted with this preparation. By recording electrical activity from electrodes along the nerve, the classical nerve action potentials can be observed and an estimate of conduction velocity could be obtained. The refractory period and summation interval of the action potentials could be analyzed by applying multiple stimuli in short intervals. All of these measurements could then be compared with variations in stimulus strength or duration. This battery of tests would demonstrate numerous properties of the tissue sample. Using the dissection previously described, any combination of the above mentioned tests can be performed.

Generic VI Development

The VIs for the neural lab all require simultaneous I/O to stimulate the nerve preparation and record from it. To that end, each is based on a **Simo I/O-SW Trig** sub-VI that accommodates the analog stimulus pulse waveform and collects the input CAP data based on a software trigger. Since the events being studied are responses to a stimulus, the stimulus itself is fed back as the trigger input. The software trigger acts as a conditional retrieval in that the incoming signal is not altered, enabling pretrigger scans to be read, if required [National Instruments (NI), 1998].

The electroneurology VIs can accept an arbitrary number of inputs; however, there are some calculations that are applied only to a single nerve recording electrode. For example, in the calculation of the fundamental relationships for the nerve as in the strength-duration curve or the restitution curve, the data involved in the measurement are only concerned with a single electrode. When conduction velocities are calculated, however, they are determined for every successive pair of electrodes, since conduction velocity can change as a function of distance along a nerve.

The graphical user interfaces (GUIs) for each of these VIs are intended to be as similar as possible, so as to minimize time spent learning the interface. As data are collected they are typically displayed in strip chart format to the right of the window, whereas output analysis is usually displayed to the left. User inputs are collected in groupings based on function. Some programmers may prefer to hide many of the setup inputs such as scan rate and channels, whereas our VIs are designed to include all this information. In this way, instructors may decide for themselves to what depth they want their students to learn about these parameters and issues associated with them. Initiation of the program compiles the software, reads the proper calibration coefficients, and waits for user input to begin collecting data.

Experiment Descriptions

This section describes the fundamental experiments that can be performed with the isolated nerve preparation. Each test description will include a brief overview of the underlying physiological phenomenon being studied and a description of the controlling VI. The VIs written for this lab include the following:

- **NrvRecruit:** Nerve Bundle Recruitment, Threshold Zone, and Conduction Velocity
- **NrvSD:** Strength-Duration Properties
- **NrvCondVel:** Conduction Velocity, Averaging Routine, and Artifact Rejection
- **NrvRefPeriod:** Restitution Properties

As outlined previously, a wide array of electroneurology experiments have been developed for use in biomedical instrumentation laboratories. The following sections provide a cursory look at some of these representative endeavors. Detailed information regarding some of these experiments will be supplied to enable readers to use or modify the appropriate VIs and sub-VIs, which are included on the accompanying CD-ROM.

Stimulus Thresholds and Recruitment

This study focuses on generating and analyzing a compound action potential (CAP) from a nerve (see Figure 3-5), particularly with regard to recruitment of nerve axons. An action potential is generated when a region of membrane is partially depolarized so that voltage-gated sodium channels open, allowing sodium ions to enter and rapidly complete the depolarization of the membrane. The extent to which the membrane must be depolarized in order to trigger an action potential is called its excitability threshold. An action potential is propagated when it partially depolarizes an adjacent region of the membrane, allowing that region to reach the excitability threshold, and so on. The stimulus itself provides a jolt that is recorded as a stimulus artifact. There is almost no delay since the artifact is conducted passively. The propagated nerve impulse follows. One can verify that the propagated pulse is not a second stimulus artifact by reversing the polarity of the stimulus and verifying that the CAP waveform doesn't change (though it will be delayed as described earlier).

A sciatic nerve is a bundle of individual axons of different types. *Control* of activity is accomplished by varying the *number* of axons that depolarize in response to a stimulus. Under the present experimental conditions, a single nerve axon will fire if the stimulus intensity is great enough, otherwise it shows no activity at all. This is what is meant by the all-or-none property of nerves. Axons of a particular type, best characterized by axon diameter, share common properties such as conduction velocity and sensitivity to a stimulus. If a large jump in compound action potential amplitude is identified over a short range of stimulus intensities, it is likely that a large number of nerves of a particular type are being sufficiently stimulated to conduct action potentials. It is also likely that a voltage range will be encountered over which little additional nerve activity can be detected. This can mean that few or no axons in the bundle have stimulus thresholds in that range. Once a stimulus is applied that exceeds the threshold for every nerve in the bundle, further increases in voltage will have no effect. The recruitment curve will reach a plateau.

VI Description

The controlling, or top-level, VI for this experiment is called **NrvRecruit** (see Figure 3-8). As mentioned in the section introduction, the backbone of this VI

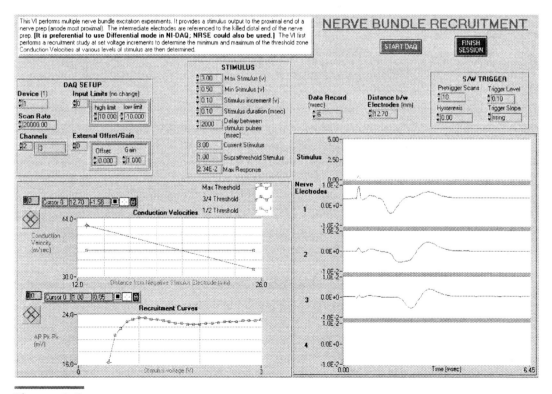

Figure 3-8.
Front panel of the **NrvRecruit** VI with typical nerve compound action potential data.

is the **Simo I/O-SW Trig** VI. This sub-VI accomplishes the short-duration stimulation (typically 0.1 msec) and records the CAPs from as many inputs as are available. (The standard used in development of this lab was three nerve electrodes.) The stimulus pulse is fed back as the first input channel and serves as the software trigger.

The user interface is a display of virtual instruments, most notably a series of virtual strip chart recorders. On these, the near real-time data from each active electrode is displayed. There is also a plot of instantaneous conduction velocity at each electrode and the recruitment curve for the first electrode. Of course, the nerve trunk being studied consists of many different size and length nerve fibers, all with different conduction velocities. Therefore, the conduction velocity being displayed is merely a representative velocity and will vary based on recruitment of individual fibers. This may also be evident in the dispersion of the action potential as the wave travels along the nerve.

Upon receiving user input via the **START DAQ** button, the VI provides a stimulus to the prep and records the data record (typically 10 msec). The data is then zeroed and any calibration coefficients are applied to it. A time column is also added to the data for post-test analysis. The peak-to-peak amplitude of the CAP is then determined. This process repeats for various stimulus levels according to the user inputs on the front panel. The CAP peak-to-peak values are then analyzed to determine the maximum response and the threshold zone of the nerve trunk. Finally, the tests are repeated at prescribed stimulus levels within the threshold zone and conduction velocity is calculated between each pair of recording electrodes. This enables the student to determine variations in conduction velocity dependent upon which nerve types are excited.

Strength-Duration Relationship

This experiment is a further exploration of the generation of an action potential and has been previously detailed in Olansen et al. (2000). This time the focus will be on the qualities of a stimulus needed to initiate a response and the properties of nerves that dictate the requirements. It is concerned with the threshold relationship for a nerve trunk, that is, the excitability properties of the nerve. We consider square-pulse current stimuli of variable amplitude (strength) and duration. In fact, several specific combinations of strength and duration can result in the production of a conducted action potential. The strength-duration curve represents a locus of critical points, which are values of strength and duration that result in the generation of an action potential. This curve separates the passive, subthreshold, nonconducting state from the active conducting state.

Since the sciatic nerve constitutes a bundle of nerve fibers of different diameters, the threshold properties of a nerve trunk will differ from those of an individual neuron. Larger fibers are excited at the lowest stimulus strengths, and smaller fibers at higher strengths. This gives rise to the concept of a *threshold zone* rather than a single threshold value as for the single fiber. The lower end of the threshold zone denotes the stimulus strength where the CAP first appears. The amplitude, duration, and overall waveshape change as stimuli of greater strength are applied (recruitment of active fibers). Finally, a stimulus strength is reached beyond which no further changes are noted in the CAP. This is labeled the maximal stimulus.

The *threshold zone* identified above detailed a change in CAP response with increasing voltage. Similar alterations in the CAP can be elicited by changing the duration of the stimulus instead of the voltage level, as demonstrated in Figure 3-9. Consider an applied voltage of a particular duration that is just

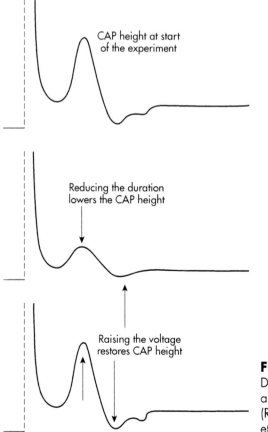

CAP height at start of the experiment

Reducing the duration lowers the CAP height

Raising the voltage restores CAP height

Figure 3-9.
Depiction of nerve CAP as strength and duration of stimulus are varied. (Reprinted with permission of Olansen et al., 2000.)

sufficient to elicit the maximal CAP for the nerve, that is, all fibers within the trunk contribute to the response. Any stimulus of a higher voltage or duration is called a *supramaximal* stimulus. Lowering either voltage or duration causes some axons to drop out, thus reducing the overall height of the compound action potential. In forming the strength-duration (S-D) relationship for a nerve trunk, we obtain a family of S-D relationships with graded stimulus intensity from threshold to maximal stimulus strength.

VI Description

The front panel of the **NrvSD** VI that performs this experiment is shown in Figure 3-10. A schematic flowchart that clearly delineates the programmatic flow of the **NrvSD** VI and a representative block diagram are shown in Figures 3-11 and 3-12, respectively. The experiment starts with the longest stimulus duration that results in a clearly defined CAP. The width should be about a millisecond. Select a stimulus voltage that gives a peak-to-peak action

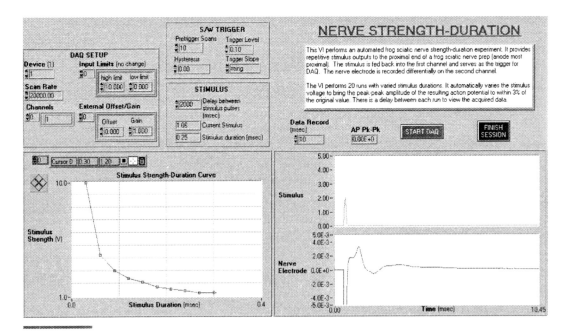

Figure 3-10.
Front panel of the **NrvSD** VI with typical nerve compound action potential and strength-duration data. (Reprinted with permission of Olansen et al., 2000.)

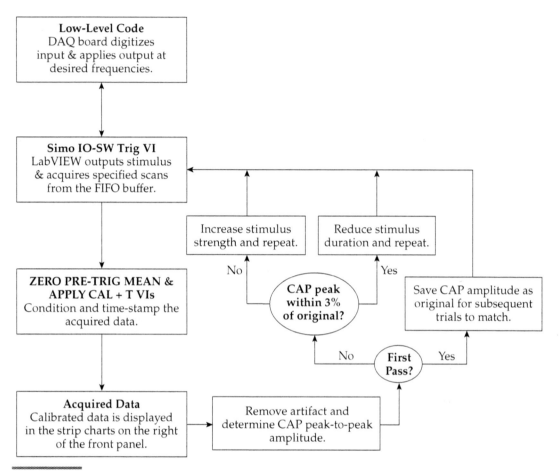

Figure 3-11.
Flowchart depicting programmatic flow of the NrvSD VI algorithm.

potential amplitude that is about one-half the maximum possible amplitude that was determined from a previous recruitment study. The algorithm then iterates through various strength-duration combinations to provide a single S-D curve that depicts the S-D relationship for a given CAP amplitude. This can be repeated for different CAP amplitudes to develop a family of curves demonstrating the global S-D characteristics of the nerve trunk. A qualifying assumption is that, as stimulus parameters are changed, a restoration to this same amplitude means that the same axons are being stimulated.

The VI automatically determines and records the peak-to-peak amplitude of the CAP for the initial voltage and duration. The stimulus duration is then

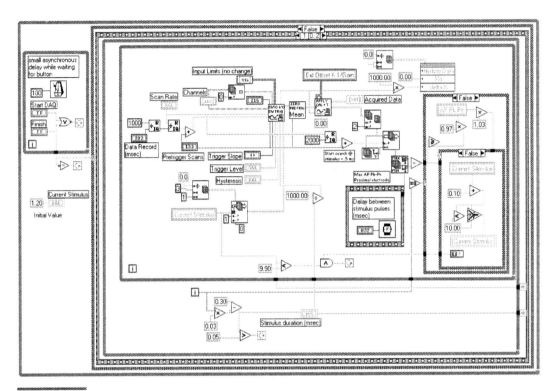

Figure 3-12.
Block diagram of the **NrvSD** VI. (Reprinted with permission of Olansen et al., 2000.)

incrementally decreased by a set amount (typically 0.02–0.05 msec). The waveform height is again determined and compared with the original. If it's smaller, as expected, the voltage level is increased and the test is rerun.

This continues until the amplitude is within 3% of the original. At that point, the voltage is recorded and the cycle is repeated at the next duration value. This continues until the duration is less than 30 sec, below which a stimulus of more than the 10 V hardware limit is usually required to generate the expected amplitude.

It should be noted that the **Apply Cal** + **t** sub-VI included is a C/C++ subroutine incorporated in a code interface node (CIN) that essentially calibrates and timestamps the recorded data. It will work as is on a Windows platform but must be recompiled on other platforms. However, this sub-VI is not at all necessary to the operation of the experiment. It is used simply to enhance charting and data storage features within these experiments, and as a simple example of code interface node (CIN) development.

Conduction Velocities

The conduction velocities determined in the recruitment study were intended to introduce the concept and demonstrate integration of various subroutines into individual VIs. In this study, the concept of conduction velocity is addressed in more detail.

Different types of nerve axons conduct an action potential at different velocities, ranging from 0.5 m/sec in unmyelinated axons to 200–300 m/sec in the fastest myelinated axons. There are well-established relationships between fiber diameter, stimulus threshold, and conduction velocity. Just as larger diameter fibers were found to require lower stimulus thresholds, they are also known to exhibit faster electrotonic conduction. This interrelationship gives some clue as to what factors affect the different properties of a nerve trunk. We refer you to any good physiology text (Berne & Levy, 1988; Ganong, 1995) for more detailed discussion of these properties. By cleverly arranging the stimulating and recording electrodes, it is possible to detect the presence of different types of nerve axons in the nerve trunk. One can also measure their actual conduction velocities and attempt to identify them.

VI Description

The **NrvCondVel** VI routine (see Figure 3-13) accepts multiple nerve electrode inputs for the purpose of evaluating the conduction velocity of the CAP of a nerve trunk. Due to the presence of multiple fiber types within each trunk, it is possible to distinguish between types via their conduction velocity. The routine discussed here allows the user to determine graphically which peaks of the waveform to consider when determining the conduction velocity. A peak detection algorithm that utilizes a polynomial interpolation scheme is then employed to improve the accuracy of the time estimate for the actual peak.

There are a number of different ways to measure the conduction velocity. One is to divide the distance between the stimulating cathode and recording electrode by the delay between the onset of the stimulus artifact and the peak(s) in the CAP. Secondly, if the CAP is recorded differentially between recording electrodes, then the compound action potential first passes over the

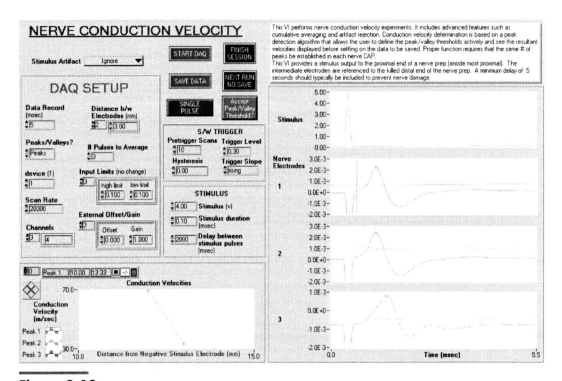

Figure 3-13.
Front panel of the **NrvCondVel** VI with typical nerve compound action potential data.

negative electrode giving a positive wave, then over the positive electrode giving a negative wave. Determine conduction velocity by dividing the physical distance between the two electrodes by the time between positive and negative wave peaks. A third means, and the one utilized here, is to record the CAP relative to a single reference point and then determine the delay in peaks between successive electrodes. These measures do not all necessarily agree. It is a good question for a student to consider which method he or she believes to be the most accurate. The student should think about what factors may contribute to the discrepancies among the types of measurement.

With the recording electrodes positioned as far away from the stimulating electrodes as possible, you may vary the stimulus intensity from below threshold until a compound action potential can be recorded and continue to increase the stimulus intensity while watching for qualitative changes in the waveform(s). It should be possible to identify groups of axons (i.e., α, β, and γ fibers) that conduct at different velocities. If you have a particularly good preparation, you may be able to identify these three or more distinct groups. It may be necessary to use the averaging function of the VI in order to see the gamma, B, and C deflections. Keep in mind that at the far end of the nerve the signal is most spread out, however the signal is attenuated because there are fewer nerve fibers at the distal end. Play around with positioning of the recording and stimulating electrodes, and try to keep track of specific bundles of fibers.

Refractory Period

No excitable cell can be stimulated continuously. There is a minimum recovery period called a *refractory period* during which the cell's ion-selective channels are readjusted and the resting potential and resting state of the membrane are restored. The phenomenon of a refractory period can be explored by applying pairs of stimuli to the nerve trunk.

The application of a stimulus pulse causes the depolarization of the axon membrane via the activation of voltage-gated sodium ion channels. This leads directly to an action potential once the threshold level is achieved. Concurrently, however, there are also voltage-gated inactivation sites on these sodium channels that prevent further sodium ion transfer until the membrane potential is restored to near its resting state. The effect of these gating channels is very well described by the Hodgkin-Huxley equations.*

While the channels are inactivated, any additional stimulus applied to the axon will have no effect on the transfer of sodium ions and, thus, will not produce an action potential. This state is known as the *absolute refractory period*. As the membrane potential of each axon returns toward its resting potential, increasing numbers of inactivation sites are disabled, allowing the sodium channels to respond to a stimulus. The amplitude of the response is directly proportional to the number of inactivation sites that have been disabled (or, the membrane potential relative to steady state). This timeframe is called the *relative refractory period* and is depicted in Figure 3-14. By studying the amplitude of the second CAP response relative to the first, the refractory period can be determined and graphically portrayed as the restitution curve.

---

*Hodgkin, A. L., & Huxley, A. F. (1952). A quantitative description of membrane current and its application to conduction and excitation in nerve. *J. Physiol.*, 117, 500–544.

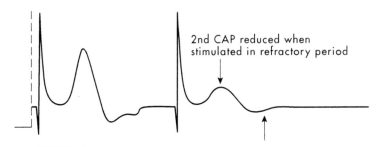

Figure 3-14.
Depiction of reduced CAP amplitude when nerve is stimulated during relative
refractory period.

VI Description

Place the recording electrodes as close as possible to the stimulating electrodes.
It may be helpful to move the stimulating electrodes to a different, more uniform section of nerve. The **NrvRefPeriod** VI algorithm automatically con-
ducts a battery of dual-stimulus tests at user-specified intervals, determines
the peak-to-peak amplitude of the second response, and creates the restitution curve (see Figure 3-15). At the smaller delay levels on the restitution
curve, it is typical to find that the compound action potential is no longer reduced continuously but drops in quantum amounts. This is due, of course, to
individual fibers dropping out and is a very good illustration of the all-or-
none property of nerves.

Another study that could easily be tacked onto this is that of summation.
Summation is a phenomenon in which two successive subthreshold stimuli,
placed close together, can elicit an action potential. Students could try setting
up the experiment to show summation in the frog sciatic. Accommodation
studies, in which a hyperpolarizing stimulus precedes the depolarizing stimulus, could also be incorporated into this experiment.

Troubleshooting the Nerve Recording

If problems occur with the nerve recordings (as is likely when dealing with biological preparations), check the obvious possibilities first, such as power applied where required and electrodes properly connected, and so on. Second,
review the areas of this book in which typical instrumentation/DAQ issues

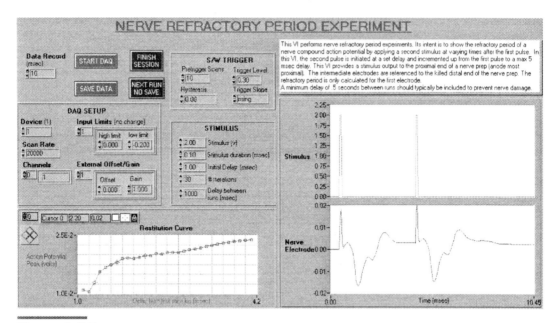

Figure 3-15.
Front panel of the **NrvRefPeriod** VI with typical nerve compound action potential and restitution data.

have been addressed for these settings. Beyond that, there are several circuits to check during recording, and you can save time by checking circuits that can be responsible for the symptom.

- **Suppose you cannot get a sweep to show up at all.**

  If you get no record on the computer screen, then somehow you are failing to trigger the recording. Check that your pulse width is at least 30 sec, and that your stimulus output is fed back as the trigger channel input. Make sure that the analog output control is set up to provide the expected stimulus and independently verify that the output is produced. (An internal short in the stimulating electrodes used in developing this lab was discovered this way).

- **You get data but you see no qualitative changes when you turn up the stimulus intensity.**

  If you see a stimulus artifact, then your recording system is working. If you do not see a stimulus artifact, then check the amplification of the channel. The height of the compound action potential and artifact varies with the quality of contact of electrodes with the nerve. You should set the amplification so that you can detect a 1 mV signal or less.

Nothing yet? Then check that your stimulus electrodes are plugged into electrodes that contact the nerve. Sometimes the electrodes are bent down, and there is air between the electrode and the nerve.

Still nothing? Check your recording system. Make sure the amplifier is not saturated (if it is due to DC offset potentials, verify that AC coupling is enabled). Make sure the amplifier is on, and in the proper mode. Check the settings. Make sure that the input is not grounded.

- **You've tried all of this and still no artifact.**

Now it is time to ask the instructor to replace your nerve chamber with one that has been tested.

- **You get an artifact but no action potential.**

You know that you are delivering a stimulus and recording properly, otherwise there would be no stimulus artifact. Again verify that the amplifier is not saturated and the nerve is resting on the electrodes. Also verify that the nerve is not unduly stretched or otherwise traumatized. If it is, it's time to use the other sciatic. If you cannot get an action potential despite all of your efforts, then ask the instructor to replace your nerve chamber with one that has been tested.

Neuromuscular Electrophysiology (Electromyography)

This study centers on the frog gastrocnemius muscle, which is innervated by a branch of a sciatic nerve. While the gastrocnemius doesn't have the fine control of the muscles of the human larynx, for example, the frog nerve-muscle preparation is nevertheless a durable and inexpensive model for muscle function. This model will enable the study of most of the important mechanisms for control of skeletal muscle function. The objectives of these laboratories incorporate fundamental animal preparation and data acquisition techniques in addition to the study of specific physiological phenomena. The properties of vertebrate (frog) skeletal muscle that can be studied include recruitment, tetanus, contractile strength, and muscle fatigue, as well as the *treppe* phenomenon (also called the *staircase* phenomenon for the initial increase and subsequent decrease in CAP amplitude during repetitive stimulus application).

All of the experiments in this section incorporate the simultaneous monitoring of the muscle compound action potential and the muscle tension as the intensity or frequency of the stimuli is varied. The VIs also include recording capability for nerve CAPs in conjunction with these neuromuscular outputs in an effort to integrate some of the lessons from the aforementioned electroneurology labs. This is only possible, of course, if the nerve is not excised as shown previously but accessed via the dissection process described later in this section.

Physiological Basis

Adult vertebrate skeletal muscle is composed of a bundle of single cells (fibers), each of which is innervated by a single motor neuron terminal. The connection between nerve and muscle is made at a single specialized site on the muscle fiber known as the *neuromuscular junction* (see Figure 3-16). The membrane at these regions of muscle fiber is modified to form a specialized structure, the motor end plate, and the motor nerve axon forms a specialized unmyelinated terminal. Each motor neuron has multiple terminals, so that it innervates from 3 to 1,000 muscle fibers (cells). The set of muscle fibers innervated by a single motor neuron is called a *motor unit* (see Figure 3-17). In general, a single muscle such as the gastrocnemius contains a large number of motor units, thus it consists of thousands of individual fibers.

The action potential conducted by a motor nerve is carried to the end plate where the neurotransmitter *acetylcholine* (ACH) is released by the nerve terminal into a space between muscle and nerve called the synaptic cleft. ACH diffuses across an approximately 20 nm gap and binds receptors on the end plate, causing a small, localized depolarization called the end-plate potential.

In the individual muscle fiber, the action potential is an all-or-none event, meaning that either it takes place or it doesn't. There are no partial action potentials. If the gross potential from the entire muscle is recorded, however, the

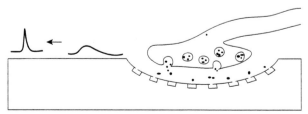

Figure 3-16.
Depiction of a neuromuscular junction.

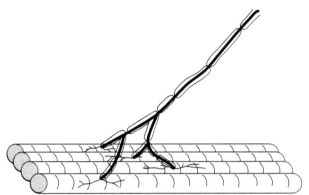

Figure 3-17.
Depiction of a single motor unit.

compound action potential is observed to vary in amplitude in accordance with the number of active motor units. The tension generated by a muscle is graded by varying the number of motor neurons recruited in a volley and the frequency with which they are stimulated.

The physiological basis for muscle contraction is given by the sliding filament theory of muscle contraction. This is based on the observation that the muscle sarcomere (the unit of force generation) contains two types of filaments, thick and thin. The thick filaments are composed of the protein myosin, and these filaments occupy the center of each sarcomere, forming the A-band (anisotropic). The thin filaments are composed of the protein actin. The actin filaments form the structures identified as I-bands (isotropic) in the light microscope (we will discuss this further in Figure 3-23).

Current theory holds that contractile force is generated in the area of overlap, or interdigitation, between the thick and thin filaments. This region of overlap can be detected within the A-band in the light microscope. In a phenomenon known as *excitation-contraction coupling*, the end-plate potential gives rise to an action potential spike that is actively propagated along the length of the muscle fiber. The action potential triggers a release of calcium from the sarcoplasmic reticulum of the muscle fiber, which in turn binds troponin, a regulatory protein associated with tropomyosin. The troponin molecules undergo a conformational change, causing the tropomyosin to expose binding sites for myosin on the actin filaments. Actin-myosin cross bridges are then made and broken, causing the muscle to contract until the calcium is re-sequestered by the sarcoplasmic reticulum. Cross-bridge binding and release continue to take place cyclically as long as ATP and calcium ions are available in sufficient concentrations, even if the muscle has fully shortened.

This equivalent of running in place can permit the muscle to maintain tension after contracting completely.

Experiment Setup

The electromyography studies are performed on an *in situ* frog nerve-muscle preparation. Nerve compound action potentials (CAPs) thus recorded are still expected to be in the 1–10 mV range, while the muscle CAPs would tend to be an order of magnitude larger, thus requiring different levels of amplification to provide input signals that cover the dynamic range of the analog-to-digital (A/D) converter. As discussed previously, performing these studies *in vivo* (i.e., nerve-muscle situated within the leg but recording made from the surface of the leg) or with the nerve encased in another type of conducting medium results in field potential recordings on the order of 10 μV.

Force Transducer Calibration

Prior to dissecting the nerve-muscle preparation, calibrate your force transducer with gram weights. The transducer used in the development of these labs is a CBSciences product, which also utilizes a preamplifier for bridge excitation and feedback. Use the FrogCal VI routine to perform a linear regression analysis on five inputs spanning the expected range of the transducer (typically 0–100 g). Hang weights up to 100 g on the transducer blade, shown in Figure 3-18, while running the FrogCal VI, shown in Figure 3-19. The VI will automatically determine the proper calibration coefficients and store them for later use. It is imperative that any external settings, such as preamp gain or offset, are not altered once calibration is complete. The experiment VIs will use these calibration coefficients to display and save the recorded data in units of grams.

Dissection

The instructor will pith a frog by inserting a dissecting needle into the spinal canal and skull. Peel off the skin on one leg from the base of the thigh. It may be helpful to refer to Figure 3-6 in the electroneurology section during the dissection. Measure and record the length of the gastrocnemius in situ, with leg fully extended then with leg fully bent. Free the gastrocnemius from the plan-

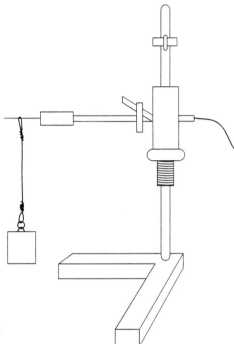

Figure 3-18.
Depiction of typical calibration
setup for force transducer.

tor surface of the foot by following the Achilles tendon to its insertion and
severing it at that point. Then free the gastrocnemius from the tibia along its
entire length. Tie a tight ligature around the upper end of the tibia just below
the kneecap. Cut the tibia just below the ligature.

Expose the appropriate sciatic nerve in the back of the thigh and free it
from the fascia and adjacent femoral blood vessel. Be careful not to damage
these vessels.

To record from muscle only: Transect the nerve at the most proximal site
obtainable.

To record from the sciatic nerve as well: To obtain both nerve and muscle
recordings, it is preferable to stimulate the nerve as proximal to the spinal column as possible. Cut underneath the urostyle (the loose bone that projects
from the end of the backbone) and cut the muscles on both sides while lifting
the end of the urostyle. You can now trace a sciatic nerve to its exit point from
the spinal column. Trace the nerve and artery through the hip region. It is important to carefully free the sciatic through the hip region to obtain access to
all descending branch nerves. These branches must be transected to eliminate
interference from the various muscles in the thigh. The nerve can be left intact

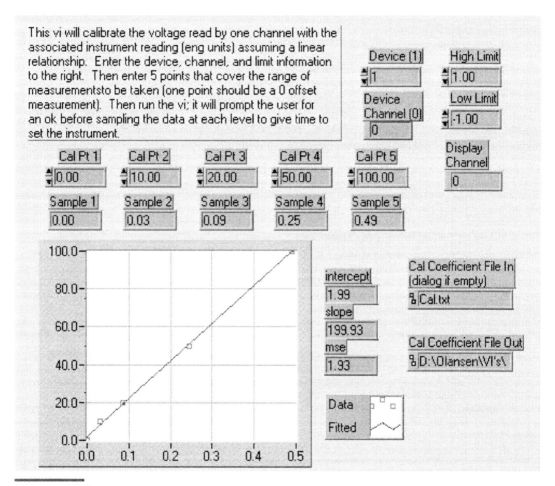

This vi will calibrate the voltage read by one channel with the associated instrument reading (eng units) assuming a linear relationship. Enter the device, channel, and limit information to the right. Then enter 5 points that cover the range of measurementsto be taken (one point should be a 0 offset measurement). Then run the vi; it will prompt the user for an ok before sampling the data at each level to give time to set the instrument.

Device (1) 1
Device Channel (0) 0
High Limit 1.00
Low Limit -1.00

Cal Pt 1 0.00
Cal Pt 2 10.00
Cal Pt 3 20.00
Cal Pt 4 50.00
Cal Pt 5 100.00
Display Channel 0

Sample 1 0.00
Sample 2 0.03
Sample 3 0.09
Sample 4 0.25
Sample 5 0.49

intercept 1.99
slope 199.93
mse 1.93

Cal Coefficient File In (dialog if empty) Cal.txt
Cal Coefficient File Out D:\Olansen\VI's\

Data
Fitted

Figure 3-19.
LabVIEW **FrogCal** Calibration VI.

or tied and cut at as high a point as possible. Keep the nerve and muscle moist with frog-Ringer's solution at all times.

Nerve-Muscle Preparation

The nerve-muscle preparation is established by inserting a parallel array of electrodes across a sciatic nerve segment and securing the proximal gastrocnemius muscle, while attaching the severed distal end to a force transducer. To accomplish this, place the frog dorsal side up on the mounting board and

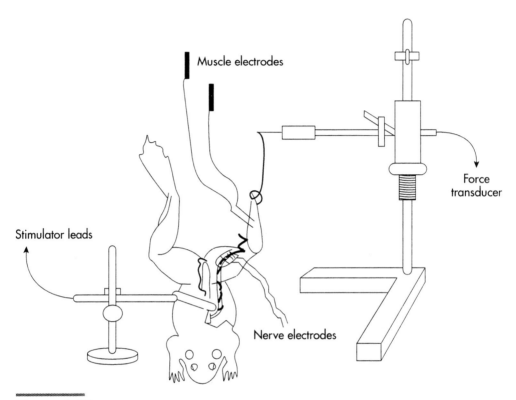

Figure 3-20.
Depiction of frog lab nerve-muscle preparation.

run a pin through the holes in the wooden columns and through the femur (proximal to the kneecap). Mount the nerve on the stimulus electrodes and the nerve recording electrodes. In developing this lab, a trough was designed of severed Tygon tubing with silver wire electrodes traversing the opening at 3 mm intervals (see Figure 3-20). The electrodes were secured and insulated and attached to a ribbon cable that was fed to the SCXI-1120. The nerve is laid across the electrodes, each of which is referenced to an electrode inserted in the thigh muscle. The muscle action potential is recorded using two lengths of fine insulated wire, each connected to a fine pin that is inserted into the muscle. The two pins should be placed about 3 mm apart near the center of the muscle and stuck well into the muscle. A ground wire is placed anywhere in contact with the animal. As soon as you verify that you can elicit a muscle twitch, protect the nerve from drying out by placing a small dab of petroleum jelly over the nerve where it contacts the electrodes.

Using surgical thread and a straight pin, connect the muscle to the transducer as you did for the weights during transducer calibration. The length of thread used should be such that with the muscle at rest length the force transducer can be vertically adjusted (e.g., using a micromanipulator) such that the muscle can be shortened to ¾ its rest length or stretched to 1½ times this length. Be sure that the thread is perpendicular to the plane of the mounting board and that the force transducer is not tilted. At rest length, the line between the muscle and transducer should be taut, but no passive tension should be observable.

Experiment Descriptions

This section describes some fundamental experiments that can be performed with the nerve-muscle preparation. As outlined previously, a variety of electromyography experiments have been developed for use in biomedical instrumentation laboratories. The following sections provide a cursory look at these representative endeavors. Detailed information regarding some of these experiments will be supplied to enable readers to use or modify the appropriate VIs and their sub-VIs, which are included on the accompanying CD-ROM. Each test description will include a brief overview of the underlying physiological phenomenon being studied and a description of the controlling VI. The VIs written for this lab include the following:

- **NMRecruit:** Muscle Fiber Recruitment, Threshold Zone
- **NMTreppe:** Muscular Stamina
- **NMTetany:** Muscle Tetanus Properties

Stimulus Threshold and Recruitment

This study focuses on generating and analyzing a compound action potential from a muscle and the resultant contraction force exerted, particularly with regard to recruitment of motor units. The phenomenon of recruitment is the result of the manner in which motor nerves are associated with muscle. As discussed previously, individual motor nerves terminate at anywhere from 3 to 1,000 individual muscle cells, each of which responds maximally in contraction to a suprathreshold stimulus. There is nothing about the quality of an individual nerve impulse that will give us variable degrees of contractile strength. The gastrocnemius is composed of thousands of muscle cells that make up hundreds of motor units. The amplitude of the muscle compound action potential is directly proportional to the number of muscle cells that are depolarized.

Recall that a sciatic nerve is composed of many individual nerve fibers, many of which are motor nerves. Recall also that different nerves had different thresholds to a stimulus. Then by varying the stimulus intensity it is possible to control how many motor nerves are depolarized with each stimulus and thus control the number of active motor units. This mechanism for regulation of tension development in skeletal muscle is called *recruitment.*

VI Description

The **NMRecruit** VI routine displayed in Figure 3-21 is used to conduct this experiment. It operates similar to the **NrvRecruit** VI described previously. The data collection is again based on the **Simo I/O-SW Trig** VI, since the recruitment study is a series of independent tests. Through the study, the stimulus is varied according to user-defined parameters, and the resulting nerve and muscle action potentials and muscle twitch tension are recorded for analysis and display. In this study, the analysis is accomplished by repeating this

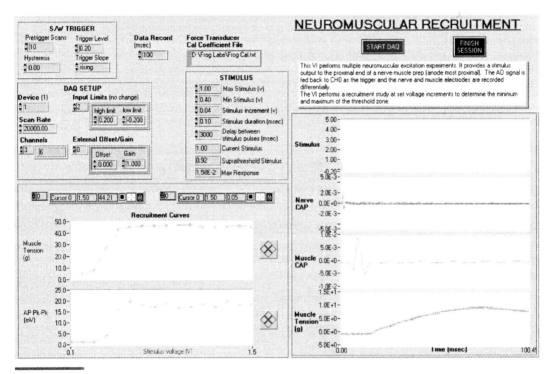

Figure 3-21.
NMRecruit VI front panel with typical data.

routine for a variety of stimulus levels and determining the maximum response in each case.

Initiation of the program compiles the software, reads the proper calibration coefficients, and waits for user input to begin collecting data. The user interface is a display of virtual instruments, most notably a set of virtual strip
chart recorders. On these, the near real-time data from the tension transducer
and the differential recordings from the muscle electrodes are displayed as
well as the nerve CAP and stimulus electrode recordings, if utilized. The user
sets the appropriate values for data collection and inputs the values desired
for stimulus duration and increment. Once this is done, the START DAQ button is pushed and the computer cycles through the entire voltage range on its
own, recording the desired data as it goes. When the experiment is complete,
the computer plots the appropriate recruitment curves on the left-hand side
of the panel. There is also a plot to correlate the recruitment of muscle fibers,
as indicated by the increasing peak-to-peak amplitude of the muscle action
potential, with the tension derived from the muscle fibers contracting.

The stimulus voltage should be increased until *both* the twitch tension *and*
the compound action potential are at maximum amplitude. Now that the
range over which the preparation can be stimulated is known, use finer increments (preferably the minimum allowable increment) to get a good number of data points covering the widest possible range of response amplitudes.
As more data is collected, the recruitment curve will become more informative. It is possible that within some ranges both the muscle compound action
potential and contractile response don't change very much. Within other
ranges, there is tremendous change over a short voltage range. That is because
within some ranges of stimulus intensity no nerve axon has a stimulus
threshold. Within other ranges, a particular group of motor nerves consists of
many axons with similar thresholds.

Staircase or Treppe Phenomenon

This lab exercise demonstrates the slow decline in output due to the lack of
relaxation time between stimuli followed by a marked increase in output due
to the buildup of Ca^{2+} in the muscle.

VI Description

The **NMTreppe** routine used to conduct this experiment is shown in Figure 3-22. Since this phenomenon requires continuous recording and stimulation over several minutes, the **Simo I/O-SW Trig** sub-VI is no longer

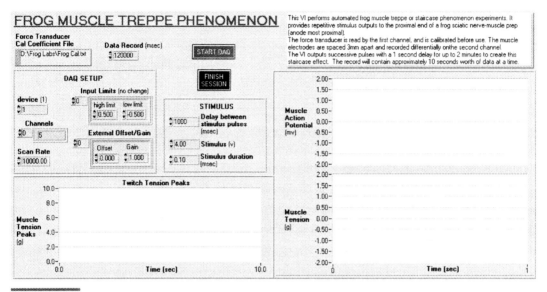

Figure 3-22.
Front panel of **NMTreppe** VI.

appropriate. Instead, a separate algorithm was developed to continuously record and stimulate according to user inputs. While this works well for these longer duration experiments, it does not include the conditional retrieval that allows for more precise recordings.

Allow the muscle to rest for several minutes. From the recruitment study, determine the stimulus intensity needed to get a near maximum twitch tension response. Set the stimulus pulse spacing to 1 sec. After a rest, stimulate the muscle continuously for a couple of minutes. What happens to pulse height? A phenomenon known as the staircase effect should be evident in which twitch tension will drop, then rise well above initial levels. This phenomenon, also known as Treppe, is related to the release and re-uptake of Ca^{2+} by the sarcoplasmic reticulum. In weaker preparations the staircase effect is masked by the decline in twitch tension due to fatigue. If it is not possible to demonstrate a staircase effect, wait until another muscle is prepared. There is a related phenomenon that will be discussed in the tetanus study.

Tetanus Phenomenon

The **NMTetany** VI automates the sequential stimulation of the nerve-muscle prep at increasing frequencies to demonstrate tetany within the muscle. Explaining why twitch tensions did not return to baseline with high-frequency stimulation may seem easy at first. The muscle simply did not have time to relax, that is, return to rest length, before the next stimulus. However, what is the *physiological* definition of muscle relaxation? Put another way, what must happen intracellularly in order for cross bridges to stop pulling, allowing the muscle to relax? How were elevated tensions obtained simply by increasing the frequency of contraction? As discussed earlier, contraction is all or none, and the same number of motor units is stimulated throughout the experiment. The explanation for tetanus is similar to that for the staircase effect. Consider that for a given twitch the concentration of free calcium in the sarcoplasm and the *duration* of elevated calcium concentration are important factors in determining the length of time the muscle cell is in the active state. The degree of tension is in fact related to duration of the active state.

VI Description

The **NMTetany** VI used to conduct this experiment is shown in Figure 3-23. Like the treppe study, this study utilizes a continuous stimulation/acquisition routine to collect the required information. Set the stimulator voltage to

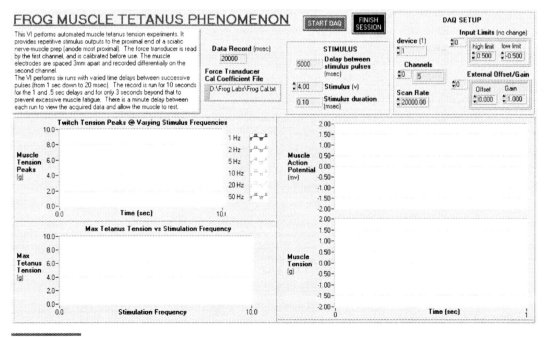

Figure 3-23.
Front panel of **NMTetany** VI.

produce a maximal *twitch* tension. The VI will cycle through various pulse spacings, from 1.0 sec through 20 msec. Each test is conducted briefly (10 sec for 1 and 0.5 sec spacings, 3 sec for the rest) while recording the sequence of twitch tensions and compound action potentials. For pulse intervals of 200 ms or less, the tests should only be allowed to run for up to 3 seconds to avoid fatigue of the neuromuscular junction. The results of the tests are displayed in plots of peak tension (in grams) obtained by the muscle and peak height attained by the compound action potential as a function of stimulus *frequency* (not pulse spacing).

(Note: The next two experiments can be conducted with the VIs given. However, they have not been automated as previous experiments have been, and more data recording and reduction is required.)

Isometric Length-Tension Relationship

This study addresses the relationship between the resting state of a muscle and the amount of contractile force it is able to produce. Muscle shortening is due to a rapid making and breaking of actin-myosin bonds, which propel the thin filaments further into the A-band. It was noted earlier that this cycle continues even when the muscle is fully shortened, that is, tension can be sustained as long as the muscle is in the active state. In this situation, the force that is generated is proportional to the number of bonds participating in the actin-myosin reaction. When the resting muscle is stretched and held in that position, the thin filaments slide out of the A-band and the region of overlap is diminished. In a completely isometric contraction this reduces the number of opposed actin and myosin reactive sites and thus reduces the tension the muscle is able to develop.

During this experiment, with each stimulus all muscle cells enter the active state, cross bridges are continually made and broken, and the muscle shortens until it cannot shorten any more. Each sarcomere contracts from the rest position (Figure 3-24A) to a position of maximum shortening (Fig 3-24C). Peak tension increases as the rest length of the muscle is increased. This is the isotonic portion of the length-tension curve, which we will study in more detail in the next experiment. At longer rest lengths the muscle does not have sufficient power to shorten completely (Figure 3-24B). It will partially shorten then hold tension until its supply of ATP is reduced, you turn off the stimulus, or something stretches or snaps.

By preventing the sarcomeres from contracting completely you prevent some of the actin-myosin cross bridges from forming. Therefore the total

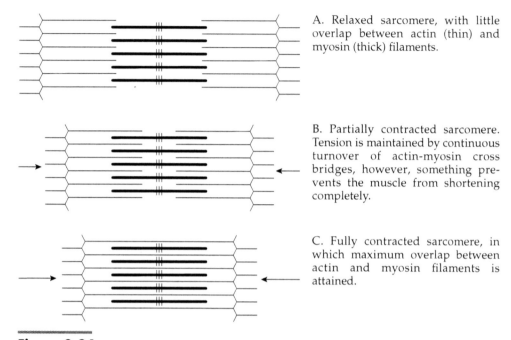

A. Relaxed sarcomere, with little overlap between actin (thin) and myosin (thick) filaments.

B. Partially contracted sarcomere. Tension is maintained by continuous turnover of actin-myosin cross bridges, however, something prevents the muscle from shortening completely.

C. Fully contracted sarcomere, in which maximum overlap between actin and myosin filaments is attained.

Figure 3-24.
Sarcomere contractile states.

number of cross-bridge interactions declines as you continue to increase passive tension on the muscle. The model suggests that isometric tension is proportional to the number of active cross bridges. You will plot passive, active, and total tension versus length in order to test this model.

VI Description

This study can be accomplished using the **NMTetany** VI modified to only operate at a single user-specified pulse interval. The data from each run can then be recorded for offline analysis. It will be necessary to accurately change resting lengths by as little as 0.5 mm. *For this experiment you must have as little distortion of the transducer as possible.* It may be wise to recalibrate the force transducer at this time. Throughout this experiment, provide a suprathreshold stimulus to ensure that all of the motor units participate in each contraction.

Stimulate continuously at 20 pulses/sec for a few seconds. Observe the tetanus tension just long enough for peak tension to develop. It is imperative that you do not fatigue the nerve-muscle junction, so as soon as tension starts to drop, stop stimulating. After a short wait repeat the previous steps, but increase the frequency to 40 pulses/sec. Continue to increase frequency until the maximum possible tetanus tension is obtained. In this state, all motor units of the muscle are in the active state and cross bridges will be continually made and broken without letting up. The muscle is working as hard as it can.

Use the vertical drive of the micromanipulator to lower the transducer so that the muscle is permitted to passively shorten to rest length. This is when the string is clearly no longer taut. Start recording, stimulate to peak tetanus tension, then stop the stimulus and allow the record to return to baseline. This generates the first data point of a length-tension curve, namely a = length in mm, b = passive tension = 0, and c = peak tension, where a is the measured rest length and c is the peak tension generated during stimulation.

Now increase the muscle length by 0.5 mm and stimulate again as previously directed. Your next set of data points will be a = rest length + 0.5 mm, b = passive tension, and c = peak tension, where b and c are the tension measurements before stimulating and at peak tension development, respectively. Continue to increase rest length in 0.5 mm increments, recording the length, passive tension, and peak tension each time. Note that for the first few data points the passive tension may remain at 0 until you begin to actually stretch the muscle. To obtain fairly consistent data, try to wait the same amount of time between application of the stimulus, so you stimulate, wait a set time, stimulate, wait, and so on.

Continue recording until the active tension generated by the muscle is dramatically reduced. At this point you will have damaged the tissue, and it will be necessary to use the other gastrocnemius or a new frog for the next experiment. For your results, prepare a single graph of passive tension, peak tension, and active tension as a function of muscle length. Active tension is total maximum tension minus passive tension. Prepare a second graph of active tension as a function of passive tension.

Isotonic Length-Tension Relationship

This study will explore yet another way in which muscle tension is regulated. It should be noted that there is no such thing as a purely isometric or isotonic contraction in muscle, *in situ*. All muscle movements are a combination of the two.

The force of contraction of a skeletal muscle pulling on the elastic elements can be likened to a piston pulling on a spring. The opposing force is directly proportional to the distance of shortening. What determines the degree of isotonic tension? Since the muscle is permitted to shorten, maximum cross-bridge overlap is achieved, therefore number of active cross bridges is not a factor. Consider that the muscle bundle is held together by connective tissue and that the ends of the muscle are anchored by tendons. Those structures, and the bending transducer leaf in the experiment, provide a passive elastic force that opposes the tension developed by contraction.

In heart muscle the isotonic length-tension relationship is referred to as the Frank-Starling Law of the Heart. More blood filling the left ventricle between beats means a greater stretch on the heart muscle, which is striated like skeletal muscle. Greater stretch means a greater force of contraction, and the additional volume is efficiently pumped. Applied to skeletal muscle, the law simply states that *within physiological limits* the greater the resting length of the muscle the greater the resulting force of contraction as the actin and myosin filaments reach a more optimal degree of interdigitation for achieving contraction. Serious weightlifters are very aware of the phenomenon (but probably not at the molecular level).

VI Description

Repeat the previous experiment with a new muscle. This time you need to obtain the maximum possible *twitch* tension, not tetanic tension. Start with the shortest rest length for which you can measure a twitch and set up to stimulate at one stimulus per second. Record about one dozen twitch tensions then stop the stimulus, allowing the muscle to fully relax. After recording the passive tension on the relaxed muscle, increase the muscle passive length as before. Again record the new passive tension before stimulating again. Under these conditions the muscle is permitted to shorten completely each time, and the active state is terminated with each twitch. This explores the isotonic portion of the length-tension curve. When active tension reaches an obvious plateau or begins to decline, terminate the experiment.

Plot the data two ways as you did for the isometric length-tension data. Note the distance over which the force of contraction continued to increase. Also note the active tension at the point at which a passive tension could first be recorded. Note how far, in mm, the muscle was stretched in order to attain maximum active tension. It is possible to have had a better isotonic portion to the curve when you performed the isometric experiment.

Additional Study: Fatigue

So, you've had a hard day. You're feeling a bit tired, not quite ready to run the 100-yard dash. If you aren't too bushed, try the following experiment to determine if it is likely that fatigue of the neuromuscular junction contributes to this weak feeling. Set up a second stimulator to stimulate the muscle directly.

With just enough tension on the muscle to measure a satisfactory twitch, stimulate the nerve at the minimum frequency previously required in order to obtain complete tetanus. Do the same for the muscle to verify that both stimuli are adequate. Stimulate the nerve while measuring both tension and compound action potential, until active tension drops to zero.

Now, stimulate the muscle without turning off the nerve stimulus. Is the action potential restored? How about the tension? What is the likely source of fatigue?

Troubleshooting the Nerve-Muscle Preparation

Here is a list of things to try just in case you run into an unresponsive preparation.

- **The muscle doesn't twitch in response to a stimulus.**

  Never mind the chart record for now. You should be able to *see* the muscle twitch when you stimulate the sciatic nerve. If it does not twitch, check that the stimulator is set up properly, as with the nerve study. The remaining possibility is that the nerve is damaged or does not make good contact with the hook electrodes. Check that the electrodes do not touch anything but the nerve (the stimulus may be grounded out). Check that the nerve directly contacts the electrodes and that they are separate. Try reversing the leads; the negative electrode should be distal (toward the muscle). Move to another part of the nerve if you can.

- **The muscle twitches but you see nothing on the recorder.**

  Watch the transducer leaf as the muscle twitches. Does it move? Gently bend the leaf with a finger. If you get no action on the record, check that everything is on including the SCXI and any preamps being used. The position control for your force transducer should move the trace on the record. If not, check that the input is not grounded. Check that the trace isn't pegged, or bottomed out. It will not respond if its rest position is far off the scale. Finally, increase the sensitivity of your recording.

- **The muscle twitches, but it will not settle down.**

  Between twitches the muscle should quickly return to baseline tension and remain still. If the muscle fails to return to baseline or appears to squirm or twitch between stimuli, a stray stimulus or bias voltage is probably reaching the sciatic. Check the ground connections throughout the system. Check that the preparation is properly grounded.

- **You do not get a compound action potential.**
  Check that the channel receiving the signal is indeed recording, and check sensitivity. Check that the amplifier is on. Check all connections. You are bound to get some kind of signal, even if the pin electrodes aren't in the muscle.

- **The signal on your action potential channel is going crazy.**
  One or both of your pin electrodes is out of the muscle. Stop the recording and secure the electrode(s) in the muscle.

- **You get a compound action potential, but it is noisy.**
  This is the most persistent problem in the muscle lab. The electrodes are small and the bare wires pick up all kinds of electrical interference. Check the filter settings on your amplifier. Check that the preparation is grounded. Try putting a ground clamp on the force transducer stand. Check that the Faraday cage (reference Figure 3-1) is properly grounded.

Cardiac Electrophysiology (Electrocardiography)

Cardiac electrophysiology is dedicated to the study of the electro-chemical activity of the heart. Studies include electrical activation of individual cells as well as the system-level activation, which results in normal or abnormal heart rhythms. The electrocardiogram (ECG) is a recording of the electrical activity of the heart. The ECG can serve in diagnosis to provide a variety of information regarding the heart's normal function, as well as predict abnormal function in disease. Similarly, it provides information about heart rhythm abnormalities. The vectorcardiogram (VCG) also provides information regarding cardiac electrical activity. In recording the VCG, two approximately orthogonal scalar ECG leads that look across the heart horizontally and from top to bottom are cross-plotted (Ganong, 1995), resulting in a loop that represents the time-varying electrical activity of the heart.

Experiments are established to demonstrate the applicability of the ECG and VCG simultaneously, in real time. These tools are a primary focus in the initial training of students in the field of cardiology. The use of signal processing tools, such as digital filtering, is also introduced in these labs as a demonstration of practical applications of fundamental engineering concepts.

Physiological Basis

In the heart, contraction of the atria (*atrial systole*) is followed by contraction of the ventricles (*ventricular systole*). The heartbeat originates in a specialized conduction system and spreads via this system to all parts of the myocardium. Figure 3-25 schematically shows the major structures that make up the heart's specialized conduction system.

The *sinoatrial node* (SA node), located at the junction of the superior vena cava and right atrium, is the normal cardiac pacemaker. The *internodal pathways*, which contain three bundles of atrial fibers containing Purkinje type fibers, connect the SA node to the *atrioventricular node* (AV node), which is located in the right posterior portion of the atrial septum. The AV node is continuous with the *bundle of His*, which branches into the *left bundle branch* (LBB) at the top of the ventricular septum and the *right bundle branch* (RBB). The LBB divides into the anterior and posterior fascicles. These branches and fascicles run subendocardially down either side of the ventricular septum and come into contact with the Purkinje system, whose fibers spread throughout the ventricular myocardium.

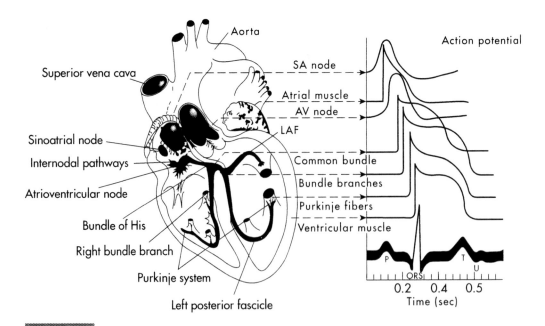

Figure 3-25.
Schematic representation of the electrical activity of the heart.

The normal heartbeat initiates at the SA node. Depolarization initiated in the SA node spreads radially through the atria, and then converges on the AV node. During atrial depolarization, which completes in about 0.1 second, the atria contract. After a delay (AV nodal delay) of about 0.1 second, excitation spreads to the ventricles. From the top of the ventricular septum, the depolarization wave spreads in the rapidly conducting Purkinje fibers to all parts of the ventricles. Depolarization of the ventricular muscle first starts at the left side of the septum and then moves to the right across the mid-portion of the septum.

The wave of depolarization then spreads down the septum to the apex of the heart. It returns along the ventricular walls to the AV groove, from the endocardial (inner) to the epicardial (outer) surface of the ventricles. The last portions of the heart to be depolarized are the posterobasal portion of the left ventricle, the pulmonary conus, and the uppermost portion of the ventricular septum. The whole sequence of depolarization of the myocardium forms the basis of the coordinated contraction of the ventricles. This coordinated contraction can be disrupted by a variety of means such as pacing (Little et al., 1982).

Experiment Descriptions

As was the case in the electroneurology and electromyography discussions presented earlier, a wide array of electrocardiography experiments can be developed for use in biomedical instrumentation laboratories. Some representative examples developed as part of a graduate research lab are briefly discussed next. Detailed information regarding some of these experiments will again be supplied to enable readers to use or modify the appropriate VIs and their sub-VIs, which are included on the accompanying CD-ROM.

This section describes some fundamental experiments that can be performed in the field of electrocardiography. Each test description will include a brief overview of the underlying physiological phenomenon being studied and a description of the controlling VI. The VIs written for this lab include the following:

- **VectorCardio:** Combined ECG and VCG Analysis

- **HRV:** Heart Rate Variability Studies

- **PRC:** Phase Response Curve Analysis

> In deference to the disclaimers annunciated at the beginning of this book, experiments conducted with these VIs should be done with appropriate protocol approvals (from the appropriate Institutional Review Board, or IRB) and sufficient electrical isolation and equipment safety. While they were all designed for use in animal studies, they are readily adaptable to human research. Proper procedures should be incorporated by knowledgeable people to ensure the safety of human subjects.

ECG/VCG

The electrocardiogram (ECG) is the primary focus of investigations in this area. As such, experiments are designed to record ECGs from a variety of animal preparations or from human subjects. Issues such as comparing open chest data with surface electrode recordings are encountered as a further extension of the volume conductor problem discussed previously. A second tool introduced in the lab is vectorcardiography, wherein two approximately orthogonal scalar ECG leads that look across the heart horizontally and from top to bottom are plotted on conventional Cartesian axes (Whitfield, 1964). The ECG leads are placed according to the standard Einthoven's Triangle depicted in Figure 3-26.

The result is a loop in the lead space (horizontal ECG vs. vertical ECG; see Figure 3-27), which represents the time-varying electrical activity of the heart. At each instant of time, the net electrical activity of the heart can be represented as an equivalent current dipole source located in the body volume conductor. In the vectorcardiograph (VCG), the electrical center of the cardiac activity is placed at the center of the coordinate system. The current dipole vector extends outward from that center and, with time, its tip traces a locus in the lead space. Since the origin of the dipole vector is fixed at the coordinate center and its orientation is variable, one may visualize the net electrical activity of the heart (see Figure 3-27). Normally net dipole vector activity is coincident with the long axis of the loop, and it is directed downward and to the right in the plot. The electrical axis of the heart, shown as an angle in Figure 3-27, depicts the orientation of the net dipole vector and is a clinically relevant parameter often utilized by physicians to diagnose electrical abnormalities in the heart.

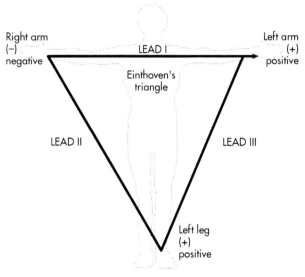

Figure 3-26.
Schematic representation of
ECG lead placement according
to Einthoven's Triangle.

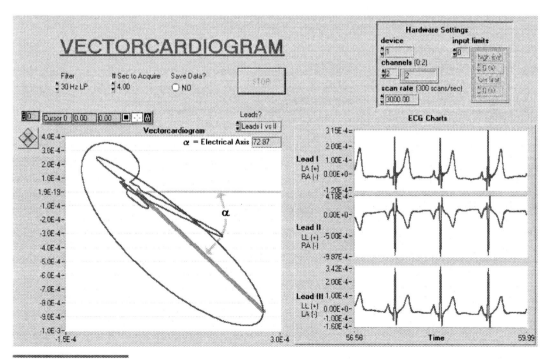

Figure 3-27.
VectorCardio VI illustrating typical ECG and vectorcardiograph waveforms.

VI Description

The **VectorCardio** VI is a relatively simple implementation of the vectorcardiogram concept. Figure 3-28 depicts the block diagram designed for a simple study of the ECG and VCG. It is assumed that the Einthoven lead system is utilized to obtain the ECG signals. The routine uses a loop to continuously read the analog input ECG waveforms and filters the data to remove noise, if desired. The **RR Intervals** sub-VI is used to isolate the data into individual beats, which provides for a clean display of the vectorcardiogram. The data is concurrently streamed to disk. This straightforward experiment readily exemplifies the ease with which LabVIEW™ can be utilized to develop biomedical laboratory applications.

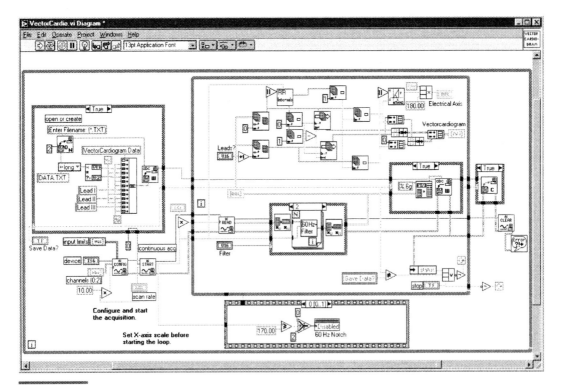

Figure 3-28.
Block diagram of the **VectorCardio** VI.

Heart Rate Variability

More advanced experiments have also been developed that analyze the potential source of the cardiac beat. One field that has received growing attention through the past 25 years is the study of Heart Rate Variability (HRV), which can be simply defined as an assessment of the *deviations* in the time interval (also known as the RR interval) between successive cardiac beats. Beat-to-beat variability of heart rate is currently the subject of significant research efforts. The research has been spurred through evidence of a significant relationship between the autonomic nervous system and cardiovascular mortality, including sudden cardiac death (Levy & Schwartz, 1994). This relationship can be exploited in the assessment of cardiovascular (and noncardiovascular) pathophysiologies and the concomitant role of fluctuating autonomic nervous control. Ongoing studies in this field should continue to "enhance our understanding of physiological phenomena, the actions of medications, and disease mechanisms" (Task Force of Eur. Soc. Cardiology and Nor. Am. Soc. Pacing and Electrophys., 1996).

Although specialized cardiac tissue, particularly the SA and AV nodes, has intrinsic pacemaker functions, the autonomic nervous system still exerts significant control over the heart rate and rhythm (Jalife & Michaels, 1994). Similar to the *neuromuscular junction* discussion earlier, the release of acetylcholine (ACH) via the vagus nerve triggers the muscarinic ACH receptors within the cardiac tissue, causing a conformational change in the various gated ion channels. This results in a slow diastolic depolarization. Conversely, the sympathetic nervous system influences heart rate through the action of epinephrine and norepinephrine, yielding an acceleration of the diastolic depolarization. Thus, the parasympathetic and sympathetic inputs cause conflicting reactions and are continuously engaged in the fine-tune control of cardiac rhythm, with vagal tone typically dominating the resting heart rate.

Time domain and frequency domain methods have been developed for the analysis of HRV. Spectral analysis of HRV has increased the understanding of neural modulation mechanisms on cardiac pacing. Clinical and experimental studies of short-term (2–5 min) HRV recordings have shown that the high frequency (HF) components of the HRV power spectrum (0.15–0.4 Hz) are modulated almost entirely by parasympathetic innervation and strongly reflect the respiratory frequency. Conversely, low frequency (LF) components are influenced by both vagal and sympathetic activity and are largely dependent on external variables such as posture (Mainardi, Bianchi, & Cerutti, 1995).

VI Description

The **HRV** VI utilizes both the aforementioned time and frequency domain techniques to readily display experimental results, as illustrated in Figure 3-29. This VI simply establishes a means for tracking and analyzing HRV and applying system identification techniques to quantitatively characterize the variations. The data acquisition and streaming capabilities are not substantially different from the **VectorCardio** algorithm detailed earlier. The use of signal processing tools, such as spectral analysis and digital filtering, is introduced in this VI as a demonstration of practical applications of fundamental engineering concepts. However, this book does not discuss these concepts in detail, as the user requires a substantive level of understanding before implementing these techniques that is beyond the current scope. Thus, we refer you to any number of signal processing texts for further information.

Phase Response/Vagal Inhibition

Another means of analyzing the electrophysiologic properties of the heart is by applying the principles of perturbation theory to the oscillatory nature of single-cell, multicell, or whole-system preparations. Contributions to the

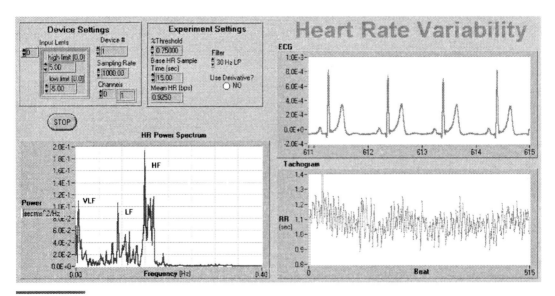

Figure 3-29.
HRV VI illustrating the typical power spectrum of a heart rate tachogram.

study of phase sensitivity with respect to the mammalian sinoatrial (SA) node date back to the 1930s when Brown and Eccles published their work on vagal inhibition in cats (1934). By inducing a perturbation stimulus (with all other external stimuli blocked), biological oscillators will often exhibit a phasic dependency of the perturbation effect on the oscillation itself. Depending on the time of the perturbation delivery, the oscillatory cycle can be either advanced or delayed.

Numerous methods exist for the analysis and display of the phasic characteristics of pacemaking cell preparations. These include transient phase response curves (PRCs), inhibition curves (ICs), phase transition curves (PTCs), and phase resetting curves (RCs). In the current study, only PRCs and vagal ICs were considered. A PRC is a measure of the normalized change in the first pacemaker period following a single vagal stimulation. An IC is a scatter plot that considers the transient response of the pacemaker cell over several succeeding cardiac periods.

VI Description

While voltage-clamp methods are typically employed for collecting data at the cellular level, the included VI was designed specifically for the *in vivo* assessment of a vagotomized animal (rat) preparation. The intent of the VI is to automate the sometimes tedious process of stimulating and recording these dispersed pieces of data. While developed for *in vivo* experiments, the basic concepts should be readily adaptable to single or multiple cellular preparations. The primary differences would exist in the physical experiment setup.

The **CardiacPRC** VI front panel is shown in Figure 3-30. The source code is included on the CD-ROM. It is necessary to take some time tracing the steps within the VI block diagram to understand the approach utilized. The algorithm essentially automates the process by triggering the stimulus pulse off the ECG, or similar, recording. Because our preparations were open chest, we found that recording the electrical activity near the SA node in the upper right atrium served as a quality signal for initiating the stimulus trigger. After recording the RR intervals for the current cycle and several beats beyond the stimulus, and allowing the heart rate to return to its rest state, the loop is reinitiated with an increase in the delay time between the QRS peak and the stimulus pulse. This process continues until vagal stimulation has occurred at discrete points throughout the cardiac cycle. Varying the stimulus delay in this manner provides a direct means of assessing the phase characteristics in response to the vagal perturbations.

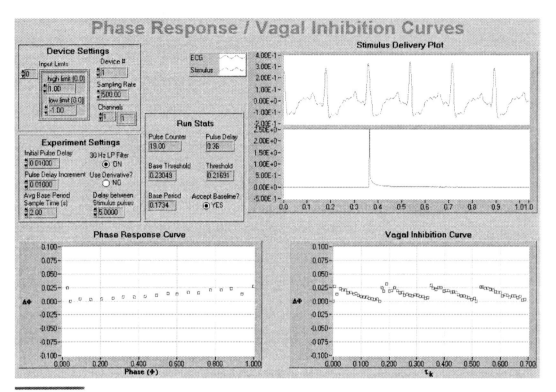

Figure 3-30.
CardiacPRC VI designed for analysis of the phasic characteristics of the cardiac sinoatrial node using *in vivo* animal preparations.

REFERENCES

Berne, R. M., & Levy, M. N. (1988). *Physiology*. St. Louis, MO: Mosby.

Brown, G., & Eccles, J. (1934). The action of a single vagal volley on the rhythm of the heart beat. *J. Physiol., 82*, 211–240.

Caprette, D. (1998). BIOS 315 lab module in physiology course notes. Rice University.

Ganong, W. F. (1995). *Review of medical physiology*. Norwalk, CT: Appleton & Lange.

Greco, E. C., & Clark, J. W. (1977). The field from an isolated nerve in a volume conductor. *IEEE Trans. Biomed. Eng.* Vol. BME-24, pp. 18–23.

Jalife, J., & Michaels, D. C. (1994). Neural control of sinoatrial pacemaker activity. In M. N. Levy & P. J. Schwartz (Eds.), *Vagal control of the heart: Experimental basis and clinical implications* (pp. 173–205). Armonk, NY: Futura.

Levy, M. N., & Schwartz, P. J. (Eds.). (1994). *Vagal control of the heart: Experimental basis and clinical implications.* Armonk, NY: Futura.

Little, W. C., Reeves, R. C., Arciniegas, J., Katholi, R. E., & Rogers, E. W. Mechanisms of abnormal interventricular septal motion during delayed left ventricular activation. *Circulation, 65*(7), 1486–1491.

Mainardi, L. T., Bianchi, A. M., & Cerutti, S. (1995). Digital biomedical signal acquisition and processing. In J. Bronzino (Ed.), *The biomedical engineering handbook* (pp. 828–852). New York: CRC Press.

McGill, K., Cummins, K., Dorfman, L., Berlizot, B., Luetkemeyer, K., Nishimura, D., & Widrow, B. (1982). On the nature and elimination of stimulus artifact in nerve signals evoked and recorded using surface electrodes. *IEEE Trans. Biomed. Eng.* Vol. BME-29, No. 2, p. 129.

National Instruments. (1998). *LabVIEW data acquisitions basics manual.* Austin, TX: Author.

Olansen, J. B., Ghorbel, F., Clark, J. W., & Bidani, A. (2000). Using virtual instrumentation to develop a modern biomedical engineering laboratory. *Int. J. Eng. Educ., 16*(3), 244–254.

Task Force of Eur. Soc. Cardiology and Nor. Am. Soc. Pacing and Electrophys. (1996). Heart rate variability: Standards of measurement, physiologic interpretation, and clinical use. *Circulation, 93*(5), 1043–1065.

Whitfield, I. C. (1964). *Manual of experimental electrophysiology.* New York: Pergamon.

LabVIEW Automates Brain Wave Experiments in the Neurophysiology Lab

by Roman Golubovski, MSc Dipl-EEng, CSEE, BME

The Challenge: Automating brain wave experiments in the neurophysiology lab.

The Solution: Building a PC-based virtual instrumentation system using a DAQ board controlled by LabVIEW.

Introduction

Brain potentials are divided into spontaneous and event-related. The spontaneous result from the regular brain activity (EEG potentials). The event-related potentials (ERP) result from external brain excitation (events) and are divided into evoked and anticipatory. Evoked potentials appear after excitation as a reflex of the brain. Anticipatory potentials appear before the corresponding event and represent an expectation of the same and usually a motor preparation process for it in the brain. The most prominent example of the expectation-related potential is the contingent negative variation (CNV) potential. It is extracted from the subject's EEG within the CNV experiment.

The CNV experiment is based on the CNV paradigm, which applies two brain stimuli (S_1/S_2) to the subject with a constant interstimulus interval. S_1 is a warning stimulus, and S_2 is an imperative stimulus to which the subject must react. The procedure is repeated tens of times, during which an ERP, produced in the EEG trace between the stimuli, shapes itself toward a specific CNV wave. The ERP after 10 to 20 trials can clearly show both components—the evoked (short) potential due to S_1 as well as the anticipatory (late, expectancy) potential together with the preparatory potential prior and due to S_2.

The dynamic CNV (DCNV) experiment is an extension of the CNV experiment. The extension involves switching S_2 on and off, which occurs automatically after fulfilling certain conditions in the experiment's environment, thus forcing a cyclic process of building and degrading of the CNV wave. The subject is not informed about the nature of both stimuli, so the expectation of appearance (absence) of S_2 during the experiment completely corresponds to the learning process.

The CNV wave (extracted ERP) can be qualified by one of its parameters, such as amplitude or slope. After the experiment, a statistical curve of the qualifying parameter is drawn across the trials. This statistical curve is denoted as the electroexpectogram (EXG) and directly presents the subject's cognitive capabilities.

Experiment Design

"AEP Research Tool," written in LabVIEW, embraces the acquisition, signal processing, and analysis of the EEG and EOG traces (latter used for validation of the EEG against artifacts) as well as reporting. The hardware used in the experiment consisted of a bandpass am-

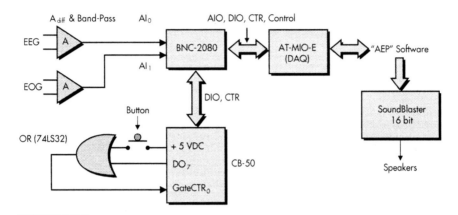

Figure 3-31.
Hardware configuration of the DCNV experiment.

plifier for mV ranges, an AT-MIO-E board, a sound card for applying the stimuli, and a push button switch with a TTL interface (Figure 3-31).

The system acquires two differential analogue channels (EEG and EOG). We use audio with S_1 being a short 1 kHz warning beep and S_2 being a longer 2 kHz imperative beep. It is essential that the subject is aware neither of the nature nor of the number of the stimuli. The acquisition lasts for 7 s. S_1 is issued in 1 second into the acquisition; S_2 is issued in 3 s if applied by the algorithm. During the experiment, the subject learns about the number, nature, and order of the stimuli, thus demonstrating the process of learning by shaping the ERP wave toward the expected CNV. The subject has to react upon hearing S_2 by pressing the button and immediately stopping it. This is to prevent falling asleep and lowering of concentration. The number of trials in the experiment is set to maximum 100 successful (120 trials total). The gap between two consecutive trials varies from 12 to 15 s to avoid timing determinism. The criterion for ERP being a CNV is defined above. After three consequent CNVs are detected, S_2 is turned off, and the subject learns to forget the imperative stimulus, thus lowering the value of the CNV-qualifying parameter. After three consequent not-CNVs, S_2 is turned on again, and so on. The EOG trace is used for automatic validation of the EEG trace against artifacts defined as voltage sequences longer and higher than preset thresholds. There is a second manual criterion applied, where the operator can reject current EEG if artifacts are recognized visually for 3 s after the acquisition. Rejection of such trials is necessary since the process of extraction of the ERP uses a cumulative iterative FIR filter that averages the acquired signal by ensemble, so every artifact that passes it will influence the extracted ERP until the end of the experiment.

Algorithm Design

The software package written in LabVIEW works under Windows NT/95. The main panel (Figure 3-32) shows the acquired EEG signal (current trial), the extracted CNV potential and

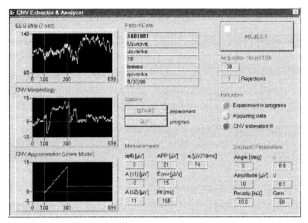

Figure 3-32.
Main DAQ panel.

its linearized model, as well as the required measurements and calculated values. A vertical marker on the CNV Morphology represents the reaction time. All of the subject's acquired and calculated data is saved in an ASCII file. The optimal filter is defined as follows—

$$CNV_i = d \cdot CNV_{i-1} + c \cdot EEG_i$$

"AEP Research Tool" is completely hardware-synchronized. Acquisition is onboard clocked. The audio stimulation is performed through a sound card. The reaction time is measured by an onboard counter, started by a digital output from the card issuing pulse at the same moment with the start of S_2 and stopped by the user pressing the button or the time-out pulse, applied again by the same digital line.

Conclusion

The EXG curve represents a cognitive wave obtained from the human brain showing the oscillatory change of the expectancy status in the human brain and is a manifestation of the expectation process and the learning process taking place in the human brain, during the DCNV paradigm.

"AEP Research Tool" successfully performs the expected task and produces an excellent nine-page report containing all required statistics. Experiments during the test phase indicated that different categories of subjects (healthy adults, groups of people with distinctive neurological disturbances, and little children) may produce quite different electroexpecto-grams, but very similar within the groups, because some are not able to raise their CNV waves above the thresholds.

The design of this experiment illustrates unlimited possibilities of LabVIEW-based systems in the medical field.

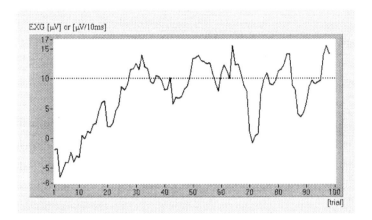

Figure 3-33.
Typical EXG curve.

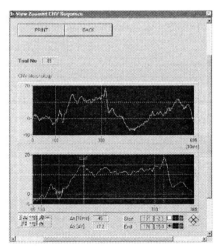

Figure 3-34.
Morphology analysis of a specified trial.

For more information, please contact
Roman Golubovski, Electrical and Biomedical Engineer, J. Sandanski 116-3/24, MK-1000
Skopje, Macedonia,
email roman.golubovski@iname.com
tel + 389 (0)91 165 367
voice/fax + 389 (0)91 165 304

Using LabVIEW for Physiological Research

By Paolo Garzella, Ph.D.

The Challenge: Automating electrical measurements used in physiological research.

The Solution: Using a virtual instrument system with GPIB and DAQ boards controlled by LabVIEW.

In a muscle physiological laboratory, several electronic instruments are required to perform physiological research. These instruments include an oscilloscope, a digitizer, an electric stimulator for the muscle fibers (both single and train of stimuli), motion controller, and other devices such as motors, force transducers, and function generators.

The instruments, which function as single, stand-alone devices capable of performing singular tasks, lack the flexibility required to perform various tests necessary for complete physiological testing and research. By replacing these singular instruments with LabVIEW and plug-in data acquisition (DAQ) boards along with a GPIB board, users can easily add more functions, such as new triggers for additional devices, new input channels, and so on. Further, the various instruments are confined to a single location—the PC.

In choosing a virtual instrument for the laboratory, several key requirements were paramount. The timer had to be extremely accurate; programming needed to be simple; and the instrument had to be flexible enough to adapt to a wide range of tests. Using LabVIEW, a PC-TIO-10 timing I/O DAQ board, a Lab-PC+ DAQ board, and an AT-GPIB/TNT GPIB board, we developed a five-function instrument for physiological testing. The instrument, termed TISTMO.VI (timer, stimulator, motor), is driven by a 486DX 66 MHz PC running Windows 3.1. TISTMO.VI is composed of a timer, an electrical stimulator, a motion controller, an acquisition controller, and an external device input.

The use of the TISTMO.VI depends on the experimental procedure required. Ordinarily, studies of the muscle tissue are conducted with either intact or skinned muscle fibers. With intact fibers, the muscle cell is activated by electrical stimulation. Skinned muscle fibers are activated by chemical stimulation. In both cases, the muscle fiber is mounted between an electromagnetic apparatus, which imposes length perturbation, and a force transducer that records muscle force.

The TISTMO.VI timing device uses the PC-TIO-10 timing I/O plug-in DAQ board to trigger the external devices of the instrument with excellent accuracy. The PC-TIO-10 triggers the oscilloscope, initiates electrical stimulation, and activates the electromagnetic apparatus to impose length perturbations to the muscle tissue. The DAQ board has four channels plus an oscilloscope trigger with a timebase of 1 ms (error 1 us in 1 ms). Each output channel can count to 10 seconds in respect to zero time.

The electrical stimulator of TISTMO.VI activates the intact fibers using either a single pulse or a train of pulses. The duration, frequency, duty time, and train of pulses are set at the discretion of the user. Electrical activation of the fibers does not affect the timer; it operates in the background of TISTMO.VI.

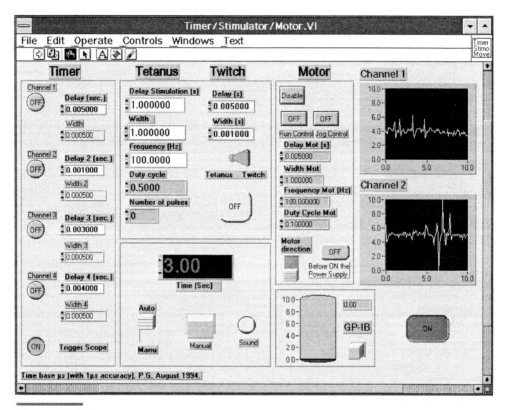

Figure 3-35.

TISTMO.VI is composed of single LabVIEW applications.

Skinned muscle fibers are activated using a chemical solution. This function is performed by motion control, which is achieved with a step motor that displaces the fiber in another chamber with the selected chemical solution. TISTMO.VI determines the direction, entity, and velocity of the motor movement.

After the muscle fibers are activated, muscle tension and length perturbation data are stored in the digital oscilloscope and then downloaded to the PC via the GPIB plug-in board for further analysis and presentation. The GPIB board maximizes throughput of data transfers from the DSO. The Lab-PC+ board is used to acquire data from external devices for slow processes, such as force transducer temperature and offset, while TISTMO.VI operates without interference.

The TISTMO.VI timer uses 8 of the 10 counters on the I/O board. Each counter is gated to an external trigger (hardware trigger) and is linked to the board clock source. The stimulator uses two counters; the motor requires two counters and four bits of the I/O port. LabVIEW.init is corrected to the millisecond.

TISTMO.VI works in both manual and automatic modes; the automatic mode has an additional timebase of seconds. When the instrument is composed of only one block, it is possible to use the timer, stimulator, and motor individually.

National Instruments provides a high-quality, low-cost solution for easily integrating hardware and software. In addition to creating a flexible instrument capable of performing a number of tests, users can use the same hardware and software for different purposes. For instance, the counters on the plug-in boards used to build the timer also drive step motors and generate stimulation patterns.

With LabVIEW, the only requirement for modifying programs and adding new tools is learning how to link them together. The object-oriented approach of LabVIEW saves hours of traditional programming without sacrificing flexibility.

Paolo Garzella, who holds a Ph.D. in physiology and a degree in biology, specializes in computer science and is conducting muscle research at Brigham and Women's Hospital, 20 Shattuck St., Boston, MA 02115 (617) 732-8248, or Via Bargagna, n42, Pisa, Italy, 011-39-50-555922

Cardiopulmonary Dynamics

4

A Cardiovascular Pressure-Dimension Analysis System
System Setup
Data Acquisition and Analysis
Clinical Significance

The capability to readily incorporate numerous measuring devices in a clinical research study has significant ramifications. Flexible, portable systems that efficiently collect and analyze data are necessary to accommodate integrated efforts commonly undertaken. While commercial equipment is essential to ongoing clinical monitoring as well as limited clinical research, it is clear that more complex research protocols require a more flexible and robust solution. At present, DAQ systems such as the LabVIEW™ systems described in this chapter provide the necessary foundation upon which portable, adaptable, and comprehensive research applications can be built. Without the type of adaptability available through a platform such as LabVIEW, comprehensive, integrated clinical research efforts will not prosper.

An extensive variety of protocols and instruments are employed in research associated with cardiovascular and pulmonary function. This chapter focuses on a subset of those ventures, as representative examples of potential applications in cardiopulmonary research. In particular, pulmonary function testing and related respiratory performance assessments will be discussed herein, as well as direct measurements of cardiovascular hemodynamics.

Typical Laboratory Workstation

Due to the diverse nature of these experiments, it is difficult to describe a typical laboratory workstation, as in Chapter 3. Research in cardiopulmonary fields often requires the use of specialized commercial equipment to interface with a patient, relegating the research endeavors to a secondary use of clinically acquired data. Additionally, collaborative efforts among clinics, hospitals, and laboratories often require a portable DAQ solution. In those instances, a mobile workstation such as the one described in Section 3.1 may be the most valuable. That setup included a laptop with a DAQCard-AI-16E-4 PCMCIA card. In other IRB-controlled or animal studies, a more permanent desktop may be more appropriate, in which case the workstation described in

Section 3.1 would be quite sufficient. It included a personal computer with an internal PCI-MIO-16E-4 data acquisition (DAQ) board. The NI Signal Conditioning eXtensions for Instrumentation (SCXI) system may not be required, as any commercial equipment utilized will typically include its own signal conditioning equipment. The unique setups for the experiments discussed herein will be addressed individually throughout the chapter.

Generic Instrumentation/Data Acquisition Issues

Commercial systems are generally sufficient for the assessment of diseases associated with pulmonary mechanics such as emphysema, asthma, or fibrosis, as well as the evaluation of cardiovascular hemodynamics in a clinical setting. While the accuracy of the outputs from these commercial systems meets industry standards, they generally do not retain the raw data used in performing the analyses, and typically use a proprietary calibration scheme that cannot be easily accessed or verified. In other words, a commercial system is virtually a black box. When conducting detailed research studies, the raw data, calibration approach/results, and details of the data processing are also typically required. A general-purpose data collection system is therefore more desirable from three standpoints:

1. The flexibility of such a system allows it to be adapted to a wide variety of studies with minimal time or additional hardware;
2. The quality of data collection can be directly controlled through hands-on calibration and filtering processes; and
3. The raw data remains available for any post-processing that may be desired.

This chapter describes the development of representative general-purpose DAQ systems and their application to clinical and laboratory research. One of the prime motivating factors in the development of the current systems was flexibility. This flexibility manifests itself in two ways. The first is the adaptability of the source code itself. LabVIEW is essentially a modular programming environment wherein VIs can be created in a hierarchical and modular fashion. Complex programs are, as a result, broken down into simpler tasks, each of which is coded into a VI. Icons representing these subroutines are embedded into the block diagram of the higher level VIs, and so on, until the top-level code is completed. This hierarchical structure permits easy modification of program elements, as well as the ability to reuse component VIs to develop completely unrelated programs.

The adaptability of the systems described in this chapter has been demonstrated through alternative pulmonary function procedures performed at John Sealy Hospital in Galveston, Texas. For example, it would be difficult at best to diagnose partial diaphragmatic paralysis using a special-purpose commercial system. Using the **Lung DAQ** algorithm (described in detail later) as a baseline, however, a quick study of suspected diaphragmatic paralysis was initiated, developed, and performed in less than six hours. Additionally, a second research study was conducted regarding the frequency dependence of lung tissue, with only minor modifications to the **Lung DAQ** system. In the partial paralysis case, a second esophageal pressure source had to be acquired. The data processing was modified to depict variations between the two esophageal pressures, recording the maximum and minimum differences. The frequency dependence study involved changes in the reference for the esophageal pressure transducer and a modified protocol. The existing VIs were adequate for obtaining and displaying the required data.

The other way in which this general-purpose DAQ system exhibits flexibility is in the portability of the system as a whole. To facilitate the routine collection of data from diseased subjects, a portable data acquisition system may be required. Because the DAQ components of the system (DAQ board, LabVIEW VIs) are contained in a PC/Mac, it is relatively simple to relocate the system to wherever it is needed. National Instruments also makes available high-quality PCMCIA DAQ cards that allow the use of a laptop, making the system truly portable. For example, this type of system would enhance the data collection capability from subjects requiring mechanically assisted ventilation.

Another primary feature of the systems presented here is the ability to utilize the collected raw data for numerous applications after the testing is complete. The data can be stored in binary or ASCII files and even tab-delimited if so desired. This makes it accessible to spreadsheet or data analysis programs. Additionally, minimizing the effects of quantization and aliasing and lightly prefiltering the acquired data greatly enhance any attempts to postprocess the data.

With the adaptability of the data collected comes the ability to find secondary uses for that data. The number of secondary uses is virtually unlimited and could include items such as an automated diagnosis sheet, or it could be used in the development/manufacture of new clinical products. Alternatively, the data collected may be used in more basic research endeavors. These may include efforts such as mathematical modeling aimed at describing the complex interactions between ventilation, perfusion, and gas transport in the lungs (see Chapter 11).

Pulmonary Function

Clinical pulmonary function testing is an established means of characterizing respiratory mechanics to screen for pulmonary disorders. Study of the physiological mechanisms underlying abnormalities in pulmonary function requires a measurement/data acquisition system that supplements commercially available pulmonary function equipment. Flexibility in adding or removing various instruments or devices based upon alterations in established protocols enables researchers to delve deeper into the physiological abnormalities associated with cardiopulmonary disorders. This section discusses representative applications relevant to the measurement and analysis of the pulmonary system. Pressures, flows, and volumes detected within the airway, chest, or lungs, as well as gas concentrations, are the typical measurement variables in this field. Measurement of these fundamental physical phenomena can also be readily accomplished with the appropriate monitoring equipment.

Pulmonary function testing (PFT) is a routine, noninvasive clinical procedure conducted to assess the functional state of a patient's respiratory system (Primiano, 1997, 1981; Conrad, Kinasewitz, & George, 1984). A typical PFT includes *spirometry*, the measure of inspiratory and expiratory flow rates, as a function of relative lung volumes during a *forced vital capacity* (FVC) maneuver. Combined with either *plethysmography* or *helium dilution*, through which absolute lung volumes can be estimated, a quantitative assessment of flow, as a function of absolute lung volume, can be conducted over the full vital capacity range. In this way, PFT results provide quantitative indices of airway/lung mechanics, such as total lung capacity (TLC), functional residual capacity (FRC), residual lung volume (RV), the volume of air forcefully expired in 1 second (FEV_1), and many others. This standard repertoire of tests evaluates an individual's lung performance and compares that information with population-specific standards. Pulmonary function tests are usually conducted to either: (1) screen for disease; (2) provide periodic checkup for chronic respiratory disease patients; (3) assess acute changes in patients with respiratory diseases; or (4) provide post-treatment follow-up. The testing can be potentially expanded to include blood gas or expired gas measurements in order to assess the effectiveness of gas exchange within the lungs.

Several brands of commercial systems, monitors, and analyzers are available for conducting these routine tests. The data so acquired is then processed on-line to yield a standard output of clinically significant parameters. These collective data are then interpreted to support a particular clinical diagnosis and course of treatment. Due to this availability of commercial equipment, pulmonary function research studies are often focused on specific, identifiable areas of interest for which established measurement techniques and commercial monitors or analyzers are available. Utilizing established protocols allows the investigator to tie specific results into established data sets to form a broader view of the system being studied. Commercial monitors or analyzers also facilitate the gathering of patient data using industry-standard equipment. Thus, the use of commercial equipment is essential to the success of current clinical research.

However, current commercial systems and established protocols may not be sufficient to support more basic research on lung function. Some investigative efforts employ measurements of both lung mechanics and expired gases to form a better assessment of lung function. Such efforts require the use of research-quality data in order to minimize errors, and they must have access to raw data rather than data manipulated by proprietary software. Research-quality data implies data obtained from signals that are thoroughly calibrated, have a substantial signal-to-noise ratio, and where discretization of these signals results in minimal quantization error and negligible aliasing (i.e., sampling is conducted at a rate significantly greater than the Nyquist frequency [Oppenheim & Shafer, 1993]). In addition, data collation and subsequent manipulation is a challenge when multiple systems or instruments are interactively involved in the problem being studied. A number of issues cannot be resolved through the use of standard equipment alone (e.g., proper synchronization of timing events, accommodation of various types of inputs, or variations in sampling frequency).

One method of addressing the aforementioned issues is to provide an external means of controlling the experiment, a separate data acquisition system. This section describes the development of this type of automated system, specifically designed for pulmonary research.

Physiological Basis

The primary function of the respiratory system is to arterialize the blood, providing O_2 and removing CO_2 for the body tissues. Essentially this is accomplished by a three-part system: (1) ventilation; (2) gas exchange; and (3) perfusion. This section focuses on ventilation. The ventilatory system is a complex network of inhomogeneous airway structures, which transport air into and out of the alveolar sacs where gas exchange occurs. Movement of the air is governed by the transairway pressure gradient created by the respiratory muscles of the chest wall, including the diaphragm. The resistance offered by the airways further modulates this airflow.

The airway divides 23 times between the trachea and the alveolar sacs within the lungs. The first 16 stages form a continuously narrowing series of conduits (*bronchi*) through which air flows. The remaining generations contain the bronchioles, alveolar ducts, and alveoli within which gas exchange occurs. These divisions increase the total cross-sectional area of the airways from 2.5 cm^2 in the trachea to 11,800 cm^2 in the alveoli (Ganong, 1995). The chest wall and the lungs are elastic structures with a thin layer of fluid in the *pleural* space between them. Contraction of the inspiratory muscles, particularly the diaphragm, increases the intrathoracic volume, thereby reducing the intrapleural pressure from its typical end-expiration value of −2.5 mmHg to about −6 mmHg. This causes the lungs to expand, reducing the pressure in the airways and causing air to flow into the lungs. After inspiration, the lung recoil force pulls the chest wall back toward equilibrium, where the recoil pressures of the lung and chest wall are balanced. This increases the pressure in the airways and air is forced out of the lungs.

The standard respiratory rate for a normal, resting human is 6–8 L/min. This amount of air is exchanged in a series of breaths (12–15 breaths/min) during which approximately 0.5 L of air is inspired and expired per breath. The amount of air that is inspired or expired in a single, resting breath is termed the *tidal volume* (see Figure 4-1). The volume of air remaining in the lungs at the end of a normal, passive expiration is the *functional residual* capacity (FRC). A maximal inspiratory effort will increase the amount of air inspired. The additional volume above tidal volume is the *inspiratory reserve volume* (IRV). *Expiratory reserve volume* (ERV) is the volume of air expired by active effort beyond the passive expiration. The amount of air left in the lungs after a maximal expiratory effort is the *residual volume* (RV).

Experiment Setup

As discussed earlier, developing a computer-based data acquisition system for conducting multifaceted research has typically required a standard programming environment such as C/C++. However, LabVIEW offers the necessary flexibility in a graphical programming and data-driven environment. The flexibility that this affords is exemplified by an automated data acquisition (DAQ) system designed for pulmonary research at John Sealy Hospital,

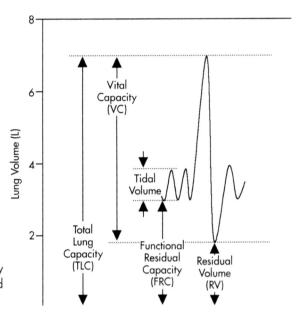

Figure 4-1.
Subdivisions of total lung capacity
(TLC) with volumes often determined
via spirometry.

Table 4-1. *Pulmonary function testing data collection components.*

	Physiological Signal	*Measurement Apparatus*
System Driving Input Pleural Pressure (cm H_2O)	Esophageal Pressure	Naso-gastric Esophageal Balloon/Pressure Transducer

Airway Mechanics
  Flow (L/sec)
  Lung Volume (L)
	Airflow at the Mouth
	Integrated Airflow
		Differential Pressure
		Pneumotachometer

Gas Exchange Expired O_2, CO_2 (%)	Expired O_2, CO_2	Datex Capnomac Gas Analyzer

University of Texas Medical Branch, Galveston, Texas (Olansen et al., 1998). There, a LabVIEW-based DAQ system was developed through which pulmonary function data from a variety of clinical instruments could be routinely collected, analyzed, and displayed. The desired measurements, namely pleural pressure, airflow at the mouth, and expired gas concentrations (particularly O_2 and CO_2), were obtained utilizing minimally invasive measurement techniques as listed in Table 4-1.

Specifically, a Sensor Medics 2800 Body Plethysmograph and associated pneumotachometer were used to establish baseline subject measurements of volume and airflow, respectively. In routine PFT, only airflow at the mouth and lung volume are utilized in the assessment of lung function. This study includes pleural pressure and expired gas (O_2 and CO_2) concentrations in an attempt to extend the capabilities of lung function assessment. Pleural pressure was measured using a latex balloon appropriately positioned in the esophagus (Milic-Emili, Mead, Turner, & Glauser, 1964) and connected to a pressure transducer in the plethysmographic system. A Datex Capnomac Ultima, using side-stream gas analysis, provided continuous measurements of expired O_2 and CO_2 concentrations.

Discrete sampling of all analog signals was, in this case, accomplished using a Macintosh Quadra 800 personal computer outfitted with a National Instruments NB-MIO-16X multifunction DAQ board and an AMUX-64T multiplexer terminal board. (The application was later ported to a Pentium PC without any modification, attesting to the cross-platform compatibility of LabVIEW.) The ensuing digital data was processed, displayed, and stored using original code developed within the LabVIEW programming environment. A depiction of the typical experimental/DAQ system setup is shown in Figure 4-2. The data acquisition system was based on that developed by Olansen et al. (1998).

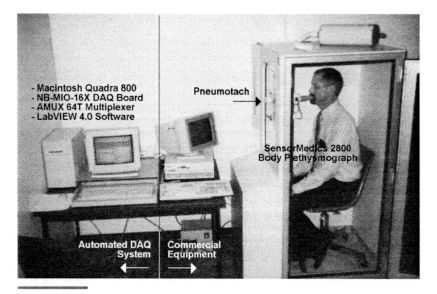

Figure 4-2.
Experimental setup for pulmonary function testing. (Reprinted with permission of Olansen et al., 1998.)

Pulmonary DAQ System Operation

The target population for this pulmonary function application was the pulmonary physicians, respiratory therapists, and other clinical personnel directly involved in the routine collection of PFT data. Hence, the graphical user interface (GUI) designed within LabVIEW enabled tests to be conducted by personnel unfamiliar with LabVIEW. As a consequence, research-quality PFT experiments could be routinely conducted in a clinical (hospital) setting, without requiring an in-depth knowledge of 'G' programming.

Lung DAQ VI

The principal DAQ procedures for this pulmonary application are incorporated into a VI called the **Lung DAQ** VI. This VI controls the complete sequence of events necessary for data collection, from acquiring the digitized samples to processing the data, including near real-time displays and data storage for future applications. Prior to performing the main tasks of data collection, the program queries the user on system readiness via a series of dialog boxes. This process ensures that (a) a thorough calibration has been performed, (b) the various instruments have been initialized, and (c) any required baseline measurements have been taken. If any of these steps are incomplete, the VI automatically invokes the appropriate software to complete the task. Upon completion of a test run, the user is able to view the full data set, determine if the test was successful, and save either the raw data, filtered data (5th order, zero phase shift Butterworth filter), or both. The primary features of the **Lung DAQ** VI program are listed in Table 4-2.

Figure 4-3 is the front panel of the **Lung DAQ** VI created using LabVIEW. Components of the front panel include both input and output. The boxes at the top and left of the panel allow the investigator to input various static data points related to the test conditions or the subject being tested. These include particulars such as room temperature; humidity; subject height; weight; and predetermined TLC, FRC, and RV pulmonary indices. Additional inputs consist of the virtual buttons, which are used to operate the system. The START DAQ and STOP DAQ buttons perform their prescribed functions, initializing acquisition of data from the First Input, First Output (FIFO) buffer on the NB-MIO-16X DAQ card, and ending it, respectively. These allow for conducting multiple test runs in one session. The CLEAR PLOT button erases all previously acquired data (in that run), without having to reinitialize the DAQ driv-

Table 4-2. *Lung DAQ and Calibration VI primary features.*

	Acquisition	Processing	Display/Storage
Lung DAQ	–Acquire continuous waveform data (default 50 Hz) from multiple analog input channels (default 4)	–Integrate pneumo-tach data to obtain instantaneous change in Lung Volume –Allow input of known FRC or calculation of FRC using mouth pressure, box pres-sure, and Boyle's law –Zero phase shift (dual-pass) Butter-worth filter (default 5th order) to lightly smooth the data	–Continuous display of all input channels and Lung Volume in a near real-time strip chart monitoring format –Continuous display of the clinically significant Flow-Volume Loop –Data is saved as a column-major, tab-delimited ASCII spreadsheet file
Calibration	–Direct, linear calibration of any transducer input –Volume-referenced calibration of the pneumotach or box pressure sensors relative to a known input volume (i.e., 3 L syringe)	–Apply the calibration to the data as it is collected on-line so that only calibrated data is directly displayed or saved	–Test parameters, including calibration coefficients and subject notes are saved after each test run for future reference

ers. The FINISH SESSION button terminates the program operation. The out-puts include the continuous display of the raw data in the strip charts on the right of Figure 4-3, a running strip chart of instantaneous lung volume on the left, and the clinically significant flow-volume loop in the lower left. The data displayed in Figure 4-3 is typical research data collected during a forced vital capacity maneuver.

The **Lung DAQ** VI is executed, as with all VIs, by mouse-clicking the opaque arrow in the upper left-hand corner of the toolbar. Responses to a series of di-alog boxes then guide the user through the calibration procedure (discussed later) and establishment of baseline measurements (FRC, RV, and TLC). With calibration complete and baseline values determined, if required, the system is initialized and awaits further input—the START DAQ button. By clicking that button, the program begins to acquire, process, and display the data.

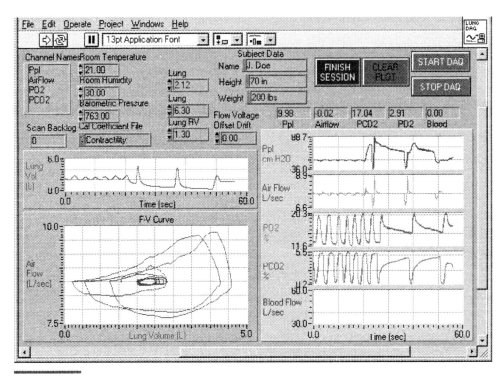

Figure 4-3.
Front panel of the **Lung DAQ** VI. (Reprinted with permission of Olansen et al., 1998.)

Lung DAQ VI Source Code

The source code necessary to implement the aforementioned VIs is developed graphically in the LabVIEW block diagram-programming format. Figure 4-4 portrays one of the main frames of the program code, within which the data is collected, analyzed, integrated, and displayed. Numerous embedded structures can be observed within this frame, beginning with the outer frame itself. This one looks like a filmstrip; it is a sequencer, the third of three. The items shown within this frame occur last in a sequence (i.e., the first two frames accounted for the calibration and FRC determination steps in the program). Everything in this frame also acts within a while loop, which is only terminated with the execution of the FINISH SESSION button.

As stated earlier, the START DAQ button begins the acquisition by terminating the while loop on the left of Figure 4-4 and activating the inner sequence. Within this sequence, the DAQ board is initialized, and data collec-

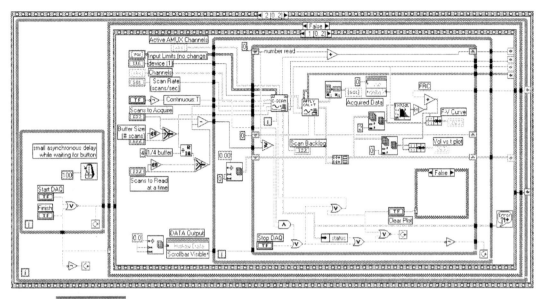

Figure 4-4.
Block diagram of the **LungDAQ** VI. (Reprinted with permission of Olansen et al., 1998.)

tion begins. The dual while-loop formation allows for the CLEAR PLOT function described earlier. Within the inner loop, the sequence of events follows the circular acquisition/process/display flow depicted in Figure 4-5.

This repetitive sequence continues until the STOP DAQ button is activated, at which time both while loops terminate and the next filmstrip frame is enabled. This next frame saves the data to the hard drive, if so desired. The data can be either saved in a raw (calibrated) data format or lightly filtered by a dual-pass (zero phase shift) Butterworth filter. A related text file is also saved with all static information that the operator entered on the front panel, as well as the calibration coefficients.

Calibration VI

All of the manipulation and presentation of data is meaningless without proper calibration of the transducers being sampled. Figure 4-6 is the front panel of one of the calibration subroutines included in the current DAQ system. This calibration scheme is volume-referenced, which is typical of plethysmograph boxes. The other subroutine is a straight linear regression

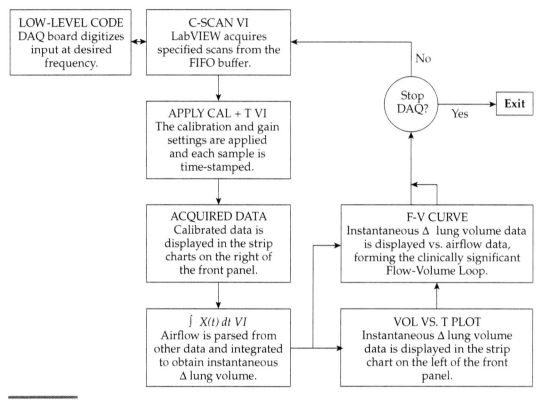

Figure 4-5.
LungDAQ VI Circular Acquisition Routine. (Reprinted with permission of Olansen et al., 1998.)

analysis based on several readings in the transducer's linear range of operation. The primary features of the **Calibration** VIs are listed in Table 4-2.

The volume-referenced calibration can be performed on the pneumotachometer or the plethysmograph box pressure sensor, if required. This method
replaces the use of a flowmeter for calibrating airflow through the pneumotachometer. In other words, instead of referencing a given pneumotachometer
output voltage to a known flow, the output is integrated to yield a volume and
the result is referenced to a known volume. For calibration purposes, a 3 L syringe is available with the plethysmograph (see Figure 4-2). The **Volume Calibration** VI reads the pneumotachometer signal, integrates it, and compares it
with the known volume change that comes from a full stroke of the syringe.
The incoming data, and the resulting calibration curve, are displayed on the
front panel. The calibration coefficients are saved for use in the **Lung DAQ** VI
discussed previously. The **Linear Calibration** VI operates in a similar manner

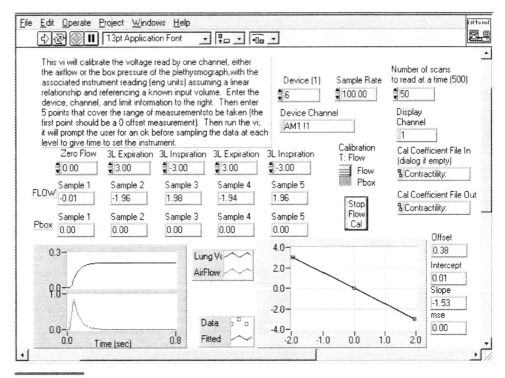

Figure 4-6.
Front panel of the **Volume Calibration** VI. (Reprinted with permission of Olansen et al., 1998.)

but compares the sampled data directly with known source values and performs a simple linear regression analysis.

With the system in place as described, care must still be taken to ensure an accurate calibration. Drift of the pneumotachometer must be considered and eliminated. Calibration must be performed over the full range of expected outputs to ensure linear operation of the transducers. And, finally, calibration should be repeated at set intervals to accommodate transducer drift or shift with time.

Lung Tissue Viscoelastance

Only the investigators' interests limit the analysis that can be applied to the data collected in pulmonary function testing. That data can be used for multiple purposes, including the original intended purpose of the **Lung DAQ** VI, which was the verification of developed theoretical models. The model addressed encompasses a study of the viscoelastic properties of lung tissue and is detailed in Chapter 11. The **Lung DAQ** VI was used in its entirety to collect the data for this experiment. Modifications were made in the hardware setup and the experimental protocol only.

The objective herein is to gain better insight into normal respiratory function for potential applications in clinical settings. One common clinical reason for studying pulmonary mechanics is the assessment of work of breathing (WOB). WOB measurement finds application in the assessment of respiratory failure and in ventilatory management, as an indicator of the metabolic energy of breathing. Recently, with the advent of new ventilatory modes for the critically ill patient, assessment of breathing effort has received revived interest among clinicians. The application of this model to the estimation of the WOB and respiration energetics is discussed in Athanasiades et al., 1999.

Experiment Setup

Instrumentation and methodologies developed for the measurement of pulmonary mechanics relied on simultaneous pleural pressure and lung volume recordings. The former is approximated by esophageal pressure, measured with a naso-gastric catheter balloon (Milic-Emili et al., 1964).Volume is measured either with spirometry or by integrating volumetric flow from a pneumotachograph. In clinical practice, the two variables (pressure and volume) are directed to a computerized monitor and plotted against each other. In this manner, work can be estimated by calculating the area enclosed in the pleural pressure loop (Armaganidis & Roussos, 1995). The setups described earlier in the sections on the typical laboratory workstation and the experimental setup are sufficient to perform this type of experiment.

Experiment Description

Measurements of airflow at the mouth and elastic recoil pressure were obtained from three volunteer human male subjects using the technique described earlier. All subjects were healthy with no current or significant history of respiratory ailments. Elastic recoil data were obtained during inspiration and expiration over the entire volume capacity of spontaneously breathing subjects. To this end, the esophageal pressure was continuously referenced to mouth pressure to enable a direct recoil pressure measurement when the airflow was zero. By instituting a shuttering technique, as shown in Figure 4-7,

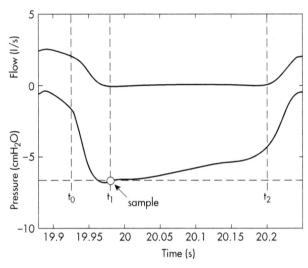

Figure 4-7.
Experimental data collected using a repetitive shuttering technique, such that elastic recoil pressure was sampled as soon as airflow ceased.

the airflow at the mouth could be repetitively halted, enabling mouth pressure to equalize with alveolar pressure. Thus, the actual measurement recorded was that of pleural (esophageal) pressure, P_{PL}, minus alveolar pressure, P_A, which equates to the elastic recoil pressure, P_{EL}. (You should note that the shuttering technique discussed herein is not necessarily valid in subjects with obstructive airways, such as asthmatic or emphysematic patients.)

Each experimental episode started with a minimum of four tidal breaths, followed by a complete inhalation to total lung capacity (TLC) and relaxed exhalation to functional residual capacity (FRC). Some episodes included forced exhalation to residual volume (RV). During these large volume excursions, a shuttering mechanism in the flow passageway was manually activated and deactivated in rapid succession (2–3 Hz). In this manner, a significant number of data points could be obtained over the entire vital capacity range.

Data were also gathered on the frequency-dependent behavior of lung tissue. Subjects executed a panting maneuver at different frequencies near FRC. During this procedure, flow shuttering was not performed, since it would alter the true frequency of breathing. Instead, only pleural pressure data and volume were collected. Pleural pressure equals the elastic recoil pressure at both end-inspiration and end-expiration (in the absence of flow, pressure drop in the airways is zero). The end-point data thus collected can be utilized to calculate the dynamic compliance of tissue. It is defined as the ratio of tidal volume to the change in pressure measured, at the respective breathing frequency (West, 1974). After each testing episode, the subject rested for a period of time that ensured full recovery of baseline levels of pO_2 and pCO_2.

Cardiovascular Hemodynamics

Cardiovascular hemodynamics describes the pumping characteristics of the heart and the associated blood flows and pressures throughout the vasculature. The circulatory system serves to transport blood and distribute essential substances to body tissues and organs, and to remove by-products of metabolism. It also plays a role in homeostatic mechanisms, such as the regulation of body temperature, humoral communication throughout the body, and adjustments of supply of oxygen and nutrients in different physiological states. The cardiovascular system that serves these purposes consists of two pumps (the right and left ventricles), each primed by an individual atrium, and a network of distributing and collecting conduits. These conduits are bridged by a vast, extensive network of thin vessels that permit rapid and efficient exchange of substances between the tissues and blood. Because the cardiovascular system plays an essential role in the functioning of the whole body, much research has been devoted to better understand its behavior under normal and pathophysiological conditions.

This section relates to the measurement and analysis of components of the cardiovascular system. Pressures, flows, and volumes detected within the heart or throughout the circulatory loop are the typical measurement variables in this field. Measurement of these fundamental physical phenomena can also be readily accomplished with the appropriate monitoring equipment.

Physiological Basis

The electrical excitation of the heart results in a wave of contraction that spreads through the heart musculature (myocardium). The synchronous nature of this contraction results in the efficient pumping of blood through the pulmonic and systemic circulations.

The Vasculature

The circulatory system can be considered in terms of the cardiac pumps and their afterloads, which are the networks of blood-transporting vasculature. The heart consists of two pumps in series. The right pump propels blood through the lungs for exchange of oxygen and carbon dioxide (the *pulmonary* circulation). Figure 4-8 schematically depicts the closed-loop nature of the circulatory system as well as the arrangement of the various series and parallel circulations in specific organs. Taking the output from the right pump and the

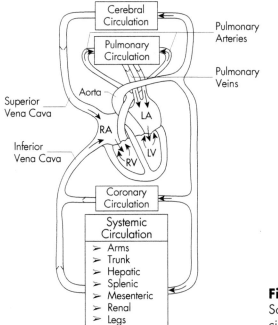

Figure 4-8.
Schematic of closed-loop circulation.

pulmonary vasculature, the left pump propels the oxygenated blood to all organs of the body (the *systemic* circulation). Oxygen and nutrients are extracted from the blood in these organs, and waste material is deposited. The deoxygenated blood is then transported through the venous system and eventually returned to the right pump, where the cycle begins anew.

The pressure afterload of the left pump (the left *ventricle*) is large relative to that of the right pump (the right *ventricle*). Approximately 20% of the total blood volume is stored in the lungs in the supine position, whereas the remainder may be considered stored in the systemic vasculature fed by the aorta. The left ventricle thus must be able to propel blood with enough energy to circulate through the relatively extensive systemic circulation. To maintain proper organ function, a continuous flow to the organs must be maintained in the face of disturbances (adequate organ perfusion). The ventricles, as the source, must maintain a reasonably constant mean flow output (*cardiac output*). During contraction (*systole*), the left ventricle generates enough pressure to exceed that in the aorta (~80 mmHg) and propel blood into the systemic circulation. From the aorta, blood moves through the systemic circulation with progressively greater decrease in pressure and velocity (Berne & Levy, 1988).

Figure 4-9 shows this progressive drop in blood pressure and flow velocity along the systemic circulation. Note that in Figure 4-9, the aorta and large

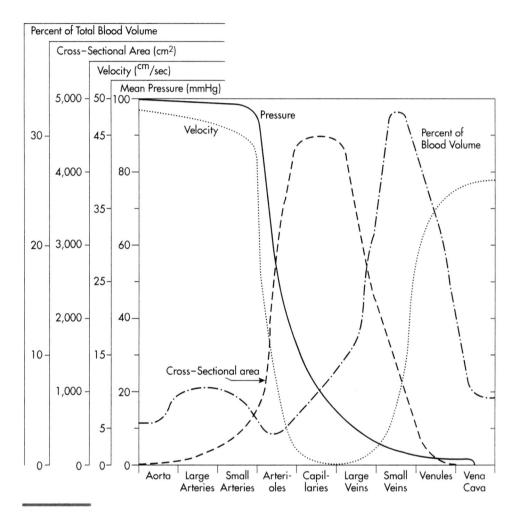

Figure 4-9.
Hemodynamic profile of systemic circulation.

arteries, which serve as major conduits for distributing blood to the regional circulations, have insignificant total cross-sectional area compared with the arterioles, capillaries, and venuoles, which comprise the dense network structures for organ perfusion. Since the capillaries consist of short tubes whose walls are very thin and the flow rate is slow, conditions in the capillaries are ideally suited for the exchange of diffusible substances between blood and tissue. Blood returns to the heart, after circulating through the capillaries, through the venous system, where the greatest portion of the circulating blood resides.

The Heart

The heart is a pump that provides the blood with energy to circulate through the body's vast and complex vascular network. As such, it is a primary component of the cardiovascular system. The anatomy of the heart and the mechanics of cardiac function are described in the following sections.

Figure 4-10 shows the anatomy of the heart. The heart is divided into four chambers: the left and right atria (LA and RA, respectively) and ventricles (LV and RV, respectively). The atria are thin-walled, low-pressure chambers that function more as large reservoir conduits of blood for their respective ventricles than as independent blood pumps. A compliant atrial (membranous) septum separates the atria, while a muscular interventricular septum separates the two ventricles. (The interventricular septum, because of its anatomical relationship to the left and right ventricles, plays an important role both in participating in ventricular contraction and mediating ventricular interaction.)

There are two types of valves in the heart, the *atrioventricular valves* and the *semilunar valves*. The atrioventricular (AV) valves separate the atria and the

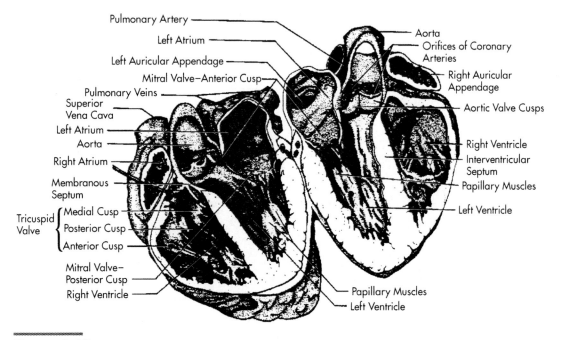

Figure 4-10.

Anatomical representation of the heart.

ventricles. The valve between the right atrium and the right ventricle is made up of three cusps (*tricuspid valve*), whereas the valve between the left atrium and left ventricle has only two cusps (*mitral valve*). Attached to the free edges of these valves are fine, strong filaments (*chordae tendineae*), which prevent eversion of the valves during ventricular systole. Semilunar valves separate the ventricles from the great vessels attached. The *pulmonary valve* between the right ventricle and the pulmonary artery and the *aortic valve* between the left ventricle and the aorta each have three cup-like cusps attached to the valve annulus.

The structure of these valves prevents regurgitation of blood into the ventricles when a brief reversal of blood flow toward the ventricles occurs at the end of the reduced ejection phase of ventricular systole. To a good approximation, these valves can be considered unidirectional and pressure-operated. That is, these valves prevent regurgitation when the pressure in the source chamber is not sufficient to open the valve, and once open, these valves maintain blood flow in a single direction under the positive pressure gradient across the valve.

The entire heart is enclosed in the pericardium, which offers structural support for the heart and normally contains a small amount of lubricating fluid to minimize friction with the epicardial surface of the heart. The pericardium is an epithelized fibrous sac and has little extensibility. As a result, it strongly resists a large and rapid over-distension of the cardiac chambers.

Figures 4-11 and 4-12 summarize the sequence of mechanical events of the heart. During *diastasis*, the ventricles fill as blood flows through the open mitral and tricuspid valves between the atria and ventricles, while the aortic and pulmonary valves are closed. After the P wave of the ECG, atrial systole

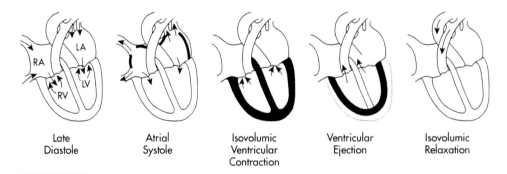

| Late Diastole | Atrial Systole | Isovolumic Ventricular Contraction | Ventricular Ejection | Isovolumic Relaxation |

Figure 4-11.
Schematic of mechanical activity in cardiac cycle.

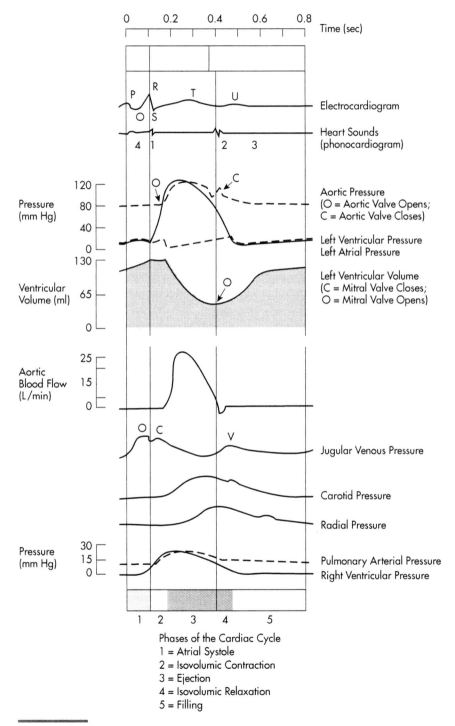

Figure 4-12.
Phases of the cardiac cycle.

begins, and the atria contract to propel additional blood into the ventricles. Ventricular volume at the end of diastole (ED) is called *end-diastolic volume* (EDV) and is dependent on the diastolic properties of the ventricle as well as hemodynamics of the input from the contracting atrium. The EDV represents the *preload* to the ventricle.

The R wave of the ECG marks the start of the ventricular systole, at which time the mitral and tricuspid valves close, in addition to the already closed aortic and pulmonary valves. This creates *isovolumic* conditions in the ventricles, as there is neither inflow nor outflow. The ventricles contract during this phase of *isovolumic contraction,* and the ventricular pressures increase rapidly until the pressures in the ventricles exceed those in the aorta (80 mmHg) and pulmonary artery (10 mmHg). When the pressure-operated aortic and pulmonary valves open, ventricular ejection (Phase 3 in Figure 4-12) begins.

Ejection is rapid at first, slowing down as systole progresses. During ejection, ventricular pressures can reach about 120 mmHg for the left ventricle and 25 mmHg for the right. At the end of ejection, the aortic and pulmonary valves close, and the residual ventricular volumes at the end of systole (ES) are called *end-systolic volumes* (ESV). The *stroke volume* (SV) is then the volume ejected by the left ventricle, or the difference between EDV and ESV. The cardiac output (CO) is then computed as the product of the stroke volume and the heart rate, in L/min. Upon valve closure, the ventricles enter the phase of *isovolumic relaxation,* during which the ventricular pressures decline rapidly (Phase 4 in Figure 4-12). When the ventricular pressures fall below those in the atria, the AV valves (mitral and tricuspid) open and the ventricles begin to fill. During *ventricular filling (diastasis),* the ventricles are passive, subject to the compliant properties of the myocardial walls and the restraining effect of the pericardium. Passive filling continues until the next atrial systole, at which time the cardiac cycle begins anew.

P-V Analysis of Ventricular Function

In the pressure-volume (P-V) plane, the complete cardiac cycle is shown as a loop, with different phases of the cardiac cycle as indicated in Figure 4-13. Note that the ventricle responds to different demands under different physiologic states (rest and exercise), and this change in the pump property of the ventricle manifests in different P-V loops. During exercise, the ventricle does more work to meet the additional demands imposed by greater physical activity, and this is indicated by a P-V loop with greater P-V area. In addition to using P-V area as a measure of work done by the ventricle, the P-V plane analysis is useful to indicate the contractile (*inotropic*) state of the ventricle.

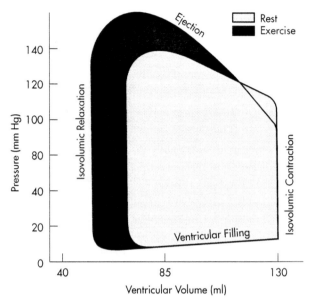

Figure 4-13.
Pressure-volume (P-V) plane analysis of the cardiac cycle.

Canine Cardiovascular Pressure Measurements

Basic experiments in hemodynamics focus on left ventricular function, blood pressure, and flow variations around the cardiovascular loop. In our studies, representative measurements are taken from cannulated arteries and veins as well as from catheter-tipped pressure transducers placed within the ventricular chamber(s) of open-chest canine preparations. Standard clinical hemodynamic indices such as stroke volume, cardiac output, and ejection fraction are also calculated.

As outlined previously, a wide array of experiments can be developed for use in cardiovascular research. The following section provides a cursory look at one representative endeavor. Additionally, this field of study is well-suited for the use of BioBench™ software. Therefore, you should refer to Chapter 2 for a discussion on BioBench basics. A demo version of BioBench is included on the accompanying CD-ROM.

Experiment Setup

Pressure measurements from selected sites in the circulatory loop were obtained from open-chest canine preparations in the Center for Experimental Cardiac Electrophysiology, Baylor College of Medicine, Houston, Texas. These anatomic sites include the left ventricle, aortic arch, descending aorta, and femoral arteries. Additional records were obtained from the inferior vena cava, the right atrium, and right ventricle. The pressure recordings from four sites were typically acquired simultaneously, and the ECG was recorded as an independent timing reference.

Pressures were obtained using solid state catheter-tip transducers from Millar Instruments (Houston, TX). Positioning of the transducers within the heart chambers or vessels was verified via X-ray imaging prior to initiating acquisition. The analog recordings were digitally sampled at a rate of 500 Hz, which was sufficient to capture all significant frequency content of the acquired waveform. Data acquisition was accomplished using a mobile computer platform consisting of a National Instruments DAQCard-AI-16E-4 PCMCIA card inserted in a laptop. Analog signal conditioning was incorporated via an SCXI-1120 module (National Instruments) in order to amplify the incoming signals (ECG x1000; Pressure x10).

The acquisition was controlled via virtual instruments (VIs) designed and developed within the LabVIEW programming environment as well as through use of the turnkey program BioBench. The data collection virtual instruments enabled (1) the raw data to be converted to engineering units (mmHg) using previously established calibration curves, (2) continuous real-time display of the acquired data for quality control, and (3) storage of acquired, calibrated data for post-processing. A sample of the data acquired is shown in Figure 4-14.

A Cardiovascular Pressure-Dimension Analysis System

The intrinsic contractility of the heart muscle (myocardium) is the single most important determinant of prognosis in virtually all diseases affecting the heart (e.g., coronary artery disease, valvular heart disease, and cardiomyopathy). Furthermore, it is clinically important to be able to evaluate and track myocardial function in other situations, including chemotherapy (where cardiac dysfunction may be a side effect of treatment) and liver disease (where cardiac dysfunction may complicate the disease).

The most commonly used measure of cardiac performance is the ejection fraction. Although it does provide some measure of intrinsic myocardial performance, it is also heavily influenced by other factors such as heart rate and loading conditions (i.e., the amount of blood returning to the heart and the pressure against which the heart ejects blood).

Better indices of myocardial function based on the relationship between pressure and volume throughout the cardiac cycle (pressure-volume loops)

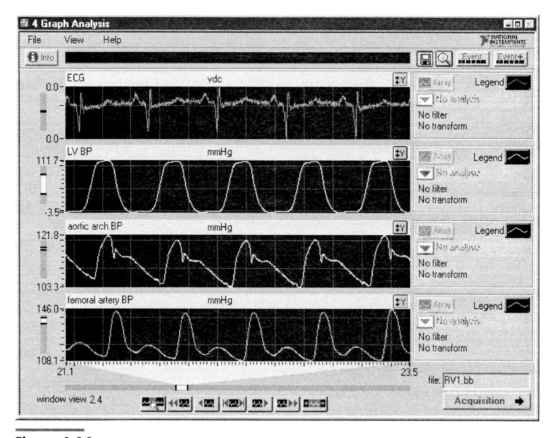

Figure 4-14.
BioBench data acquisition and analysis software illustrates collection of canine cardiovascular hemodynamic data.

exist. However, these methods have been limited because they require the ability to track ventricular volume continuously during rapidly changing loading conditions. While there are many techniques to measure volume under steady state situations, or at end-diastole and end-systole (the basis of ejection fraction determinations), few have the potential to record volume during changing loading conditions.

Echocardiography can provide on-line images of the heart with high temporal resolution (typically 30 frames per second). Since echocardiography is radiation-free and has no identifiable toxicity, it is ideally suited to pressure-volume analyses. Until recently, however, its use for this purpose has been limited by the need for manual tracing of the endocardial borders, an extremely tedious and time-consuming endeavor.

System Setup

Biomedical and software engineers at Premise Development Corporation (Avon, CT), in collaboration with physicians and researchers at Hartford Hospital, have developed a sophisticated research application called the Cardiovascular Pressure-Dimension Analysis (CPDA) System. The CPDA system acquires echocardiographic volume and area information from the acoustic quantification (AQ) port, in conjunction with ventricular pressure(s) and ECG signals, to rapidly perform pressure-volume and pressure-area analyses. This fully automated system allows cardiologists and researchers to perform on-line pressure-dimension and stroke work analyses during routine cardiac catheterizations and open-heart surgery.

The system has been designed to work with standard computer hardware. Analog signals for ECG, pressure, and area/volume (AQ) are connected to a standard BNC terminal board. Automated calibration routines ensure that each signal is properly scaled and allow the user to immediately collect and analyze pressure-dimension relationships.

The development of an automated, on-line method of tracing endocardial borders (Hewlett Packard's AQ Technology; Hewlett-Packard Medical Products Group, Andover, MA) has provided a method for rapid on-line area and volume determinations. Figure 4-15 illustrates this AQ signal from a Hewlett Packard Sonos Ultrasound Machine. This signal is available as an analog voltage (-1 to $+1$ volt) through the Sonos Dataport option (BNC connector).

The diagram in Figure 4-16 illustrates the measured parameters and the specific hardware used for this application. Although this application was initially developed on a Macintosh platform, LabVIEW's cross-platform compatibility enables the system to be run on multiple platforms, including Windows 9X/Me/NT/2000. The CPDA system also takes advantage of the latest hardware developments and form-factors and can be used on either a desktop or a laptop computer.

Data Acquisition and Analysis

Upon launching this application, the user is presented with a dialog box that reviews the license agreement and limited warranty. Next, the Main Menu is displayed, allowing the user to select from one of six options as shown in Figure 4-17.

When conducting a test, an automated calibration sequence for the pressure and ultrasound signals (acoustic quantification) is generally performed. If the

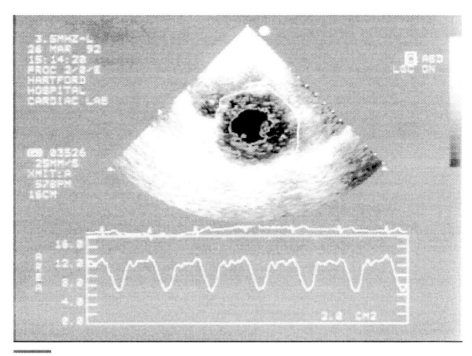

Figure 4-15.
The Acoustic Quantification (AQ) signal.

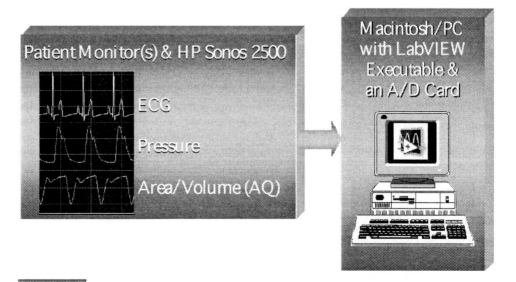

Figure 4-16.
Schematic diagram of the Cardiovascular Pressure-Dimension Analysis System.

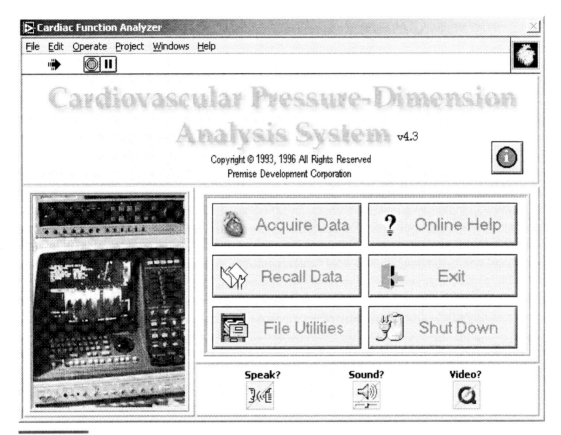

Figure 4-17.
Cardiovascular Pressure-Dimension Analysis Main Menu.

user elects not to perform a calibration, the most recent calibration values are accessed from the integrated calibration log. Pressure calibration data is retrieved from a lookup table containing specific gain and offset values for a variety of different manufacturers' pressure monitors. The AQ calibration procedure involves scaling and mapping the display image signal over the –1 to +1 volt output range. Figure 4-18 illustrates the front panel for the Hewlett Packard Sonos calibration procedure. Sequential instructions in the form of a scrolling string indicator, as well as dialog boxes, are also available.

The default sampling frequency for each channel is 200 Hz. Data is typically collected for 20 to 60 seconds. The user is presented with a Pre-Scan panel to ensure that each signal is calibrated and tracking appropriately.

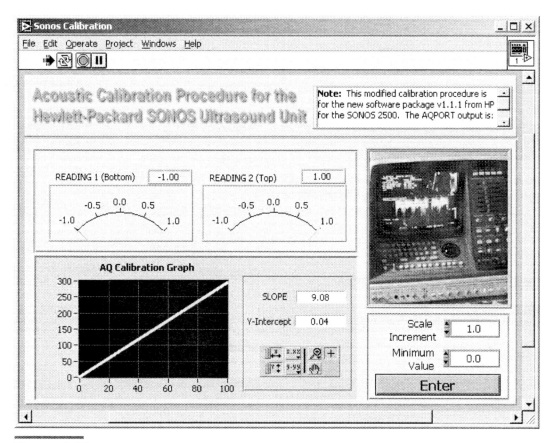

Figure 4-18.
Hewlett Packard's Sonos Ultrasound Machine Calibration front panel.

When the user is ready to collect and store data, the Cardiac DAQ instrument is called (see Figure 4-19). This sub-VI uses double-buffering to collect and display each channel for the predefined time. In order to maintain high temporal resolution, data is displayed in 10-second sweeps. An indicator is provided to display the instantaneous and total collection time.

Once data is collected, the user is presented with the Data Selection sub-VI to define a particular range of data to save to a file. This option allows the user to store only the portion or subset of data that is useful among the entire collected data set (i.e., the last 25 seconds of a 60-second array). The default setting will store the entire data set. Interactive cursors are used to interactively

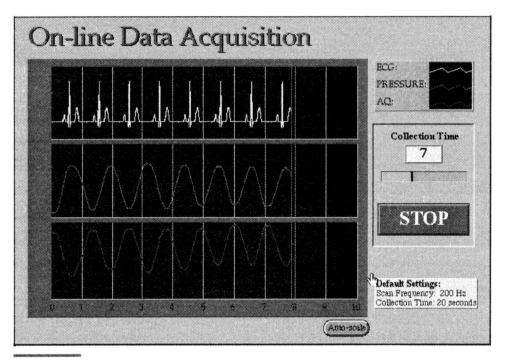

Figure 4-19.
The **Cardiac DAQ** front panel.

set the initial and final indices of the data subset for analysis as illustrated in
Figures 4-20 and 4-21.

Clinical Significance

Several important relationships can be derived from these signals. Specifically, a parameter called the *End-Systolic Pressure-Volume Relationship (ESPVR)*
describes the line of best fit through the peak-ratio (maximum pressure with
respect to minimum volume) coordinates from a series of pressure-volume
loops generated under varying loading conditions. The slope of this line has
been shown to be a sensitive index of myocardial contractility that is independent of loading conditions. In addition, several other analyses, including

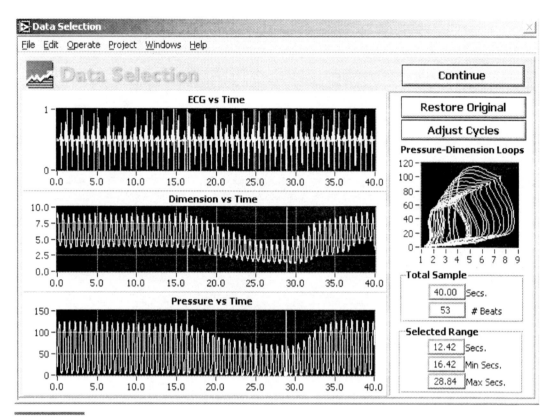

Figure 4-20.
The **Data Selection** front panel.

time-varying elastance (E_max) and *stroke work,* are calculated. Time-varying elastance is measured by determining the maximum slope of a regression line through a series of isochronic pressure-volume coordinates. Stroke work is calculated by quantifying the area of each pressure-volume loop. Statistical parameters are also calculated and displayed for each set of data. Figure 4-21 illustrates the pressure-dimension loops and each of the calculated parameters along with the various analysis options. Finally, the user has the ability to export data sets into spreadsheet and database files and export graphs and indicators into third-party presentation software packages such as Microsoft PowerPoint™.

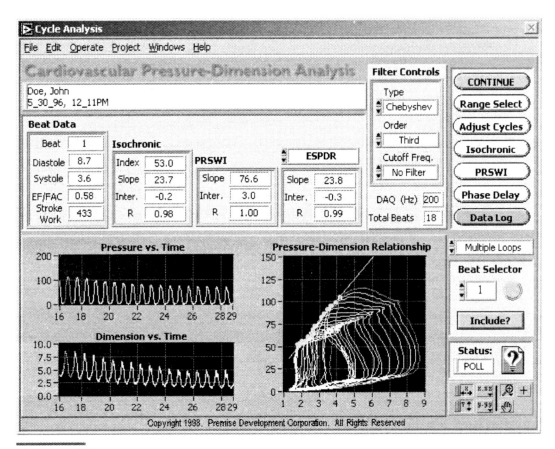

Figure 4-21.
The **Cardiac Cycle Analysis** front panel.

REFERENCES

Armaganidis, A., & Roussos, C. (1995). Measurement of the work of breathing in the critically ill patient. In C. Roussos (Ed.), *The thorax* (2nd ed.). New York: Marcel Dekker.

Athanasiades, A., Ghorbel, F., Clark, J. W., Jr., Niranjan, S. C., Olansen, J., Zwischenberger, J. B., & Bidani, A. (2000). Energy analysis of a nonlinear model of the normal human lung. *J. Biol. Systems, 8,* 115–139.

Berne, R. M., & Levy, M. N. (1988). *Physiology* (2nd ed.). St. Louis, MO: Mosby.

Conrad, S. A., Kinasewitz, G. T., & George, R. B. (Eds). (1984). *Pulmonary function testing—Principles and practice.* New York: Churchill Livingston.

Ganong, W. F. (1995). *Review of medical physiology* (17th ed.). Norwalk, CT: Appleton & Lange.

Milic-Emili, J., Mead, J., Turner, J. M., & Glauser, E. M. (1964). Improved technique for estimating pleural pressure from esophageal balloons. *J. Appl. Physiol., 19*(2), 207–211.

Olansen, J. B., Ghorbel, F., Clark, J. W., Jr., Deyo, D., Zwischenberger, J. B., & Bidani, A. (July/August 1998). An automated LabVIEW-based data acquisition system for analysis of pulmonary function. *J. Clin. Eng. 23*(4), 279–287.

Oppenheim, A. V., & Schafer, R. W. (1993). *Discrete-time signal processing.* Englewood Cliffs, NJ: Prentice Hall.

Primiano, F. P., Jr. (1981). A conceptual framework for pulmonary function testing. *Ann. Biomed. Eng., 9,* 621–632.

Primiano, F. P., Jr. (1997). Measurements of the respiratory system. In J. G. Webster (Ed.), *Medical instrumentation: Application and design* (3rd ed., pp. 372–439). New York: Wiley.

West, J. B. (1974). *Respiratory physiology—The essentials.* Baltimore: Williams & Wilkins.

Independent Solution Articles

LabVIEW: Our Choice for Cardiothoracic Research

by Robert M. Wise, Department of Surgery, Thoracic and Cardiovascular Division, School of Medicine, University of Maryland

The Challenge: Having resident physicians perform cardiothoracic research without extensive programming.

The Solution: Using LabVIEW, residents can bypass software engineers and get closer to the data.

As part of the Thoracic Surgery Resident Physician's training, a year or more is usually spent in a research environment where the resident physician is required to learn how to conduct research and how to produce publishable results. Most physicians entering the program have minimal engineering background and programming experience. This is unfortunate, because acquiring and analyzing data from experiment models is an essential part of this regimen—understanding the data acquisition trail is essential to good research.

In the past, data in the laboratory was either tediously copied from chart recordings or acquired by hideously long programs written by programmers who usually had little knowledge of the reasoning behind the experiment. In addition, finished programs could not be modified by the residents; any alterations, which often proved painstaking and time-consuming, were performed by software engineers. Today, using LabVIEW, we can bypass the software engineers and get closer to our data. Residents not only perform complex data analysis on their own, they can also rapidly alter the experiments as needed. We are currently conducting experiments on the perfused beating heart.

Data such as pressure, flow, length, and electrocardiogram readings are recorded daily. Using a Gould recording system and transducers, we acquire pressure and electrogram data. We obtain flow data through in-line flow probes from Transonic Systems Inc. Length measurements are taken from a Triton Technology sonimicrometer. These analog signals are fed to an I/O interface box connected to an AT-MIO-16X data acquisition (DAQ) board and digitized using the easily modifiable virtual instruments (VIs) supplied with LabVIEW.

Our current analysis program started from the Graph Acq'd File VI provided by LabVIEW; we modified it significantly to provide the kinds of data we wanted to record. The original voltage signals fed to the analog channels are transformed into aortic and left ventricular pressure in millimeters of mercury. Aortic and left atrial flow are transformed into milliliters per minute. Length measurements are fed to elliptical volume equations embedded in sub-VIs—from this data, ventricular volume is obtained.

Subsequently, we obtain pressure volume loops, which provide a classic method to access the contractility and efficacy of muscle tissue. We can display mean values of end diastolic pressure and end systolic pressure, the first-order differential of pressure (dP/dt), coronary flow, and ejection fraction of several beats. All of the data is easily obtained from the front panel by means of a ring control that selects visible and invisible attributes of the waveform charts needed. We can then send the data to other software, such as spreadsheets.

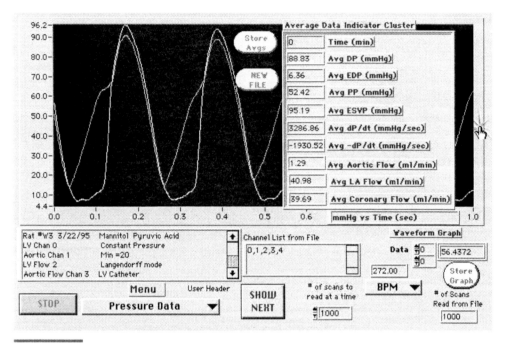

Figure 4-22.
With our analysis program, we can inspect experimental data while the calculations are performed.

With LabVIEW, we easily modify and continue to expand our system. In the near future, we plan to incorporate a waveform analysis package for frequency analysis of the cardiac electrogram. LabVIEW has aided our research efforts and stimulated ideas for future research experiments.

For more information contact Robert Wise, MSTF Bldg Room 400,
10 S. Pine St., Baltimore, MD 21201,
tel (410) 706-0504, fax (410) 706-0311,
e-mail: rwise@surgery2.ab.umd.edu

SpiroPro: A Novel Spirometer Software Application for Lung Function Testing

By Mikael Fagerholm, Keijo Leinonkoski, and Mika Holmström (Tecono Oy); and
Mikko Eloranta, Timo Haikonen, and Pekka Heikkinen (Medikro Oy)

The Challenge: To develop a sophisticated Windows software product for analyzing ventilation of the lungs. This product should take care of communicating with measuring device, analyzing the measured signal, comparing the results with predicted values, and storing analyzed data to database. One essential task was to make the UI very intuitive and easy to use.

The Solution: The system is based on the following software components: main program, serial driver, result and reference group databases, HTML-report templates, and measuring device. The driver is written with MSVC++ and the main program, which consists of about 190 VIs, is written with LabVIEW.

Tecono Oy/Logos Test Engineering department together with Medikro Oy has developed a sophisticated medical application for analyzing the ventilation of human lungs. SpiroPro is a Windows application software for the Medikro spirometer system. The measuring system consists of a very small-scale spirometer developed by Medikro Oy. The spirometer includes a disposable flowhead (Fleish-type pneumotachometer), tubing, and spirometer unit (pressure sensor and A/D conversion unit) attached to a serial port (Figure 4-23). The system is commercial and sold to hospitals and private clinics by Medikro Oy.

Typical Use of Spirometer

A spirometer is used to measure accurately the ventilation of the lungs as well as the nature and the degree of functional disorder such as obstruction or restriction. Typically, the test includes measuring lung volume as well as the flow of forced inspiration or expiration. The results are expressed in universally agreed upon parameters, such as vital capacity, peak flow, forced expiratory volume at 1 second, and maximum voluntary ventilation. These and several other parameters are compared with a reference value, which is based on research conducted on a large group of people. The reference value depends on the age, gender, size, and ethnicity of the patient.

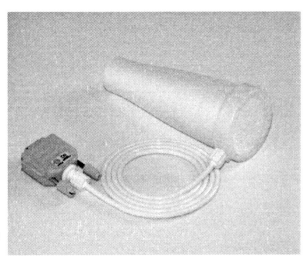

Figure 4-23.
Spirometer unit: flowhead with tubing.

SpiroPro Software

The SpiroPro software not only had to have an easy-to-use graphical interface but also had to meet several objectives:

- The measurement had to be real time.
- The signal rate was 100 Hz.
- The signal needed to be linearized.
- All necessary parameters were to be calculated.
- A new patient database needed to be created for flexible storage of measured signal and patient data.
- Reference groups were to be incorporated and flexibly stored in the database.
- A new graphical user interface for the measurement was to be developed.
- A Windows driver was to be developed.
- A novel HTML format was to be developed for final reports.
- Localization issues were to be included (dimensions, languages, local reference values).
- Users needed to be able to send results to a consulting doctor via e-mail (and only specified results related to desired patient session, not the whole database).

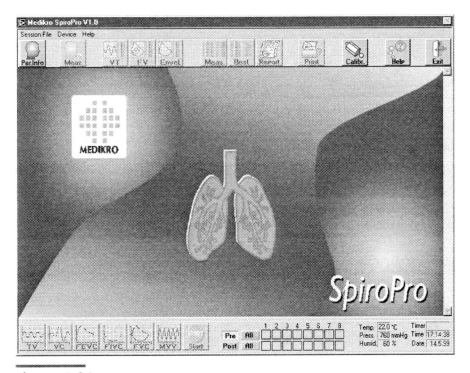

Figure 4-24.
SpiroPro's Initialize view.

All these objectives were met, and early tests with the Pulmonary Lung Function Generator indicate that the new software meets or exceeds the spirometer standards outlined by the American Thoracic Society (1994 update).

Typical Use of SpiroPro

After the user has logged into the program, there are two possible actions: start interpretation with patient and patient data or execute the volume calibration. In Figure 4-25 we have chosen to retrieve an existing session from the Microsoft Access database. This session is understood as a document that consists of one or several different measurements from one patient during one day.

From the Personal Information view (Figure 4-25), the physician will most likely start a new measurement with the patient or analyze the existing session and write interpretation text based on the measured values. In either case, the next step is to get familiar with the Measure view (Figure 4-26).

From the bottom of the window the user gets valuable information of ambient parameters and a rough picture of how many signals have been acquired for different phases (pre- and

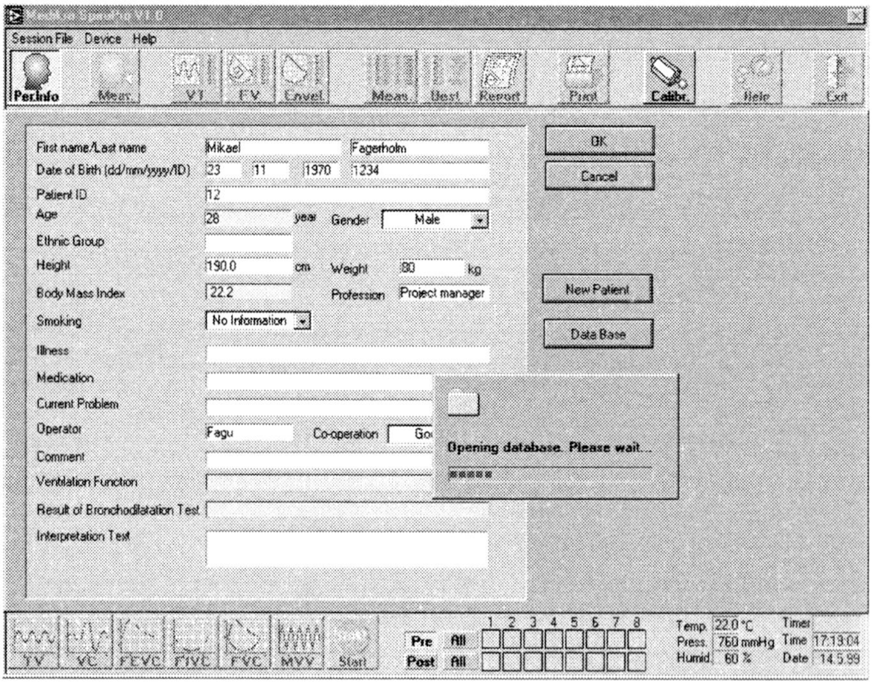

Figure 4-25.
Retrieving a session from the patient database.

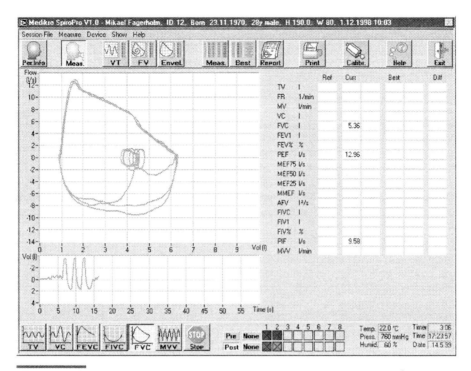

Figure 4-26.
Measure view during the acquisition.

post-phase). New measurements are made easily by pressing the Start button (Figure 4-26) or pressing one of the measure type buttons (TV, VC, FEVC, FIVC, FVC, or MVV). Different measure types can be established in one signal!

During the measurement some result variables, depending on the selected measure type, are calculated during the signal acquisition (Figure 4-26: PEF, FVC, and PIF). Other measure type variables are calculated after the acquisition has terminated, if the user accepts the signal.

Acquiring and displaying the flow values is just half the truth. What is perhaps even more important is analyzing the signal: calculating result variables from all signals belonging to the current session and comparing variables with predicted values. To get different graphical presentations of session signals, the physician might browse signals in Volume-Time graph (VT button), Flow-Volume graph (FV button), or have an Envelope presentation in Flow-Volume graph (Envel button). To get more detailed feedback, the user can view calculated variables in a spreadsheet-like presentation (Figure 4-27). This also includes a centered bar graph. Markers represent the calculated variables in proportion to predicted value range (green bar graphs).

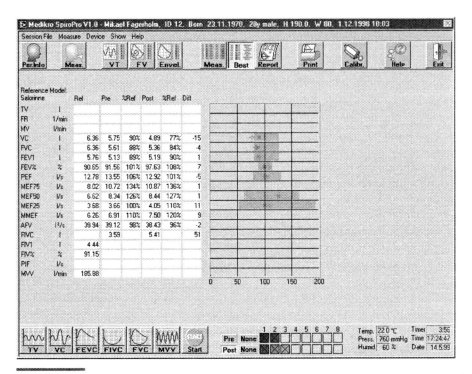

Figure 4-27.
Representative results and Histogram view.

The system's open design provides an easy upgrade path to different kinds of other measurements (e.g., cardiological). Future demands to monitor ventilation function remotely in real time will see the incorporation of DataSocket. This technology will probably be included in the next release version.

One key element in the development project has been a very carefully designed user interface. The interface makes advanced use of LabVIEW and uses very dynamic HTML report generation (Figure 4-28) to meet the varying needs of hospitals. Each end-user can easily modify the contents and the appearance of the final report. This state-of-the-art software includes a total of 190 VIs.

The final reports can be edited with common HTML editors such as Microsoft FrontPage. Contents of the report are selected during the execution of SpiroPro. For example, the contents of the graphs in the full report are selected by using the Full Report Wizard (Figure 4-29).

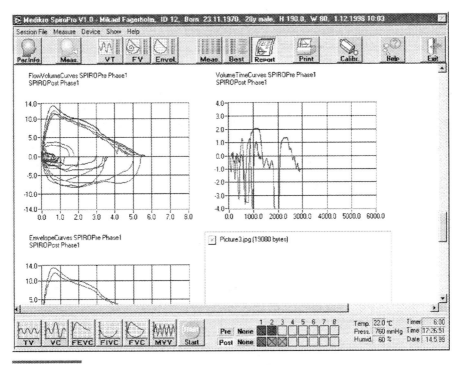

Figure 4-28.
Report view—Full report edited in HTML.

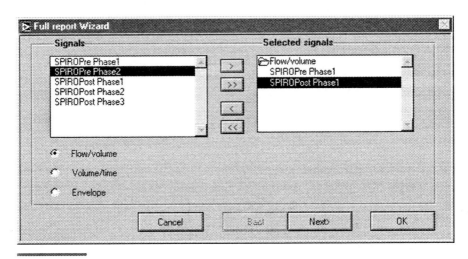

Figure 4-29.
Full Report Wizard—Easy report configuration.

Part III
Clinical
Applications

Cardiopulmonary
Applications

5

IntelliVent™ Data Logger: A Real-time Research Tool for Data
Acquisition, Analysis, and Display of Ventilatory Parameters
       Applications
       Hardware

Virtual instrumentation (VI) allows hospitals and research and educational institutions to conceive, develop, and implement biomedical applications that both achieve enhanced clinical interpretations and result in significant cost reductions. Premise Development Corporation, together with Hartford Hospital, developed several innovative applications that demonstrate significant opportunities for the use of VIs in the clinical environment.

In a collaborative approach, physicians, researchers, and biomedical and software engineers developed data acquisition and analysis systems that successfully integrate virtual instrumentation principles in a noninvasive manner. This chapter describes several LabVIEW™-based research applications that quantify and assess cardiac and pulmonary function.

The Cardiopulmonary Measurement System (CMS)

The Cardiopulmonary Measurement System (CMS) is an easy-to-use data acquisition and analysis application that interfaces with a mass spectrometer, or any rapid responding multigas analyzer, and a flow/volume measurement device. The LabVIEW-based system is fully automated and can run on a standard computer.

The CMS offers pulmonologists, cardiologists, exercise physiologists, and researchers a choice between the single-breath and rebreathing method. The capabilities of this system offer a unique and powerful research tool for the study of pulmonary diffusion and cardiac parameters in a variety of populations ranging from patients with chronic pulmonary disease to elite athletes.

Background

Pulmonary blood flow, which is proportional to cardiac output in normal subjects, may be determined by rebreathing a gas mixture containing a foreign soluble gas (C_2H_2 or N_2O) for periods of time not exceeding 30 seconds. If no diffusion limitation is present, the rate of uptake of the foreign gas is a function of (1) the solubility coefficient of the gas, and (2) pulmonary blood flow (i.e., cardiac output).

A custom application using LabVIEW was developed to measure the carbon monoxide diffusing capacity of the lung, a function of the lung's ability to transfer oxygen to the blood. In addition, this system is also able to noninvasively measure lung volumes, lung tissue volume, oxygen consumption, and cardiac output. The ability to measure cardiac function in conjunction with pulmonary function is desirable because (1) a change in cardiac output affects measured diffusion capacity, and (2) it can be determined whether a patient suffers from a pulmonary disorder, a cardiac disorder, or both.

A unique feature of this system is that it has the ability to measure diffusing capacity by either a single-breath or a rebreathing technique. The system has been independently tested and validated against commercially available products (Warren E. Collins, Inc., Braintree, MA) and the standards set forth by the American Thoracic Society (*American Review of Respiratory Disease*, 1975).

The System

An integrated LabVIEW-based Cardiopulmonary Measurement System was developed to measure oxygen consumption, total lung capacity, diffusion capacity, membrane diffusing capacity, and cardiac output.

Figure 5-1 illustrates the system's inputs (indicators) and outputs (controls). Specifically, gas concentrations are measured with a Random Access Mass Spectrometer (RAMS™; General Electric Medical Systems, Milwaukee, WI), flow and volume are measured via a Hans Rudolph's RSS 100HR Research Pneumotach (PNT) System (Hans Rudolph, Inc., Kansas City, MO).

The CMS typically employs a rebreathing technique that is based on the rebreathing procedure described by Sackner and associates in 1975. The custom data acquisition and instrument control software was written in LabVIEW 5.1. The system can acquire up to 16 differential analog inputs at a sampling frequency of 50 to 200 Hz. A 12-bit analog-to-digital (National Instruments' DAQ-6062E) conversion board is used to digitize each channel. A digital input/output board (National Instruments' DAQ-DIO 24) is also used to control a series of relays that in turn control the mass spectrometer and operate pneumatic lines to inject test gas and evacuate residual gases from the rebreathing

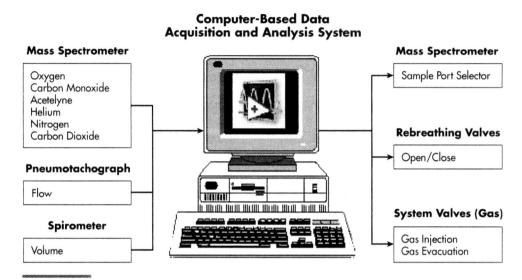

Figure 5-1.
System diagram: This image schematically illustrates the specific input parameters and output controls involved in the Cardiopulmonary Measurement System.

chamber. In addition, digital controls are also used to momentarily switch the patient from breathing room air to the rebreathing chamber.

A sub-VI was also developed to test the integration of all individual system components. Specifically, it validates the data acquisition board's ability to acquire analog input readings from the RAMS and PNT systems. In addition, the sub-VI tests digital outputs, which are output to the Electromechanical Relay System (National Instruments' ER-16). The ER-16 programmatically orchestrates the triggering of the Pneutronics' Digital Valves (Parker Instrumentation, Pneutronics Division, Hollis, NH) that in turn control the flow of gas through a pneumatic manifold system into the pneumatic lines. The digital valve and manifold system automate the process of gas multiplexing between different pulmonary function test gases and the vacuum/suction port. The digital valve also controls the toggling between the RAMS capillary sample line and pneumatic air lines that open and close the three-way pneumatic valve in the mouthpiece.

Finally, all the aforementioned system's components were assembled into a META 4 Metabolic/Pulmonary Test System (General Electric Medical Systems, Milwaukee, WI), hence making this system highly mobile and compact

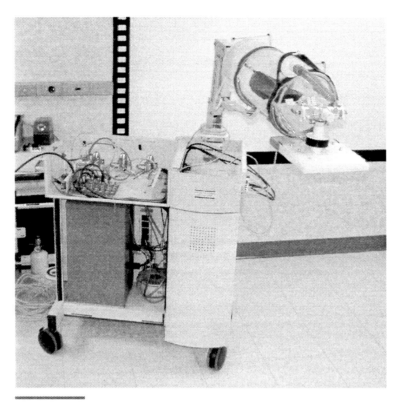

Figure 5-2.
Photograph of the CMS.

enough to move around the hospital setting with ease. The portable CMS is shown in Figure 5-2.

Figure 5-3 illustrates the analog inputs and data acquisition hardware of the CMS.

Digital outputs from the DAQ-DIO 24 are conditioned through a National Instruments' ER-16 Electromechanical Relay Board to control the mass spectrometer and to activate pneumatic solenoids, which in turn drive both patient and system valves. Both the RAMS and the Hans Rudolph's RSS 100HR Research Pneumotach (PNT) System have been configured to communicate serially through the RS-232 protocol and have been integrated into the LabVIEW application. Figures 5-4 and 5-5 illustrate both the digital control logic as well as the flow of the various gas, pneumatic, and vacuum lines.

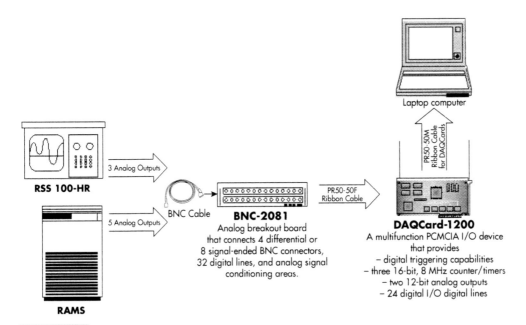

Figure 5-3.
Schematic diagram of analog inputs and data acquisition hardware.

This CMS's LabVIEW software program performs three primary functions:

- Instrument control
- Data acquisition
- Data analysis and presentation

These primary functions are executed through a series of sub-VIs. A block diagram of one of these sub-VIs is shown in Figure 5-6. This diagram illustrates how LabVIEW's case structure embedded within a while loop can function as a state-machine. In this case, the sub-VI or state-machine is used to configure the mass spectrometer, the vacuum and gas valves, and the pneumotachograph.

The LabVIEW software enables an intuitive graphical user interface (GUI) based on the principles of a virtual instrument. In effect, the computer is transformed into a VI designed specifically for cardiopulmonary assessment.

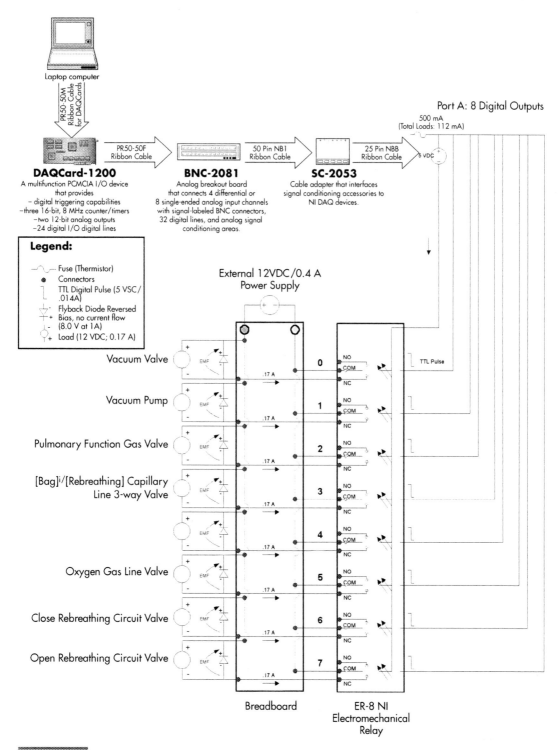

Figure 5-4.
Digital output control diagram of the Cardiopulmonary Measurement System.

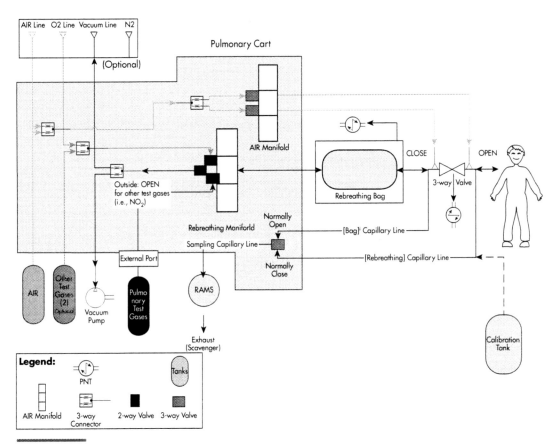

Figure 5-5.
Schematic diagram of the Cardiopulmonary Measurement System's pneumatic circuits.

Some of the program's unique features include:

- Option for single-breath, rebreathing, or dual technique
- Noninvasive measurement of cardiac parameters (cardiac output)
- Digital I/O (control of mass spectrometer, solenoids, relays, valves)
- Automated calibration of volumes, flow rates, and gas concentrations
- Data acquisition of signals from mass spectrometer, pneumotachs, thermocouples, and differential pressure transducers
- Automated gas injection and evacuation

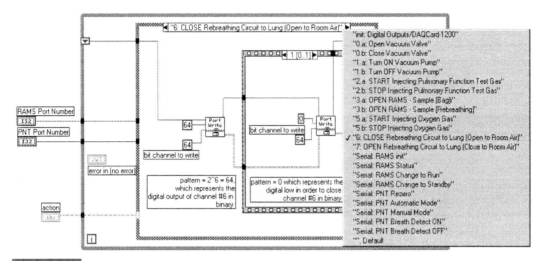

Figure 5-6.
Sub-VI (or state machine) that controls the mass spectrometer, the vacuum and gas valves, and the pneumotachograph.

- Automated detection of end of expiration and regression slopes
- Customized reports for physicians
- Remote accessibility and control of system via modem
- Interactive multimedia: audio (voice instructions, warnings, and error messages), animation and video (via Microsoft ActiveX technology such as interactive Agents to instruct patient(s) or staff)

How the Cardiopulmonary Measurement System Works

Upon launching this application, the user is presented with a top-level VI and is given four options from which to choose: performing a test, recalling a previous test, system calibration, or exiting the program (see Figure 5-7). In addition, several multimedia options are available to provide support for both the user and the patient in the form of error messages, warnings, and audio and video instructions on specific test procedures.

When performing a new test, the user is presented with the Test Selection screen shown in Figure 5-8. This screen allows the user to perform a test using

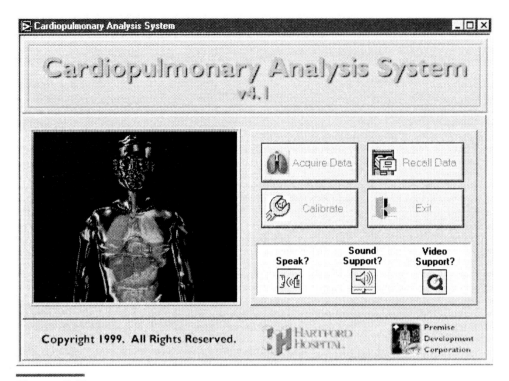

Figure 5-7.
The Main Menu allows the user to select options to calibrate system components, perform a new test, recall a previous test, or exit the application.

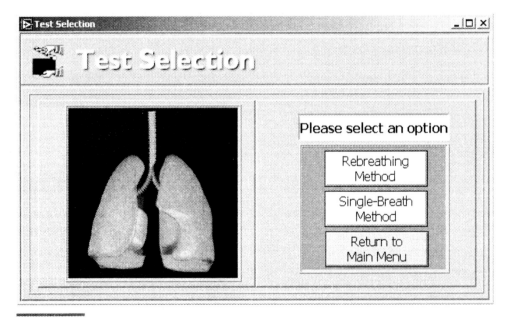

Figure 5-8.
Test Selection screen.

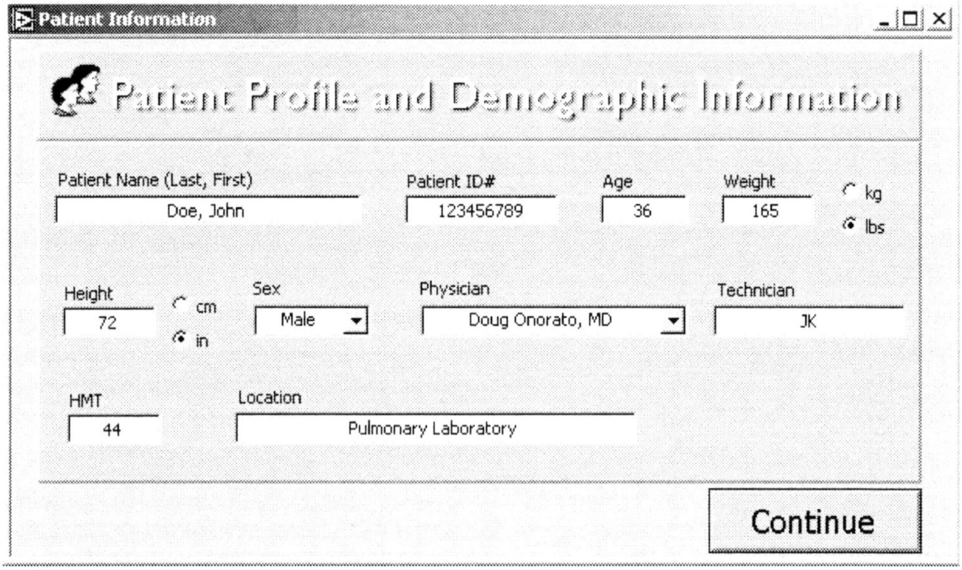

Figure 5-9.
Patient Information screen.

either the rebreathing or the single-breath protocol. Alternatively, the user may return to the Main Menu.

When a new patient is tested on the CMS, the user enters the patient's data in the Patient Information screen (see Figure 5-9). The user can select and verify previously tested patient data in the Patient Information screen as well. All patient profiles and demographic information are stored in a relational database.

After patient information has been entered, the CMS would automatically record the current environmental conditions through pressure, temperature, and humidity sensors that provide analog outputs. As shown in Figure 5-10, the current environmental conditions (barometric pressure, temperature, and humidity) are automatically inserted into sliders to be confirmed by the user.

The patient's inspiratory capacity is either entered or calculated into the Setup Module so that the CMS can fill the rebreathing chamber with the appropriate amount of test gas. The inspiratory capacity is either determined in a previously performed spirometry test or estimated from an algorithm based upon the patient's age, sex, height, and weight. The slider control on the

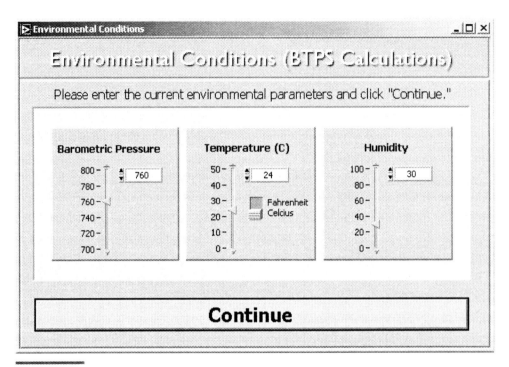

Figure 5-10.
Environmental conditions (barometric pressure, temperature, and humidity).

left side of the Setup VI (see Figure 5-11) allows the user to adjust the inspiratory capacity, and the Bag Volume indicator on the right side of the Setup VI displays the volume of gas entered into the rebreathing chamber. Ideally, the two numbers should be within 3% of each other.

When performing a study, the patient breathes from a two-way valve that can switch from room air to a special gas mixture containing 10% helium (He), 0.5% acetylene (C_2H_2), 0.3% carbon monoxide ($C^{18}O$), and a balance of oxygen (O_2). In order to achieve accurate pulmonary measurements, it is very important to switch the patient into the breathing circuit when the patient is at end-expiration or when the patient has fully exhaled (to his or her residual volume). Most commercially available systems typically rely on the patient to determine this point and signal the pulmonary technician that he or she is at end-expiration (often by a hand signal), at which point the technician manually switches the patient into the breathing circuit. In the case of the CMS, however, the system is able to automatically determine end-expiration by employing some simple logic on both the flow and the carbon dioxide channels

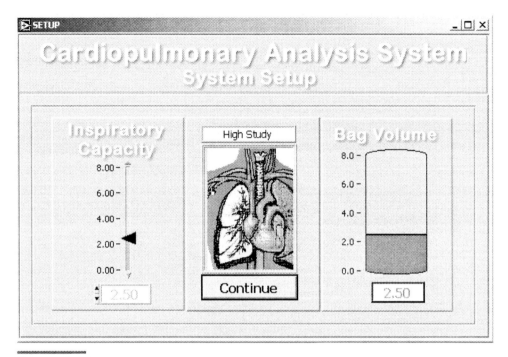

Figure 5-11.
Setup Module for bag-in-a-box chamber.

across the patient's rebreathing valve. Flow will be momentarily equal to zero at the end of inspiration and expiration, but the carbon dioxide level will greatly exceed ambient conditions (typically 0.03 to 0.05%) only during expiration. Therefore, the CMS can accurately and consistently determine end-expiration when

$$Flow = 0$$

$$Carbon\ dioxide > 2\%$$

These two values are continuously measured in the End-Tidal Breath Determination VI (see Figure 5-12). The user typically clicks on the Exhalation button while the patient is exhaling, and the system will automatically determine end-expiration. Alternatively, the user can invoke the manual valve control to instantaneously switch the patient into the test circuit.

During a test sequence, the system represents a closed circuit consisting of the patient's lungs and airway, the valve apparatus, and the airtight anesthesia bag containing the test gas. A sample port near the valve's mouthpiece allows the mass spectrometer to measure changing gas concentrations during

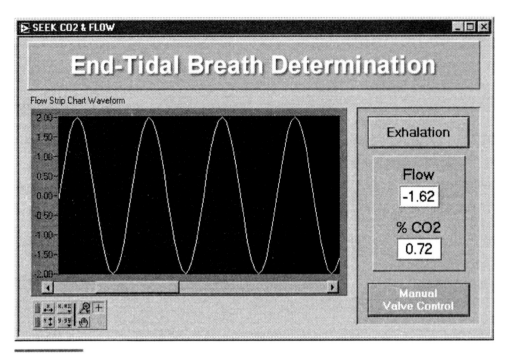

Figure 5-12.
End-Tidal Breath Determination.

the study, along with the respective flow and volume signals. The sampling rate on all channels is typically 50 Hz (but can be up to 200 Hz). Figure 5-13 illustrates a typical rebreathing pattern for a rebreathing protocol.

Lung Volumes: One of the simplest and most widely used methods of measuring lung capacities is with a spirometer. However, a spirometer can only measure ventilated volumes and flows, it cannot measure residual volumes (the amount of gas remaining in the lungs at the end of maximal expiration) or lung volumes that include residual volumes (i.e., total lung capacities and functional residual capacity). The advantage of the diffusion system is that it can measure these absolute volumes via a helium or methane dilution technique.

Helium (He) and methane (CH_4) are inert gases that are unable to diffuse through the lung tissue. As the patient breathes, the gas eventually equilibrates in the lung and the bag. Since no helium or methane has been lost or absorbed by the body, the amount of gas present before equilibration and after equilibration must be equal. This can be expressed in the equation

$$C_1 * V_1 = C_2 * V_2$$

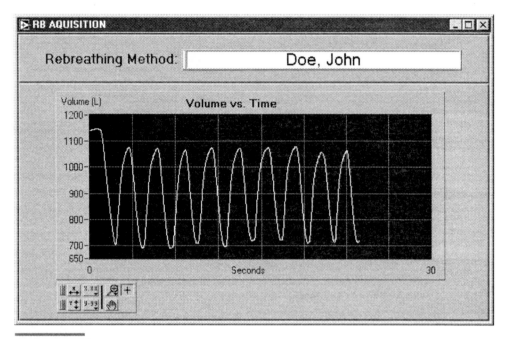

Figure 5-13.
Rebreathing Data Acquisition Module.

where C_1 and V_1 are the initial concentration and volume, and C_2 and V_2 are the final concentration and volume, respectively. This is illustrated in Figure 5-14. Therefore, by knowing the initial concentration of helium (or methane), the initial volume of the bag, and the final helium (or methane) equilibrium concentration, the system volume (i.e., volume of the bag plus the lung volume) can be calculated from the equation

$$(Fa_{He_{initial}})(V_{initial}) = (Fa_{He_{final}})(V_{system})$$

where

$$Fa_{He_{initial}} = \text{the initial concentration of helium (or methane)}$$
$$V_{initial} = \text{the initial volume of the rebreathing bag}$$
$$Fa_{He_{final}} = \text{the final concentration of helium (or methane)}$$

and

$$V_{system} = V_{dead\ space} + V_{initial} + V_{lungs}$$

where $V_{dead\ space}$ is the mechanical and physiological dead space, and V_{lungs} is the initial volume of the lungs (based on protocol, usually the residual volume or functional residual capacity).

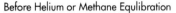

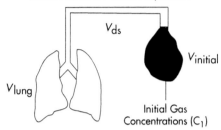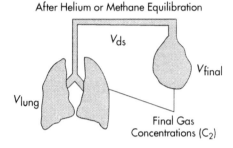

Figure 5-14.
Helium/methane equilibration of the lungs.

From the system volume, the lung volume is easily calculated knowing the initial bag volume and dead space. However, the volume of the lung that is actually calculated depends on the patient's maneuver prior to the initial inspiration from the bag. For example, if the patient breathes normally (at tidal volume) and then inhales from the bag, V_{lung} will be approximately his or her functional residual capacity (FRC). Based on current protocol, the patient is instructed to expire to residual volume before beginning the test. Therefore, the volume that is calculated is the patient's residual volume (RV). The exact equation is

$$RV = V_{system} - (V_{inspired} + V_{dead\ space})\ [BTPS*]$$

At the end of the test, the patient is instructed to inspire deeply in the bag. This measures his or her inspired volume and is added to the residual volume to calculate the patient's total lung capacity:

$$TLC = RV + V_{inspired}$$

where $V_{inspired} = V_{initial}$ (initial bag volume) $- V_{residual}$ (the residual volume in the bag after a maximal inspiration).

To simplify the calculation, the previous two equations are combined and expressed in the following equation:

$$TLC = V_{system} - (V_{dead\ space} + V_{residual})\ [BTPS]$$

Vital Capacity: Vital capacity will be defined as total lung capacity less residual volume. Alternatively, it can be measured by having the patient maximally inhale followed by maximal exhalation. The total exhaled volume will be equal to vital capacity (forced vital capacity).

The front panel in Figure 5-15 illustrates a data set from a rebreathing diffusion study. In this case, the helium gas array is displayed over a 30-second rebreathing interval. This VI measures the initial and final helium concentra-

---

*Barometric Temperature and Pressure Saturated

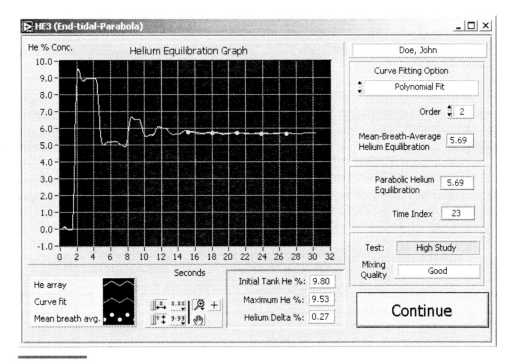

Figure 5-15.
Helium equilibration curve: This module profiles the rate at which helium (and the other soluble gases) is absorbed into the lungs.

tions (to calculate system volume and total lung capacity); it is also used to assess the mix quality of the test and to determine if there were any leaks during the rebreathing interval. Based on these criteria, a lookup table of descriptive sentences is accessed and the appropriate ASCII strings are concatenated to form a computer interpretation of each study.

Once the system volume has been determined, the operator proceeds to the next set of regression analyses by clicking on the Continue button. The user is presented with a module that determines the time-0 intercept value.

The time-0 correction is based on the premise that since different gases behave differently in the lung compartment, a delay would be introduced in the calculations in order to let all the gas contents in the rebreathing bag compartment equilibrate with the lung compartment. This was accomplished by observing the time at which helium (He) equilibrated between the two compartments (as shown in Figure 5-15). Once He has equilibrated in both compartments, it was assumed that the other gases in the mixture, that is, acetylene (C_2H_2) and carbon monoxide ($C^{18}O$), have equilibrated as well. At this point, the $C^{18}O$ and C_2H_2 disappearance curves were forced to intercept at time-0

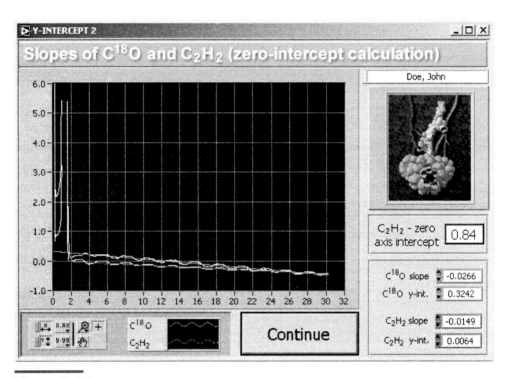

Figure 5-16.
Time-Zero Determination Module.

and have the same origin. From this moment on, the rate of disappearance of each gas was observed and calculations drawn. Figure 5-16 illustrates the determination of the time-0 correction concept.

Historical Note: Although the technique of rebreathing of C_2H_2 has been used for many years to estimate cardiac output, it was not until the advent of the rapid mass spectrometers in the second half of this century that quick and reliable results were obtained with several gas rebreathing methods. Marvin Sackner et al. (1975) proposed one such rebreathing method, based on procedures by Cander and Forster (1959). Sackner and his colleagues developed the time-0 correction where a mixture of gases was rebreathed for 20 to 30 seconds in a closed circuit into a bag. By sampling the rate of disappearance of each gas in the rebreathing circuit using a rapid mass spectrometer and by normalizing the data according to the time-0 technique, the diffusing capacity (DLCO), membrane diffusing capacity (DM), capillary blood volume (VC), pulmonary tissue volume (Vt), and cardiac output (Qc) could be estimated repeatedly and reliably.

Researchers have demonstrated that the slope of a linear regression through the natural logarithm of the acetylene, carbon monoxide, and oxygen disappearance curves will correspond to cardiac output (Qc), diffusion capacity of the lung (DLCO), and oxygen consumption (VO_2), respectively. These values are displayed in Figure 5-17. Note that the linear regressions through the acetylene and carbon monoxide arrays are using the last two-thirds of the mean-breath average value for each breath. The mean-breath average is the average concentration value of each breath, represented by a single point within each breath cycle. The linear regression through the oxygen array is based on the end-tidal value for each breath.

Cardiac Output: Cardiac output (Qc) is equal to the product of stroke volume (ml) and heart rate. Two methods will be used to determine cardiac output: thermodilution using a pulmonary artery catheter (PAC) and the acetylene rebreathing technique. The rate of C_2H_2 disappearance is directly

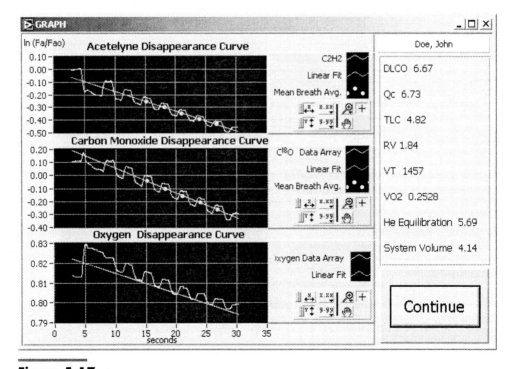

Figure 5-17.
Soluble gas disappearance curves: This module quantifies the rate at which acetelyne, carbon monoxide, and oxygen are absorbed into the lungs. Each disappearance curve (which is exponential) is linearized by plotting it as a function of a natural logarithm against time.

proportional to cardiac output. Cardiac output can therefore be calculated using the following modification of Fick's equation (Cander & Forster, 1959):

$$\dot{Q}_c = \left[\frac{\left(V_{system} \frac{760}{PB - 47} \right) + \alpha_t Vt}{\alpha_b(t_1 - t_2)} \right] * \left[\ln \frac{Fa_{(t_2)}}{Fa_{(t_1)}} \right]$$

where:

$\dot{Q}_c$ = cardiac output
V_{system} = system volume (alveoli volume + bag volume)
$PB - 47$ = barometric pressure minus water vapor pressure
α_t = Bunsen solubility coefficient of C_2H_2 in pulmonary tissue
α_b = Bunsen solubility coefficient of C_2H_2 in blood
Vt = pulmonary tissue volume

$\left[\ln \frac{Fa_{(t_2)}}{Fa_{(t_1)}} \right]$ = slope of the C_2H_2 regression line

Diffusion Capacity: Diffusion capacity of carbon monoxide will be defined as the rate at which carbon monoxide is transferred from the alveolar to the pulmonary capillary per minute for each mmHg of partial pressure difference. The technique requires the patient to rebreathe a gas mixture of known concentrations (specifically, carbon monoxide, helium or methane, oxygen, and acetylene) for 20 to 30 seconds. By measuring the disappearance curves of each gas, diffusion capacity (DLCO), total lung capacity (TLC), cardiac output (Qc), lung tissue volume (VT), and oxygen consumption (VO2) can be determined. The rate of disappearance of carbon monoxide is related to the following equation:

$$\frac{dV_{co}}{dt} = \frac{D_{L_{co}}}{P_{a_{co}}}$$

where:

dV_{co} = the rate of disappearance of CO
$D_{L_{co}}$ = the diffusion capacity of the lung
$P_{a_{co}}$ = the partial pressure of CO within the alveoli

Oxygen Consumption: Oxygen consumption (VO_2) is the amount of oxygen the body uses per minute. It is directly related to cardiac output and ventilation. The measurement of O_2 consumption is simple in principle. The amount of oxygen consumed is equal to the amount of oxygen inspired minus the amount of oxygen expired (Cotes, 1979):

$$\frac{d\,\mathrm{VO_2}}{dt} = \left(\frac{V_1}{dt}\mathrm{FiO_2} - \frac{V_E}{dt}\mathrm{FeO_2}\right)$$

where

$\dfrac{d\,\mathrm{VO_2}}{dt}$ = consumption of oxygen per minute

$\dfrac{V_1}{dt}$ = volume of gas inspired per minute

$\mathrm{FiO_2}$ = inspired concentration of oxygen

$\dfrac{V_E}{dt}$ = volume of gas expired per minute

$\mathrm{FeO_2}$ = expired concentration of oxygen

However, the actual measurement of $\mathrm{FiO_2}$ and $\mathrm{FeO_2}$ is unnecessary. During the rebreathing maneuver, the patient breathes from a known volume of mixed gases, which includes oxygen. The rate at which the oxygen disappears is equal to the rate at which it is consumed. Therefore oxygen consumption can be calculated directly using the slope of the oxygen regression line:

$$\dot{V}_{\mathrm{O_2}} = (V_{\mathrm{system}}) \left[\frac{d\,\mathrm{FaO_2}}{dt}\right]\left[\frac{60}{\Delta t}\right]$$

where

(V_{system}) = system volume (alveoli volume + bag volume)

$\left[\dfrac{d\,\mathrm{FaO_2}}{dt}\right]$ = slope of oxygen regression line

$\left[\dfrac{60}{\Delta t}\right]$ = conversion factor to uptake per minute

Upon completing a DLCO study, a final report is generated. This report compares the actual measured values with a set of predicted values based on the patient's age, height, and sex. In addition, a detailed computer interpretation is displayed as a string indicator. By using Microsoft Agent Technology, or any other text-to-speech engine, it is even possible to read or speak the ASCII string directly from the computer. A sample report is shown in Figure 5-18.

System Validation: This system was tested in a NASA-supported study at The Institute for Exercise and Environmental Medicine in Dallas, Texas. The

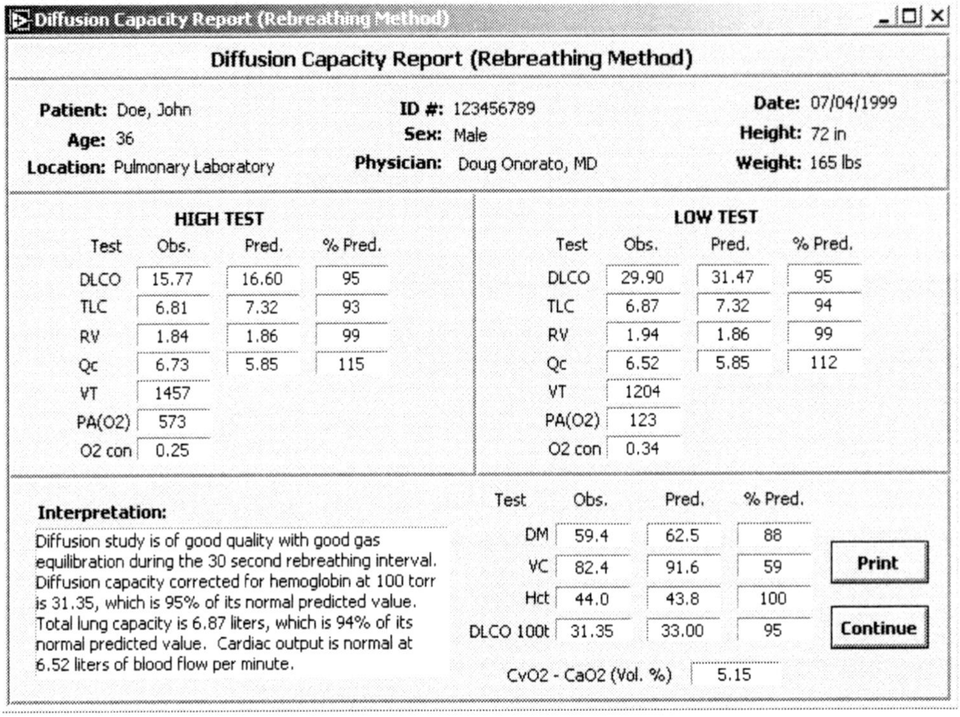

Diffusion Capacity Report (Rebreathing Method)

Patient:	Doe, John	**ID #:** 123456789	**Date:** 07/04/1999
Age:	36	**Sex:** Male	**Height:** 72 in
Location:	Pulmonary Laboratory	**Physician:** Doug Onorato, MD	**Weight:** 165 lbs

HIGH TEST

Test	Obs.	Pred.	% Pred.
DLCO	15.77	16.60	95
TLC	6.81	7.32	93
RV	1.84	1.86	99
Qc	6.73	5.85	115
VT	1457		
PA(O2)	573		
O2 con	0.25		

LOW TEST

Test	Obs.	Pred.	% Pred.
DLCO	29.90	31.47	95
TLC	6.87	7.32	94
RV	1.94	1.86	99
Qc	6.52	5.85	112
VT	1204		
PA(O2)	123		
O2 con	0.34		

Interpretation:

Diffusion study is of good quality with good gas equilibration during the 30 second rebreathing interval. Diffusion capacity corrected for hemoglobin at 100 torr is 31.35, which is 95% of its normal predicted value. Total lung capacity is 6.87 liters, which is 94% of its normal predicted value. Cardiac output is normal at 6.52 liters of blood flow per minute.

Test	Obs.	Pred.	% Pred.
DM	59.4	62.5	88
VC	82.4	91.6	59
Hct	44.0	43.8	100
DLCO 100t	31.35	33.00	95

CvO2 - CaO2 (Vol. %)   5.15

Print

Continue

Figure 5-18.
Sample report for a Rebreathing Protocol Test.

main objective of this study was to compare cardiac output determination via the rebreathing technique with direct Fick and thermodilution methods. Eighteen volunteer subjects participated in this validation study. Four rebreathing measurements were performed at five-minute intervals during supine rest, standing rest, and 100-watt cycle ergometer exercise. The results of this study revealed a strong correlation between the soluble gas rebreathing technique and the direct Fick and thermodilution methods ($r=0.92$). Figures 5-19 and 5-20 illustrate the correlation among the methods used to determine cardiac output (Pawelczyk et. al., 1995).

Clinical Significance

Noninvasive diffusion capacity and cardiac output measurement by this soluble gas absorption technique can be of tremendous clinical importance. Other noninvasive techniques for assessing cardiopulmonary performance,

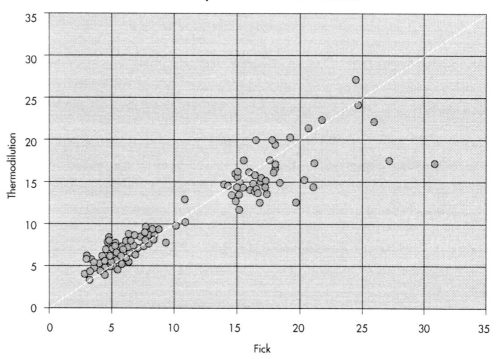

Figure 5-19.
Comparison of cardiac output measurements (L/min): Thermodilution vs. direct Fick method.

specifically cardiac output, include echocardiography, bioimpedance cardiography, arterial pulse analysis, magnetic resonance imaging (MRI), and ultrafast computed tomography (CT). These techniques, however, are limited by impracticality (MRI and CT), high cost (echocardiography), imprecision, and inaccuracy.

Invasive techniques, such as the pulmonary artery catheter (PAC), are most frequently used for assessing ventricular performance by thermodilution measurement of cardiac output. Placement of the PAC, however, is extremely invasive, requiring percutaneous cannulation of large central veins, such as the internal jugular or subclavian veins. This can result in significant morbidity to the patient; namely, injury to the carotid or subclavian vessels, thrombosis, hemorrhage, pneumothorax, and nerve trauma. Other reported complications from routine use of the PAC include arrhythmias, infection, pulmonary infarction, and pulmonary artery rupture (PAR).

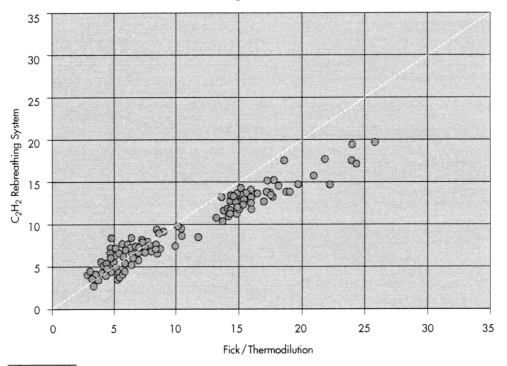

Figure 5-20.
Comparison of cardiac output measurements (L/min): Soluble gas technique vs.
Fick/thermodilution.

The Fick method is the method currently used for cardiac output determination in most cardiac laboratories. The arterial-venous oxygen saturation is
measured by sampling blood from the pulmonary artery and from the aorta.
Because of the technical difficulties and the time required for accurate collection of expired gas in a collecting or Douglas bag, the value for the oxygen
consumption is assumed based on standard nomograms. This is a standard
practice in many catheterization laboratories. Recently this practice has been
called into question and researchers have concluded that the oxygen consumption should be measured in order to accurately assess the cardiac output.

The accurate determination of cardiac output and oxygen consumption is
essential in a variety of cardiac conditions. Cardiac output measurements
help clinicians make appropriate decisions regarding diagnosis or alter therapy in a number of conditions. Clinically, it can provide hemodynamic characterization essential for the design of an appropriate medical regimen and

prognosis in a critically ill patient. Experimentally, it can provide an assessment to quantitatively examine the effects of physical training on maximal aerobic power (VO_2max) in an athlete.

The CMS has the advantage of being completely noninvasive in that a right heart catheter does not have to be inserted for the purpose of obtaining an arterial-venous oxygen difference. Thus, the CMS has potential widespread use throughout the hospital in such settings as intensive care units, general cardiology units, and exercise laboratories. The availability of the information provided by this technique may allow for the optimization of therapy in patients with a wide variety of cardiac pathology, including, but not limited to, congestive heart failure, ischemic heart disease, and valvular heart disease.

Summary

Virtual instrumentation allows the development and implementation of cost-effective biomedical applications. As the healthcare industry continues to respond to the growing trends of managed care and capitation, it is imperative for clinically useful, cost-effective technologies to be developed and utilized. Thus, it is clear that virtual instrumentation will continue to play a growing role in medical instrumentation, clinical assessment, and treatment.

Validation of a LabVIEW-based Vibrotactile Stimulation System to Treat Apnea of Prematurity

Human gestation consists of continual growth, including physiological and neurological development. Infants born preterm (prior to 37 weeks) are more susceptible to impairments as a result of their incomplete development. Respiratory impairment is a particular risk, including pauses in breathing (apnea), a direct result of physiological and neuronal immaturity of the respiratory control system. Incidences of apnea have increased in recent years as the neonatal survival rate has increased due to improved medical care.

The incidence of apnea is particularly high in premature infants of low birth weight and inversely related to age. In preterm infants between 30 to 35 postconceptional weeks, an apnea incidence rate of 71% has been reported. It has been estimated that 24% of infants weighing less than 2,500 g at birth and 84% of infants under 1,000 g may experience apnea during the neonatal period.

Currently, treatment of apneic episodes is achieved with mechanical ventilatory assistance, pharmacological treatments, or physical stimuli. Medical personnel use physical stimulation, usually via hand contact, to terminate apneic events. This hand contact is often delayed and may present infection risks, due to inadequate hand washing between infants.

The concept of automatic mechanical stimulation via a vibratory stimulator was devised and patented by Dr. Leonard Eisenfeld (1996), a neonatologist at the Connecticut Children's Medical Center in Hartford, Connecticut. The objective of this project (called BabySave™) was to develop a software system to collect and analyze data to assess the efficacy and safety of mechanical vibrotactile stimulation (VTS) as a means to interrupt an apneic episode. The hospital's Institutional Review Board (IRB) has approved this study. Parental consent is also required.

Applications

Current monitoring technologies allow for trending of data, that is, storing waveforms for review of events of interest for further diagnosis. However, these systems are extraordinarily expensive, as they typically network multiple monitors from various patients. In addition, these trending systems are hardwired to record signals that originate from the patient monitors and cannot be adapted to record custom waveforms, such as those driving the vibrotactile stimulator.

For this reason, the BabySave data acquisition system was developed to acquire analog signals from a neonatal apnea monitor. In addition, analog signals were simultaneously acquired to indicate if stimulation was being applied and whether the stimulation was manual or from the VTS system. Software algorithms were developed to analyze recorded data and provide an output that would allow analysis of the efficacy and effectiveness of vibratory stimulation.

The VTS System

Figure 5-21 illustrates the main components of the VTS system. These include a computer, a LabVIEW software application, a patient monitor (model 511, CAS Medical Systems, Branford, CT), a vibrotactile stimulator (VBW32 transducer, Audiological Engineering Corporation, Somerville, MA), a foot pedal, a buffer/filter module, and an infrared module.

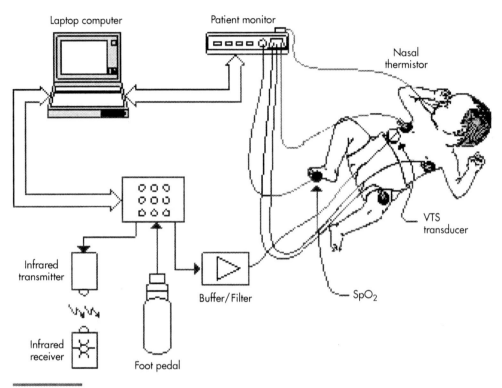

Figure 5-21.
System diagram of the vibrotactile stimulation system.

The computer records EKG, heart rate (from EKG), pulse rate (from SpO_2), thoracic impedance, nasal airflow, and SpO_2 waveforms from the monitor using a data acquisition card. The nurse may trigger a three-second, 250 Hz, square wave vibrotactile stimulus that will go to the VTS transducer. The nurse will step on the foot pedal when hand stimulation is being provided so the system can record that event.

An infrared transmitter has been developed to send the monitor's alarms to a receiver carried by the nurse. The VTS buffer/filter module limits the amount of current that can be drawn from the computer. It also provides electrical isolation for patient protection when no pulse has been applied by means of a relay. The signal sent to the transducer is generated in the computer. Different waveforms can be selected: sine, square, triangular, or sawtooth. Other variable parameters are the duration of the pulse and the amplitude and frequency of the signal.

Study Protocol

The goal is to study 20 apneic neonates for a minimum of 24 hours. The study is divided into VTS and hand stimulation periods. In the case of VTS, a transducer will be placed on the baby's thorax. At the beginning of the study the patient demographics are entered. The system was modified to allow random selection of the stimulation type (hand or VTS) that is to be used initially.

Figure 5-22 illustrates the Data Collection screen. The system has an apneic event counter. After five events the software will visually indicate to the nurse that the stimulation type has to be switched. If 12 hours pass and five events haven't occurred, the software will also indicate to the nurse to switch the stimulation method.

When an apneic event occurs and hand stimulation is being used, the nurse will step on a foot pedal, while the stimulation is applied. After three seconds

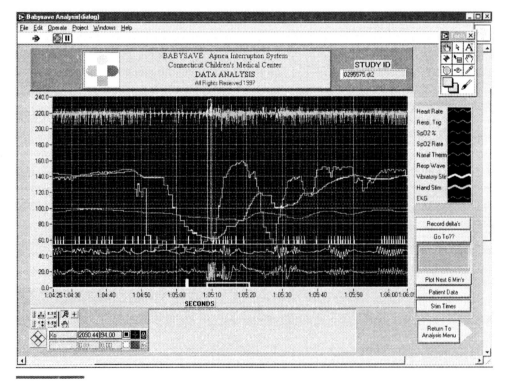

Figure 5-22.
Data Collection screen.

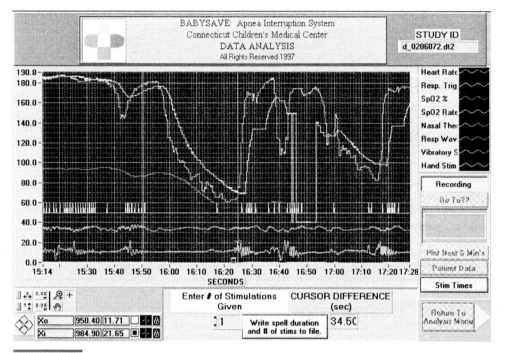

Figure 5-23.
Record Stimulation Information screen: User follows prompts to place cursors around areas of interest (spell duration, stimulation duration, etc.). Data is automatically calculated and exported to a spreadsheet.

of stimulation the computer will voice-prompt the nurse to stop. Then, after three more seconds, the software will voice-prompt the nurse to assess if the stimulus was or was not successful.

During the VTS period, the nurse will trigger a three-second pulse by either hitting the F1 key on the keyboard or clicking a button on the computer screen. Three seconds after the end of that pulse, the software will voice-prompt the nurse to assess if the stimulus was successful. Every 12 hours the system will indicate to the nurse to check for redness or skin breakdown in the stimulation site.

If breathing has not resumed after the computer prompts for assessment (during either hand or vibrotactile stimulation) the nurse will restimulate by hand or ventilate as is clinically appropriate. The nurse will mark this additional intervention by stepping on the foot pedal. Figures 5-23 through 5-27 illustrate the user interface and functionality of this research tool.

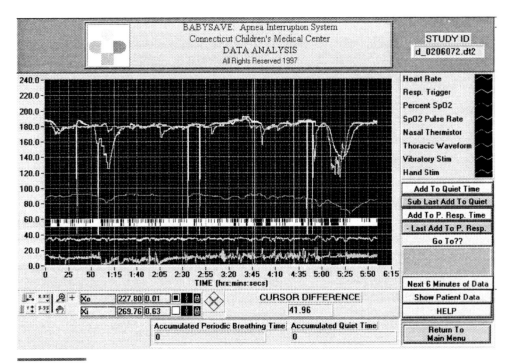

Figure 5-24.
Record Quiet Time and Periodic Respiration screen: User places cursors surrounding quiet time or periodic respiration and clicks on Add To Quiet Time or Add To P. Resp. Time. Output is in spreadsheet format. Yields percent quiet time and percent periodic respiration for entire study and total time of quiet time and periodic respiration for study.

Results

A function generator virtual instrument was developed with LabVIEW to drive the VTS transducer directly from the computer. No problems were encountered by driving an inductive load (the transducer) with the computer since the buffer/filter was designed and implemented. The previous system used a freestanding function generator. Voice prompts, visual indicators, and software-controlled timers now guide the nurses through the study.

Recording the EKG signal resulted in a larger data file. Secondary storage drives are recommended. The sampling rate for the waveforms is 50 Hz. The use of a higher sampling rate is required if diagnostic EKG is needed since a preterm baby has an average heart rate of 150 beats per minute. This sampling rate can be used in the future to derive information about respiration from the EKG signal and also to quantify cardiogenic artifact in the thoracic imped-

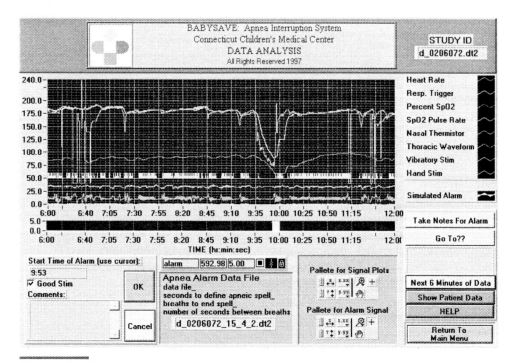

Figure 5-25.
Apnea Detection Algorithm screen: Lower plot displays simulated alarm. Upper plot contains study data. Notes field in lower left allows the user to annotate each simulated alarm signal.

ance waveform. The infrared notification is affected by interference from the fluorescent lighting in the rooms. Limitations have been encountered with IR (infrared) module transmitting distance. To date the system has been successfully tested on a single infant.

Discussion

In the previous study only one baby had apnea. The VTS was as successful as hand stimulation. However, hand stimulation was noted to take twice as long (six vs. three seconds). This current study has important additional value. A larger number of patients will provide data as to the efficacy and safety of this system.

Regarding the infrared alarm notification, better results can be achieved by using filters, lenses, and a higher power IR LED. The use of radio frequency as wireless alarm notification can also be explored in future studies. This

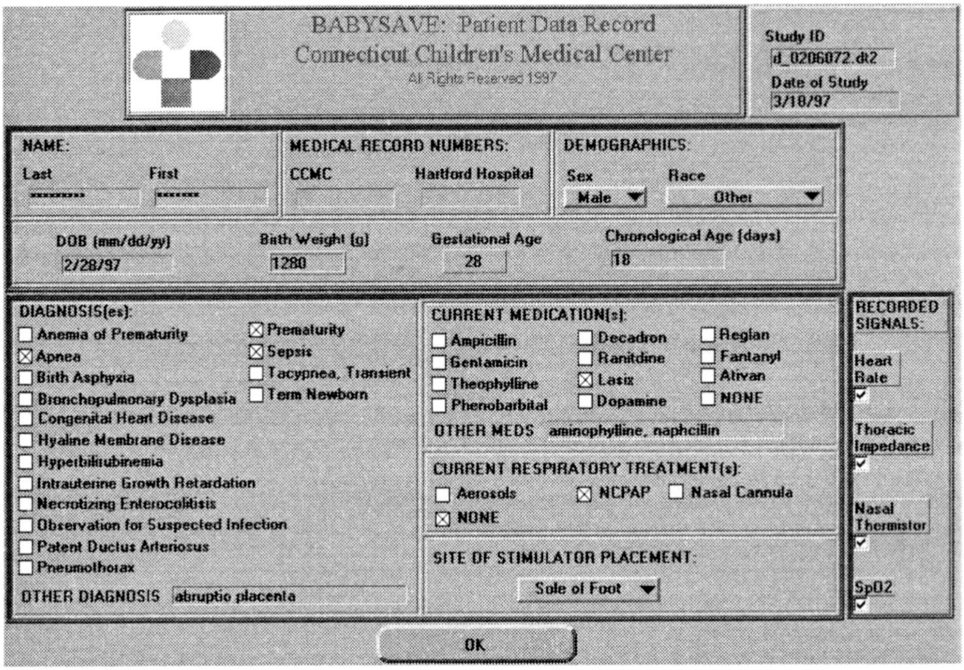

Figure 5-26.
Patient Data screen: User enters clinical data pertaining to new study patient. All data is maintained in a database.

implementation will help to reduce the noise level in the NICU, which has an adverse impact on preterm babies. More work is needed in apnea detection algorithms to try to close the loop so the system can automatically trigger the VTS pulse.

IntelliVent™ Data Logger: A Real-time Research Tool for Data Acquisition, Analysis, and Display of Ventilatory Parameters

IntelliVent™ is an easy-to-use data acquisition and analysis product that interfaces with the Puritan-Bennett 7200 or 840 series ventilator. IntelliVent acquires and analyzes both patient and ventilator parameters. Physicians, biomedical engineers, and software engineers have designed this interactive research system.

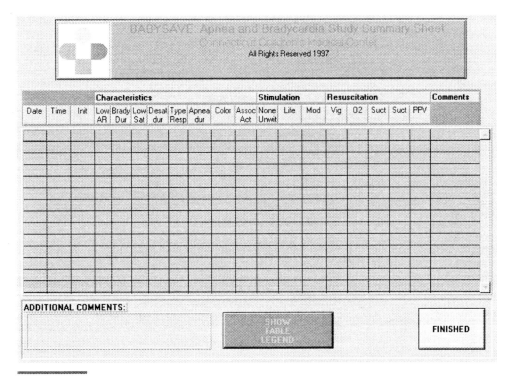

Figure 5-27.
Apnea and Bradycardia Data sheet: Identical to paper records maintained by nursing staff. Computer form is copied from paper record at end of study. Only entries during the time of study are recorded.

Applications

Microprocessor-controlled mechanical ventilation systems utilize continuous data measurements to monitor and maintain oxygenation and ventilation in critically ill patients. Although a microprocessor-based ventilator continuously provides data, clinical restraints often result in decision analyses, which are based upon intermittent data sets.

Minute Ventilation (Ve), the product of respiratory rate (RR) and tidal volume (Vt), is the physiologic integrator of the carbon dioxide production, oxygen utilization, and the mechanical performance of the respiratory system. Collectively, these parameters define the relative need for ventilatory support. Vt and RR are not stereotypically constrained, however, and they vary instantaneously with respect to metabolic demands and external stimuli. Frequently, these parameters demonstrate inherent periodicity, confounding accurate intermittent analysis.

An on-line data acquisition and analysis system capable of trending the coefficients of variation and mean Ve, Vt, and RR would rapidly identify and quantify trends in respiratory periodicity. In addition, Vt and Ve effectively define and quantify the intensity of pressure support ventilation. Collectively, these parameters provide insight, and enhance decision making with respect to the relative need for ventilatory support. The acquired datalogs indicate that continuous data collection and statistical analysis of respiratory parameters is reliable and enhances insight into mechanical support systems. Future applications of predictive equations and computer-ventilator articulation will provide an opportunity to expedite decision making and maximize patient-ventilator synchrony.

Hardware

The system has been designed to work with any computer. Data is acquired from the ventilator's digital communications interface (DCI) through the computer's serial communication port. Future versions of IntelliVent will also support the IEEE-1073 standard (Medical Information Bus). An optional plug-in module allows for simultaneous acquisition of analog signals and waveforms. Figures 5-28 through 5-32 show the IntelliVent Data Logger screens at work.

IntelliVent™ is dedicated to the memory of Douglas J. Onorato—physician, scientist, and friend. Doug was a caring physician, a wonderful teacher, and a visionary researcher. He will be missed by all who knew him.

REFERENCES

American Thoracic Society. (1987). Single breath carbon monoxide diffusing capacity (transfer factor): Recommendations for a standard technique. *Am. Rev. Respir. Dis., 136,* 1299–1307.

Cander, L., & Forster, R. E. (1959). Determination of pulmonary parenchymal tissue volume and pulmonary capillary blood flow in man. *J. Appl. Physiol., 14*(4), 541–551.

Cotes, J. E. (1979). *Lung function: Assessment and application in Medicine,* 4th ed. Oxford: Blackwell Scientific Publications, 203, 329.

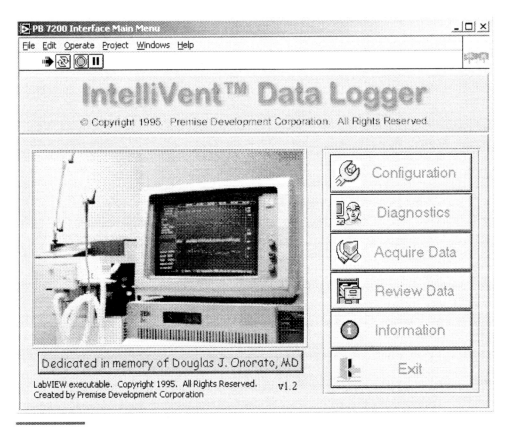

Figure 5-28.
IntelliVent™ Main Menu.

Eisenfeld, L. (September 1996). Patent Number: 5,555,891, United States Patent.

Pawelczyk, J., Levine, B. D., Prisk, G. K., Shykoff, B. E., Elliot, A., & Rosow, E. (1995). Accuracy and precision of flight systems for determination of cardiac output by soluble gas rebreathing. Presented at the 1995 American Institute of Aeronautics and Astronautics Conference.

Sackner, M. A., Greeneltch, D., Heiman, M. S., Epstein, S., & Atkins, N. (1975). Diffusing capacity, membrane diffusing capacity, capillary blood volume, pulmonary tissue volume, and cardiac output measured by a rebreathing technique. *Am. Rev. Respir. Dis., 111*, 157–165.

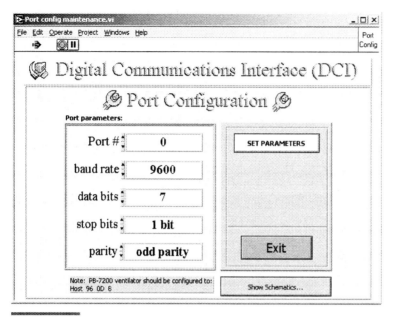

Figure 5-29.
Port configuration utility: A serial port configuration utility allows you to easily configure any serial (com) port to communicate with the digital communications interface (DCI) on any Puritan-Bennet 7200 or 840 series ventilator.

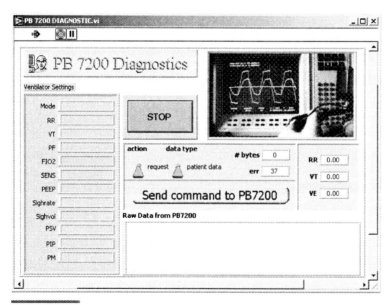

Figure 5-30.
IntelliVent provides a Diagnostic Module to verify data transfer of both ventilator settings and patient parameters (as measured by the ventilator). In addition, the raw data stream (ASCII text) is also available.

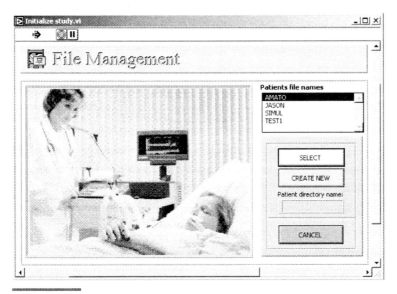

Figure 5-31.

File management utilities allow you to select previously acquired data sets or create new records. In addition, data sets may also be exported into standard spreadsheet format for further analysis by third-party software applications.

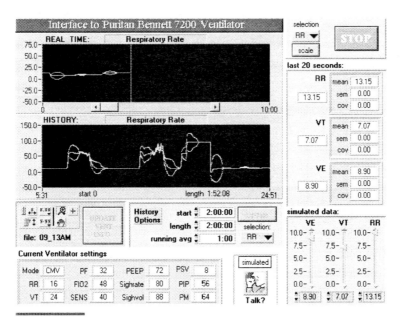

Figure 5-32.

Real-time analysis: IntelliVent provides real-time trending of several ventilatory parameters. By displaying the coefficients of variation and mean Ve, Vt, and RR, it is possible to rapidly identify and quantify trends in respiratory periodicity and the relative need for ventilatory support.

Independent Solution Articles

A Virtual Instrument for Critical Care Monitoring of Newborn Patients with Hypoplastic Left Heart Syndrome Following Norwood Operation Using LabVIEW

By Mark J. Schroeder, Ph.D., Thomas Yeh, Jr., M.D. Ph.D., Erle H. Austin, M.D., and Steven C. Koenig, Ph.D., Jewish Hospital Heart and Lung Institute

The Challenge: Provide pediatric cardiac surgeons with a virtual instrument to monitor the delicate balance of flow from a single ventricle through the systemic and pulmonary vasculature for the post-operative care of neonates born with hypoplastic left heart syndrome recovering from a Norwood procedure.

The Solution: We created a user-friendly interface enabling physicians to monitor the balance of flow through the parallel systemic and pulmonary vasculature using a laptop PC-based clinical data acquisition system with a National Instruments 16-channel A/D board (model: DAQCARD-AI-16XE-50) and LabVIEW software.

Abstract

Hypoplastic left heart syndrome occurs in newborns in which the left ventricle is unable to support the systemic circulation. It has the highest mortality among cardiac defects in newborns. Untreated, none will survive beyond four months. Pediatric surgeons treat these patients with the Norwood operation, a complex surgical procedure in which a shunt is connected from the right subclavian artery to the pulmonary artery so that the single (right) ventricle can pump blood to both the systemic and pulmonary vasculature. Post-operative care is managed by measuring arterial and venous oxygen saturations, which are used to estimate the systemic (Qs) to pulmonic (Qp) flow ratio. In order to optimize survival, blood flow through these parallel circulations must be balanced (Qp/Qs ratio $= 1$). National Instruments A/D hardware and LabVIEW software allowed us to integrate these multiple measurements into a real-time virtual instrument to monitor flow ratio trends and other physiologic signals, which otherwise would not be possible.

Background

Optimal post-operative survival following the Norwood procedure requires that one ventricle pump blood to the lungs and to the body in a balanced fashion. The Norwood operation is the first of three surgical procedures and is usually performed within the first two weeks of life. To ensure survival, particularly during the critical first 48 hours after surgery, balanced flow through the systemic and pulmonary circulations must be maintained to ensure adequate oxygenation of blood and perfusion of end-organs. The standard technique for determining the balance of systemic (Qs) and pulmonary (Qp) flows is the calculation of a recently reported flow ratio (Qp/Qs) based upon oxygen saturation measurements that include systemic arterial (SaO_2), systemic venous (SvO_2), and pulmonary venous ($SpvO_2$) shown in equation 1. Additionally, an oxygen excess factor (Ω), which is a function of systemic arterial (SaO_2) and systemic venous (SvO_2) oxygen saturations (equation 2), has been proposed as a helpful tool in managing infants with critical cardiovascular problems.

$$Qp/Qs = (SaO_2 - SvO_2)/(SpvO_2 - SaO_2) \qquad \text{Equation 1}$$
$$\Omega = (\text{Oxygen Delivery})/(\text{Oxygen Consumption}) = (SaO_2)/(SaO_2 - SvO_2) \quad \text{Equation 2}$$

Pediatric surgeons rely heavily upon these oxygen saturation measurements from multiple instruments, which then must be processed manually to determine the flow ratio (Qp/Qs), oxygen excess factor (Ω), or both in order to select an appropriate treatment and to monitor post-operative recovery. The current problem lies in the delay of small, yet critical, changes in blood flow ratios and trend information that are used for expeditious management decisions. National Instruments A/D hardware and LabVIEW software allowed us to develop a virtual instrument that enables pediatric surgeons to monitor real-time flow ratio trends and other physiologic signals that would otherwise not be possible, thereby improving a surgeon's ability to manage this life-threatening condition.

Virtual Instrument Design

The pediatric surgeons requested that the virtual instrument be physically small, monitor and acquire data continuously in real time, and display long-term history of pertinent parameters. Specifically, the virtual instrument needed to acquire and display the flow ratio, oxygen excess factor, and other physiological parameters (aortic pressure, central venous pressure, aortic flow, and ECG). The requirements were met by developing a virtual instrument using a DAQCARD-AI-16XE-50 16-channel A/D board and LabVIEW 5.1 software (both from National Instruments, Austin, TX) housed on a laptop computer (Gateway Pentium III, Sioux Falls, SD). Oxygen saturation and other physiological measurements are made from multiple clinically approved instruments (i.e., HP model 66, pulse oximeter, etc.).

The virtual instrument is initiated by completing front panel documentation (Figure 5-33), in which the user enters the study name, subject number, DAQ operator, and the base filename. The base filename is automatically incremented and appended to include the correct date and time saved.

Once documentation data have been appropriately entered, the user selects the Continue button to initiate the monitoring and data acquisition virtual instrument (Figure 5-34). The virtual instrument supports five primary functions: (1) *tank indicator* for monitoring systemic and pulmonary oxygen saturation levels, (2) flow ratio and oxygen excess factor *meters*, (3) continuous real-time physiological parameter *waveform charts*, (4) flow ratio and oxygen excess factor four-hour *waveform chart*, and (5) save options *control buttons*. The tank indicator and meters are updated every second and the real-time waveform charts are updated every 0.5 seconds. The four-hour waveform chart provides the surgeon with a history of flow ratio and oxygen excess factor parameters from meters, which allows him or her

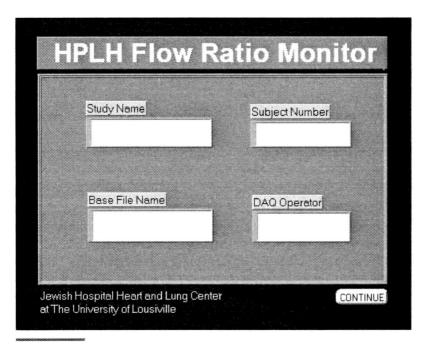

Figure 5-33.
Front panel of virtual instrument for documenting data files.

to observe the long-term trends associated with post-operative recovery or therapeutic interventions. The save control buttons provide the user the option of saving data in three different ways: (1) continuous recording, (2) data epoch save, where the length is specified by the user, or (3) automatically save the specified data epoch at a specified time interval. Data is saved to a new time- and date-stamped spreadsheet file for every save. The saved data can be easily imported into numerous spreadsheet programs or the user can view the data in the custom-designed program HPLP_Viewer.VI, which allows the user to easily access and view any of the previously saved data.

Clinical Application

Our virtual instrument provides pediatric surgeons with a valuable tool for post-operative management of newborns following a Norwood procedure for the treatment of hypoplastic left heart syndrome (Figure 5-35). Specifically, the surgeon can now continuously track changes in the flow ratio and oxygen excess factor in real time, which he or she was previously unable to do. Surgeons can also observe trends in the flow ratio and oxygen excess factor over a four-hour period, which provides extremely important information in deducing the effects of therapeutic interventions for maintaining a balanced flow ratio ($Qp/Qs = 1$). For example, if the flow ratio is out of balance (i.e., $Qp/Qs > 5$), the surgeon would be

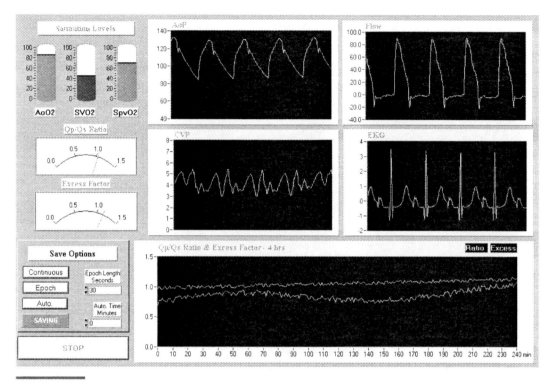

Figure 5-34.
Virtual instrument for continuous real-time monitoring and acquisition of oxygen saturation, flow ratio, oxygen excess factor, and physiological parameters.

interested in the effects of vasodilators/vasoconstrictors that adjust the balance of flow but may take an extended period of time to have an effect (e.g., 20 minutes). The surgeon may also be interested in how stable the flow ratio is over a four-hour period, where large fluctuations in the flow ratio may indicate an unstable system that may require aggressive treatments and close monitoring. Finally, the data acquisition feature allows surgeons and researchers to post-process these data and identify promising therapeutic treatments, develop postoperative management strategies, and elucidate critical physiologic control mechanisms.

Conclusion

A medical instrument for monitoring the flow ratio and oxygen excess factor or acquiring these data does not exist. National Instruments A/D hardware and LabVIEW software provided the platform to construct a virtual instrument that integrated oxygen saturation and physiological measurements from multiple medical devices for calculating, displaying, and saving these vital parameters. We believe this integrated virtual instrument will allow pediatric surgeons to better treat these patients and improve clinical outcomes.

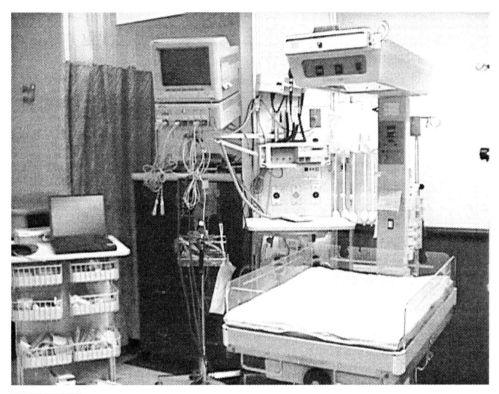

Figure 5-35.
Post-operative care instrumentation and unit for monitoring post-operative recovery of new-borns following Norwood procedure for treatment of hypoplastic left heart syndrome.

Biomedical Patient Monitoring, Data Acquisition, and Playback with LabVIEW

By Guy A. Drew and Steven C. Koenig, Ph.D., Jewish Hospital Heart and Lung Institute, Louisville, Kentucky

The Challenge: Real-time patient monitoring, data acquisition, and post-data collection review during intra-operative surgical procedures and post-operative recovery.

The Solution: Combined DAQ system (AT-MIO-16E-10 16-channel A/D board) with PC-based LabVIEW software to create a physician-friendly real-time patient monitoring, data acquisition, and playback interface.

Investigators at the Jewish Hospital Heart and Lung Institute include cardiovascular surgeons, cardiologists, physiologists, and biomedical engineers. They actively conduct medical research designed to characterize new surgical techniques, test innovative cardiovascular devices, and evaluate pharmacological agents. During intra-operative procedures and post-operative recovery, physicians rely upon real-time physiologic measurements of cardiac function that include heart rate, blood pressure, and cardiac output to monitor surgical procedure outcomes. Further, investigators require accurate and precise data collection of physiological measurements of cardiac, systemic, and pulmonary function for post-processing analyses. Many of these experiments must also be conducted in compliance with Good Laboratory Practice (GLP) for submission of study data to the Food and Drug Administration (FDA) to receive approval for clinical trials.

Design Challenge and Solution

Our challenge was to design a sophisticated, multi-feature data acquisition (DAQ) program that would be flexible enough to meet a broad range of experimental needs yet simple enough for nonengineers (physicians, cardiologists, and physiologists) to operate. Additionally, the DAQ program had to meet FDA guidelines for GLP compliance. We developed a data acquisition system with the ability to measure up to six channels of pressure, two channels of flow, ECG, and seven auxiliary inputs. Specifically, amplifier analog outputs are conditioned by a signal distribution system containing buffering, gain, offset, and low pass anti-aliasing filters, and then digitally converted using a 16-channel A/D board (AT-MIO-16E-10, National Instruments, Austin, TX). This digital data can be continuously displayed in real time on a PC monitor or saved in a data file for post-processing analyses or playback. Investigators simply complete a worksheet (Figure 5-37) that provides the documentation required to configure the DAQ program according to their study protocol.

In order to provide numerous DAQ features and meet our physician-friendly requirements, the DAQ program was configured to run under four modes of operation. First, a DAQ

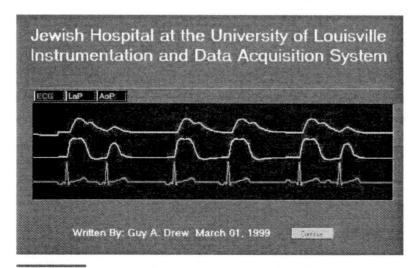

Figure 5-36.

Wizard (National Instruments, Austin, TX) is used to assign channel names (i.e., Aortic Pressure), identify analog input channels (i.e., channel 1), and convert physical input values to physiologically equivalent units (i.e., 0–2 V = 0–200 mmHg). Next, the main DAQ program is selected and the data recorder Program Profile Menu screen is invoked (Figure 5-38).

Support documentation for experimental data sets is entered into the Data File Parameters and File Header Information panels. Individual channel names assigned in the DAQ Wizard can then be mapped in the DAQ Channel Setup panel. Other important user-selectable options include the ability to simultaneously record two separate data sets, select fixed data collection epochs or continuous data recording, and load a previously defined program file. Finally, a data file into which experimental data will be stored must be identified. Indicators and pop-up warning menus provide real-time feedback to ensure no errors have been inadvertently made during configuration. Upon successful completion of these setup requirements, Run Profile is selected.

Following Run Profile, an automatically fitted continuous waveform chart display of up to 16 channels (one subject) or 8 channels (two subjects) is invoked (Figure 5-39). Selected patient parameters can be monitored continuously by physicians or stored in automatically incremented data files. Additional documentation for each data set can be logged in a Notes indicator. An indicator also identifies whether continuous or epoch data function has been selected and illuminates when invoked. A split screen feature enables simultaneous monitoring and data collection of separate patients. An overlay feature allows multiple waveforms to be displayed in the same graph (i.e., overlay Aortic and Left Ventricular Pressure). A freeze indicator allows the user to stop continuous display and study individual waveform characteristics without interrupting data collection. An exit indicator allows the user to return to the Program Profile Menu.

A Data Viewer option allows the user to retrieve previously recorded data sets. The header information and ASCII data for each channel are displayed. The entire data set or individual epochs can then be replayed through data display graphs. Indicators located below the data display graphs identify starting and ending data points and length of data set displayed. ASCII data sets can easily be imported into a spreadsheet (i.e., Excel) or loaded in a data analysis package (i.e., HiQ or Matlab).

Application

Our DAQ system and software have been successfully used in support of an ongoing preclinical study of the AbioCor Implantable Replacement Heart (ABIOMED, Danvers, MA; Figure 5-40). Cardiovascular surgeons, post-operative care attendants, and ABIOMED engineers rely on this system for subject monitoring and hourly recordings of physiological parameters for assessment of device performance and subject status.

Conclusion

Previously, physicians and investigators relied on expensive medical grade monitoring systems with limited ability to store digitized data and with low frequency content (< 50 Hz).

Data File Parameters

Base File Name	
Data Set Size (sec)	
Stating Sequence	

File Header Information

Study Title	
Subject Number	
Actual Surgery Date	
Medical ID Number*	
DAQ Operator(s)	

DAQ Channel Setup

Channel # 1**		Channel # 9	
Channel # 2		Channel # 10	
Channel # 3		Channel # 11	
Channel # 4		Channel # 12	
Channel # 5		Channel # 13	
Channel # 6		Channel # 14	
Channel # 7		Channel # 15	
Channel # 8		Channel # 16	
Sample Rate (Hz)			

Graph Layout

Graph 1

Channel # 1	
Channel # 4	
Channel # 7	

Graph 4

Channel # 9	
Channel # 12	
Channel # 15	

Graph 2

Channel # 2	
Channel # 5	
Channel # 8	

Graph 5

Channel # 10	
Channel # 13	
Channel # 16	

Graph 3

Channel # 3	
Channel # 6	

Graph 6

Channel # 11	
Channel # 14	

* Medical ID Number typically provided by RRC

** Channel names as defined in Channel Wizard

Figure 5-37.

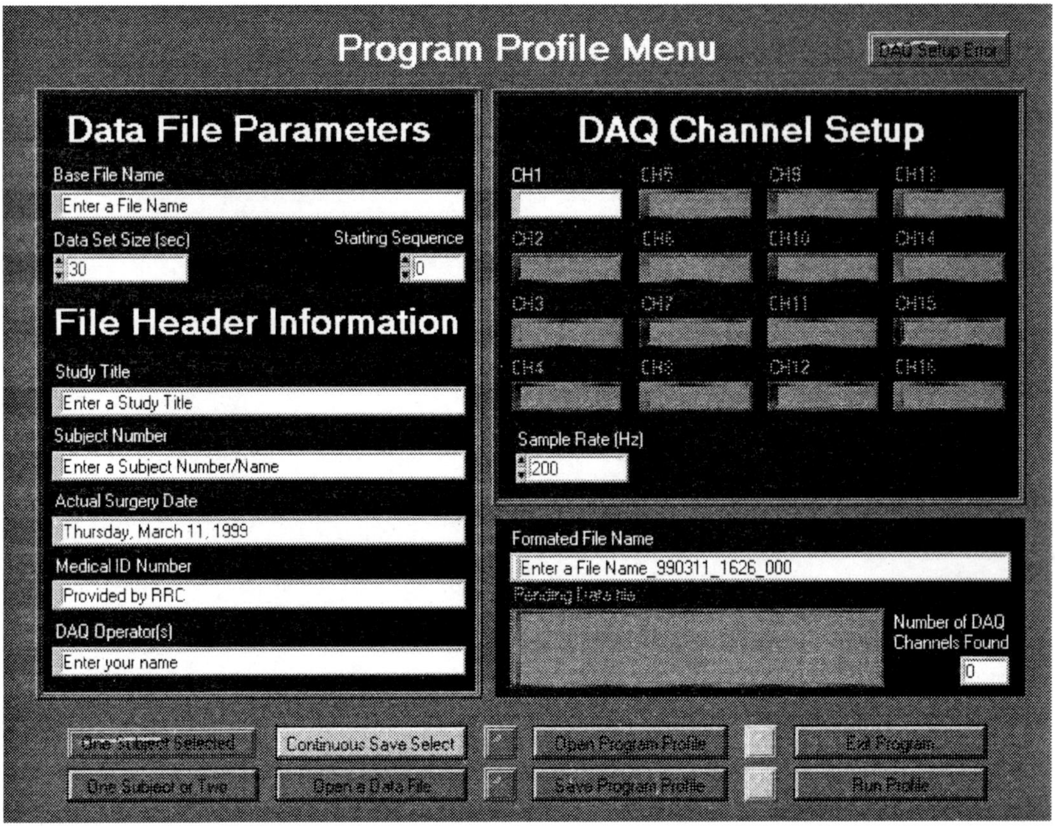

Figure 5-38.

Additionally, investigators relied primarily upon data chart recorders for collection of experimental data sets. These hardcopy data sets limited analysis to calculation of simple cardiac performance parameters that took an extreme amount of time to complete. We selected LabVIEW as the ideal platform for developing a physician-friendly DAQ program that could provide a low-cost alternative for real-time patient monitoring, provide more sophisticated parameter calculations for characterizing cardiovascular function, and substantially reduce post-processing data analysis time.

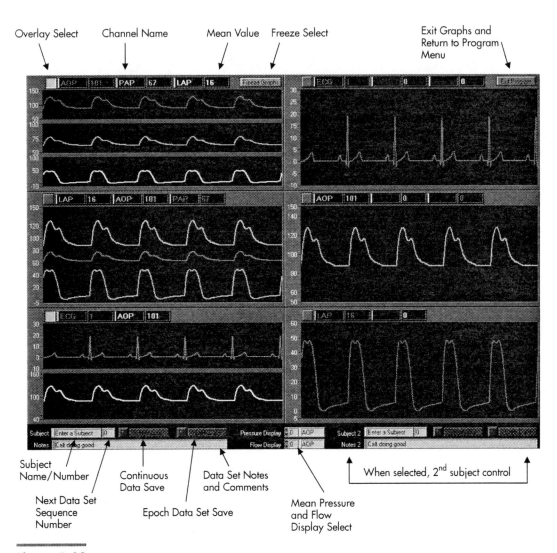

Overlay Select   Channel Name   Mean Value   Freeze Select   Exit Graphs and Return to Program Menu

Subject Name/Number

Next Data Set Sequence Number

Continuous Data Save

Epoch Data Set Save

Data Set Notes and Comments

Mean Pressure and Flow Display Select

When selected, 2nd subject control

Figure 5-39.

Figure 5-40.

For more information, please contact:

Guy A. Drew,
Sr. Electronics Engineer,
US Army Institute of Surgical Research,
3400 Rawley E. Chambers Avenue,
Building 3611,
Fort Sam Houston, TX 78234-6315
tel (210) 916-5992
e-mail guy.drew@cen.amedd.army.mil

Steven C. Koenig, Ph.D.,
Assistant Professor of Surgery,
Jewish Hospital Heart and Lung Institute,
500 South Floyd Street, room 118,
Department of Surgery,
University of Louisville,
Louisville, KY 40202
tel (502) 852-7320
e-mail sckoen0l@athena.louisville.edu

Funded by a grant from the Jewish Hospital Heart and Lung Institute (Louisville, KY).

Machine Vision and Motion Control Applications

6

Overview
Machine Vision Defined
       Step 1—Conditioning
       Step 2—Acquisition
       Step 3—Analysis
Computer Technologies for Machine Vision:
Making Machine Vision Easier
       Machine Vision Functions
       Machine Vision Application Development
       Application of Machine Vision Functions
       Color Analysis in Machine Vision
       Color Pattern Matching
       Color Pattern Matching Features
       Color Location
       Cameras and Interfacing

Overview

The human observer uses a wide range of cues, drawing from color, perspective, shadings, and a vast library of particular individual experiences. Visual perception relies on a uniquely human capacity to make judgments. However, a machine vision system does not have an experience base from which to make comparative decisions. Everything must be specifically defined. Simple problems, like locating an object, determining top from bottom, distinguishing a light artifact from a part of the interesting object, becomes a complicated task in machine vision. In order to reduce the number of variables, the vision system must be provided with the best image possible, meaning the image that makes it easiest for the machine vision computer system to do its task.

It is our objective in this chapter to provide useful information on the three steps required for successful machine vision applications—conditioning, acquisition, and analysis. In addition, we will present several real-world examples of machine vision (and computerized motion control) to illustrate the practical applications of these technologies within the biomedical field.

Machine Vision Defined

It is always helpful to specifically define a new subject of study. Machine vision can be defined as "the acquisition and processing of images to identify or measure the characteristics of objects."

This section will introduce you to three simple steps that help you find success with machine vision (see Figure 6-1).

- **Conditioning:** the process of preparing your imaging environment. You need to consider factors such as lighting and motion, as well as the specific feature(s) of the objects at which you are looking.

- **Acquisition:** involves selecting image acquisition hardware, a camera, and a lens. You also use software to capture and display the image.

- **Analysis:** includes all of the image processing steps you use to answer questions about your image. You can answer questions such as "How big is this object?" and "Is this object present in the image?"

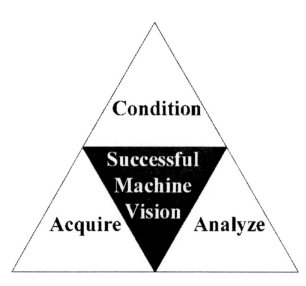

Figure 6-1.
Three steps required for successful machine vision applications.

Step 1—Conditioning

Conditioning your imaging environment is a critical first step. By properly conditioning, you can focus your development energy on the application rather than problems from the environment. Further, you save precious processing time at run time.

Lighting is probably the most important part of conditioning. Because poor lighting can cause shadows and glares that degrade the performance of an image processing routine, time spent on selecting and setting up lighting is well spent.

Motion in an image is both beneficial and detrimental to a machine vision application. Properly used, motion control equipment can aid you in imaging an object from different angles or positions. Conversely, if you don't take motion into account prior to acquiring your first image, your images may prove difficult to process because of blur and other factors.

Finally, it is important to take note of your image features. For example, how big is the object you are inspecting? Can you use its shape to expedite your processing? The following discussion will address these questions.

Lighting Considerations

The type of lighting you select can determine the success or failure of your application. Illuminating the scene and part is a critical step when developing an inspection system. The objective is to separate the feature or part you want to inspect from the surrounding background by as many gray levels as possible. If the feature you want to inspect is difficult to separate visually from the background or the defects do not stand out, then the inspection task will prove difficult. Common types of lighting are halogen, fluorescent, laser, incandescent, and LED.*

Maximum image contrast between features observed relative to features ignored is of critical importance. Often, you want the observed feature to have the maximum energy, which increases the signal-to-noise ratio at the camera. Another consideration is minimum sensitivity to object/lighting position and

---

*Additional information on lighting products and accessories can be obtained from the following companies (each of which is National Instruments' lighting partner): Graftek (www.graftek.com), Edmund Scientific (www.edsci.com), Fostec (www.fostec.com), and Stocker & Yale (www.stkr.com).

orientation. The image quality should be independent of these factors. Highly directional light sources, for example, increase the sensitivity of specular highlights (glints). Minimum sensitivity to environmental and ambient lighting is also critical. Changes in ambient illumination, for example, sunlight varying with the weather and time of day, can cause problems in your machine vision system.

Many other factors go into determining the lighting for an application. The factors include the camera, lens, and image processing technique. Beginning with an image that has proper lighting makes image processing software development easier.

Motion Considerations

The next part of conditioning your imaging environment is motion. Because of the method that some cameras use to acquire images, moving image objects can appear blurry. For instance, if the charge-coupled device (CCD) is exposed to a moving image for a long duration, blurring occurs. You can use a strobe light in motion applications to get very crisp images. The strobe light is turned on for only microseconds to limit the time of exposure of the image to the CCD. The strobe light requires accurate timing with the camera. It has a very long lifetime but personnel must be shielded from the strobe. Many frame grabbers (including National Instruments image acquisition hardware devices) have up to four standard digital input/output lines that you can use to send pulses to control strobe lights. In addition, you can use data acquisition hardware to control strobe lights. Figure 6-2 shows how a blurry image can be adjusted with special equipment.

Size Considerations

Before you buy equipment, such as cameras and lenses, for your application, consider the size of the object you want to examine. Many applications require resolution of small particles. The challenge is to find the correct lens and camera to meet your requirements. In order to see greater detail over a large field of view, you need a camera with high resolution.

Resolution can be defined as how many pixels are within a given image and *field of view* is what the camera can see. For example, the overlaid images of a circuit board in Figure 6-3 illustrate the difference in resolution for the same field of view.

Figure 6-2.

An example of how blur can be eliminated with special cameras and lighting. (Reprinted with permission of National Instruments.)

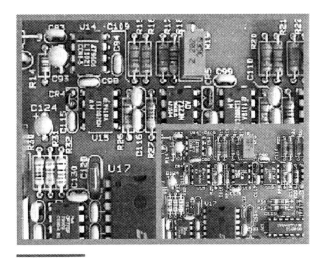

1280 × 960 =
1.23 million pixels

640 × 480 =
310,000 pixels

Figure 6-3.

Greater detail of the same field of view is obtained by using a camera with higher resolution. (Reprinted with permission of National Instruments.)

Surface Considerations

When objects in your image are covered by shadows or glare, it becomes much more difficult to examine the image. Some objects reflect large amounts of light because of the nature of their external coating or due to their curvature. Poor lighting setups in the imaging environment can cause shadows to fall across the image. Figure 6-4 illustrates the impact of reflective and curved

Figure 6-4.
Example of the impact of proper lighting for reflective or curved surfaces. (Reprinted with permission of National Instruments.)

surfaces. When possible, position your lighting setup and your imaged object such that glare and shadows are reduced or eliminated. If this is not possible, you may need to use special lighting filters or lenses. These are available from a variety of vendors.

Shape Considerations

It is important to pay special attention to the shape of an object in an image. Before setting up a camera and lighting equipment, determine if you can solve your application by looking only at the shape of the object. If so, you may want to create a silhouette of the object using a technique called *backlighting*. Finding the edges and measuring the distance between, or area within, a particular region of interest is quick and simple. Operations that depend only on the object shape execute quickly and efficiently. This kind of operation will be demonstrated when we discuss the Wound Measurement System later in this chapter.

Step 2—Acquisition

The next step in the formula for machine vision success is the acquisition of an image. In this section, we discuss camera and lens selection. Figure 6-5 shows a typical setup of a computer-based measurement and automation system.

Analog Cameras

Analog cameras are those that output a video signal in an analog format. You may wonder how to generate the analog signal. Additionally, you may have heard the term *CCD*, and have wondered how this relates to the analog video signal.

A charge-coupled device (CCD) is an array of thousands of interconnected semiconductors. Each CCD sensor is a pixel-sized, solid-state, photosensitive element that generates and stores an electric charge when it is illuminated. The sensor is the building block for the CCD imager, a rectangular array of sensors on which an image of the scene is focused. In most configurations, the sensor includes the circuitry that stores and transfers its charge to a shift register, which converts the spatial array of charges in the CCD imager into a time-varying video signal. Timing information for the vertical position and horizontal position plus the sensor value combine to form the video signal.

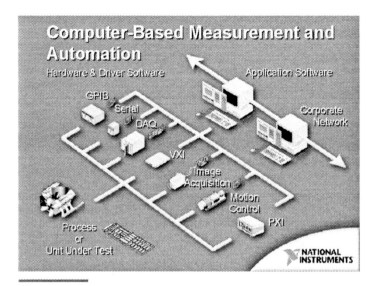

Figure 6-5.
Schematic diagram of a typical machine vision system. (Reprinted with permission of National Instruments.)

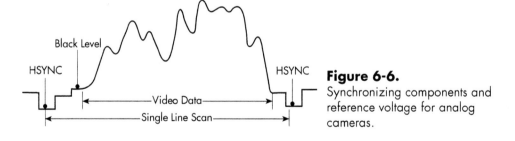

Figure 6-6.
Synchronizing components and reference voltage for analog cameras.

The horizontal sync (HSYNC) identifies the beginning of each line; several lines make up a field. An additional pulse, called the vertical sync (VSYNC), identifies the beginning of a field. For most low-end cameras, the fields are interlaced to increase the perceived image update rate. Two fields make up one frame. Electronic Industries Association (EIA) RS-170 and NTSC cameras update at 30 frames per second with 640 × 480 lines. CCIR and PAL cameras update at 25 frames/s with 768 × 576 lines. Finally, you see in Figure 6-6 that one point is labeled the Black Level. This level is a certain reference voltage to measure pixel intensities.

Digital Cameras

Digital cameras have several advantages over analog cameras (see Table 6-1). Analog video is more susceptible to noise during transmission than digital video. By digitizing at the CCD camera level rather than the image acquisition board, the signal-to-noise ratio is typically higher, resulting in better accuracy. Digital cameras now come with 10- to 16-bit gray levels of resolution as a standard for machine vision, astronomy, and thermal imaging applications.

Important Camera Technologies

For standard analog cameras, the CCD is divided into two fields; the odd field (rows 1, 3, 5, and so on) and the even field (rows 2, 4, 6, and so on). The fields combine to form a full image frame. Each field acquires every 1/60 of a second, resulting in 30 frames/s. Standard analog cameras with interlaced scanning can solve most applications. For progressive scan cameras, the CCD sensor array is exposed at the same time rather than in two steps. Progressive scan cameras have less blurring because the CCD sensor array is exposed for one duration, which makes them especially useful in applications where the object being imaged is moving. National Instruments IMAQ™ 1407, 1408, and 1409 boards work with progressive scan and standard analog cameras.

Unlike an area scan camera, a line scan camera acquires an image that is only one or a few pixels wide. Line scan cameras are usually chosen for applications where the item is moving, such as a conveyor or stage on a production system. Use the IMAQ 1424 and 1422 series boards with line scan cameras.

Table 6-1. *Comparison of digital versus analog cameras.*

Digital Cameras	Analog Cameras
Faster data rates	Slower data rate
Higher spatial resolution	Lower spatial resolution
Advanced features	Simpler but easier
Greater pixel depth	Typically 8-bit
Higher signal-to-noise ratio	Lower signal-to-noise ratio
Thicker cabling	Simple BNC cabling
Larger footprint	Smaller footprint
More expensive	Less expensive

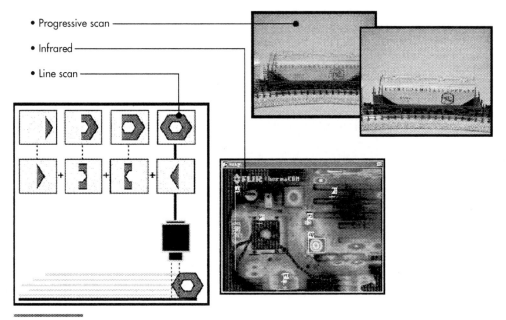

Figure 6-7.
Examples of progressive scan, infrared, and line scan camera technologies.

Infrared or thermal cameras are needed for many scientific and industrial applications. You can see many defects easier in the infrared spectrum. Other advantages include seeing through opaque objects. For example, you can see oil level in a compressor when the oil temperature differs from the casing temperature. IMAQ Vision software includes special analysis functions for analyzing thermal images. You can easily acquire images from infrared cameras from FLIR, Raytheon, and Indigo because they use standard analog output. Some high-resolution cameras output digital images and work with the IMAQ 1424 and 1422 series boards. Figure 6-7 shows examples of progressive scan, infrared, and line scan technologies.

Camera Lenses

The lens and camera sensor size and image contrasts determine the spatial resolution often noted in pixels/mm or line pairs/mm of your vision system. Physical dimensions, such as the working distance, the distance between the front of the lens and the object under inspection, and field of view (FOV)—the area of the object that the camera sees—are important considerations when selecting a lens. Telecentric lenses are often useful because they work off a wide depth of field (DOF). With a telecentric lens, you can image objects of different distances from the lens and the objects stay in focus. National Instruments partners such as Edmund Industrial optics and Graftek Imaging are good resources for lenses.

Image Acquisition Hardware

Features that you should consider for machine vision or scientific imaging applications include fast data movement to PC memory, integration with data acquisition and motion control hardware, advanced triggering, and onboard preprocessing functions.

An NI custom ASIC, the MITE direct memory access (DMA) chip for imaging, offers superior image data throughput. The MITE chip performs the interlacing and copying to PC memory instead of the CPU. Standard DMA controllers cannot do this. Unlike other frame grabber offerings, using the MITE does not result in the need to lock down a larger block of memory (such as 20 MB) in the driver for image acquisition.

Standard and Nonstandard Multichannel Boards

The IMAQ 1408 boards offer easy-to-use driver and camera configuration software for up to four standard and nonstandard cameras. You can use the 1408 to acquire data from nonstandard cameras that have variable pixel clocks from 5 MHz to 20 MHz. You can configure monochrome acquisition from RS-170, CCIR, NTSC, PAL, RGB, and progressive scan cameras. You can configure color image acquisition from NTSC, PAL, and RGB cameras using the StillColor™ feature. StillColor acquisition of the image quality is better, but the image acquisition rate is slower than full frame color video.

Low-Cost Standard Monochrome Boards

The 1407 series boards offer low-cost single-channel monochrome accuracy. You can configure the 1407 for standard RS-170 and CCIR analog monochrome cameras. Unlike other low-cost machine vision image acquisition boards, the 1407 series offers advanced features such as partial image acquisition, onboard decimation, look-up table processing, programmable gain, and triggering.

Low-Cost Standard Color

The 1411 series boards are configurable for color image acquisition from standard NTSC, PAL, and S-Video cameras. You can also acquire images from monochrome RS-170 and CCIR cameras. Unlike other color machine vision boards, the 1411 series offers fast color conversion to hue, saturation, and luminance (HSL) image data. This is especially useful for high-speed color matching and inspection applications, even in varying illumination.

Digital Cameras

With a digital camera IMAQ board, you can acquire images at thousands of frames/s with greater grayscale resolution and with more spatial resolution (up to 8,000 × 8,000 pixels). National Instruments 1424 series boards for PCI and PXI/CompactPCI™ are some of the fastest digital image acquisition boards available, with a 50 MHz pixel clock and 32-bit wide digital input (four 8-bit pixels). With 80 MB of onboard memory, the NI 1424 can acquire data at a top rate of 200 MB/s. The 1422 series boards feature a 16-bit input and a 40 MHz pixel clock for lower cost digital image acquisition applications.

National Instruments' latest offering includes IMAQ hardware for the digital camera, low-voltage differential signaling (LVDS) standard. LVDS is a device that extends the performance of the commonly used digital camera RS-422 differential data bus. RS-422 limits the frequency to the 20 MHz range. However, LVDS cameras can clock data out at 50 MHz using the IMAQ PCI-1424 LVDS board. Plus, you can use the LVDS of the IMAQ board to transmit data as far as 100 feet. LVDS also reduces noise significantly.

You can use the digital IMAQ boards for area and line scan cameras. Unlike an area scan camera, a line scan camera acquires an image that is only one or a few pixels wide. Line scan cameras are usually chosen for applications where the item image is moving, such as a conveyor or stage on a production system. Line scan cameras are useful in high-resolution applications because you can acquire lines at a fast rate as the part moves. IMAQ Vision software can process each line individually or process several lines that have been stitched together. You can use the 1424 and 1422 series boards with line scan cameras.

NI-IMAQ Driver Software

Whether you are using LabVIEW™, Measurement Studio™, Visual Basic™, or Microsoft Visual C++, NI-IMAQ gives you high-level control of National Instruments image acquisition devices. NI-IMAQ is a complete and robust application programming interface (API) for image acquisition. NI-IMAQ performs all the computer- and board-specific tasks for straightforward image acquisition without register-level programming. NI-IMAQ is compatible with NI-DAQ and all other National Instruments driver software for integrating imaging into any National Instruments solution.

NI-IMAQ is an extensive library of functions that you can call from your application programming environment. These functions include routines for video configuration, image acquisition (continuous and single-shot), memory buffer allocation, trigger control, and board configuration. NI-IMAQ performs all functionality required to acquire and save images.

NI-IMAQ internally resolves many of the complex issues between the computer and IMAQ hardware, such as programming interrupts and DMA controllers. NI-IMAQ provides the interface path between LabVIEW, Measurement Studio, and other programming environments and the hardware product.

National Instruments Intelligent Configuration Software

You can easily configure standard and nonstandard video capture with Measurement and Automation Explorer (MAX), which is shipped with NI-IMAQ. This interactive tool sets the camera type (RS-170, CCIR, NTSC, PAL, and nonstandard), region of interest, auto-exposure, and antichrominance filter. Plus, you can use this interactive utility to set up acquisition from noninterlaced progressive scan cameras and to create your own camera configurations for nonstandard video by setting the VSYNC, HSYNC, and other timing information.

You can use the external lock feature to set variable scan acquisition for microscopes and other sources that generate their own PCLK, HSYNC, and VSYNC. NI-IMAQ saves you development time because you can use these settings automatically in your application development environment. Plus, you can use the utility to snap and grab images and then save the images to

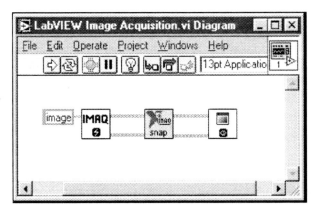

Figure 6-8.
Example LabVIEW diagram for image acquisition.

file. In addition, you can display the histogram to verify the contrast, dynamic range, and lighting conditions.

Image Acquisition in LabVIEW

NI-IMAQ has both high-level and low-level functions for maximum flexibility and performance. High-level functions include single-shot and continuous-mode image acquisition. Low-level functions include imaging sequence setup.

It is evident from Figure 6-8, which shows how to acquire an image in Lab-VIEW, that using NI-IMAQ is quite simple. With only three icons in LabVIEW, you can snap and display an image. You can build more complex configurations using this as a foundation. Additionally, once you write code for use with a particular image acquisition board or camera, this code does not change if you need to change your hardware. NI-IMAQ is robust enough to work identically across National Instruments' entire product line of image acquisition devices and across a broad range of cameras.*

Step 3—Analysis

The final step to machine vision success is the image analysis. Using National Instruments IMAQ Vision software, you can apply image-processing techniques to your images to make decisions based on software analysis. Using these tools, which plug into the programming environment of your choice,

---

*The accompanying CD-ROM contains a folder of IMAQ Vision evaluation software (VBI CD-ROM/sections/vision/IMAQvision).

you can take measurements, locate objects, and perform other tasks through an easy-to-use application programming interface.

Image Processing Techniques

IMAQ Vision software from National Instruments adds high-level machine vision and image processing to LabVIEW, Measurement Studio, and other programming environments. IMAQ Vision includes an extensive set of MMX-optimized functions for gray-scale, color, and binary image display; image processing, including statistics, filtering, and geometric transforms; and pattern matching, shape matching, blob analysis, gauging, and measurement. Users, systems integrators, and OEMs can use IMAQ Vision to accelerate the development of industrial machine vision and scientific imaging applications.

The possibilities and range for image processing and machine vision are numerous, if not overwhelming. So many algorithms exist for you to select from you might ask which one is right for your application and where to begin. In this section we introduce you to image processing and the algorithm requirements for different applications.

To start with, in many applications you need a quantitative or statistical description of your image or region of interest. Statistical functions are calculated quickly; you can solve many inspection applications using simple functions such as average and standard deviation. Pattern matching functions are key for machine vision applications for locating features in the image. Particle analyses are important for counting and sorting applications. Edge detection is useful in applications where you want to find the position of an object. Gauging works hand in hand with edge detection in applications where you want to measure the distances or angles between points in an image.

Statistics

Statistical image processing functions simply give a representation of the image, such as the average pixel value or the standard deviation of the gray-scale values. In an inspection application, under strict lighting conditions, you can calculate the average intensity of a region of interest (ROI) to quickly determine if a chip is present. The average gray-scale value varies depending on whether the chip is present. In less constrained environments, you can use the standard deviation of intensity in an ROI the same way. You can use the ROI control to draw a line on the image; the pixel on the line is the line profile. Line

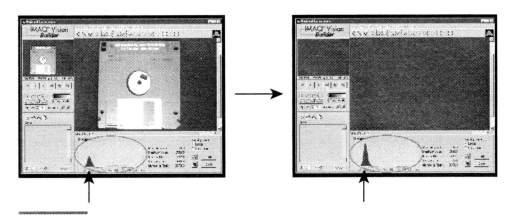

Figure 6-9.
The Histogram function can be used to provide information about lighting, focus, and the presence of an object—in this case a floppy disk. (Reprinted with permission of National Instruments.)

profiles are useful in making a 2D graph of the pixel intensity values or for finding distances between objects and edges. Figure 6-9 illustrates how the histogram is useful in providing information about lighting, focus, and the presence of an object.

Pattern Matching

Pattern matching provides information about the presence or absence, number, and location of the model within an image. For example, you can search an image containing a printed circuit board for one or more alignment marks called *fiducials*. You can use the positions of the marks to align the board for chip placement from a chip-mounting device. In addition, you can use pattern matching to locate key components in gauging applications. In gauging applications, key components are located—then, the distance or angle between these objects is gauged. If the measurement falls within a tolerance range, the part is good; if it falls outside the tolerance, the component is rejected. In many applications, searching and finding a feature is the key processing task that determines the success of the application.

- **Positioning**—Determining the position and orientation of a known object by locating features. The features are used as points of reference on the object.

- **Inspection and Examination**—Detecting simple flaws, such as missing parts or illegible printing.
- **Gauging**—Measuring lengths, diameters, angles, and other critical dimensions. If the measurements fall outside a set of tolerance levels, then the component is rejected.

Particle Analysis

You can use particle analysis functions to find dozens of parameters for the blob, such as the orientation, area, perimeter, center of mass, and coordinates of each blob. You can use these unique parameters to identify parts in a sorting application. You can also use them to ensure that the part meets quality standards. Subsets of the particle analysis functions deal with morphology. These functions are extremely useful in counting and inspection applications. With these functions, you can change the shape of the image particles so you can count them. For example, you can use the erosion function to erode the perimeter of objects so that you can count two particles that are close to each other or overlapping. In addition, the morphology functions can remove small particles (an unwanted blob) caused by reflection in the image. Figure 6-10 illustrates the particle analysis function.

Edge Detection

Edge detection is an important image processing and machine vision function. This function clearly determines the boundary or edge of an object and

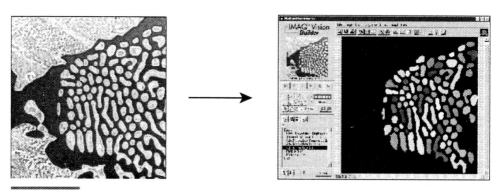

Figure 6-10.
Particle analysis. (Reprinted with permission of National Instruments.)

Figure 6-11.
Edge detection analysis.
(Reprinted with permission of
National Instruments.)

shows up as intensity discontinuities in an image. It is used often in gauging or calipering applications. In gauging applications, you measure lengths, diameters, angles, and other critical dimensions of edges. If the edge measurements fall outside of set tolerance levels, then the component is rejected.

Gauging sometimes is used inline with the manufacturing process as well as offline. If the measurements are made offline, you can use a sampling of components to determine the quality of a sample or batch of manufactured components. Edge detection can help identify a crack in an object (as will be illustrated in a case study at the end of this chapter). Plus, parallel edges in a soldering application can help determine the quality of the solder application. The edge information in pattern recognition is very important; you can often recognize an object from only a crude outline. IMAQ Vision includes many different edge detection functions, such as Sobel, Sigma, Roberts, Prewitt, Gradient, and Differentiation. Figure 6-11 illustrates an example of edge detection analysis.

Gauging

You can use gauging functions to automatically measure distances and angles between edges of an object. You measure from point to point using the line profile and interactive display window. You can also specify markers and calculate measurements between markers. In a production application as shown

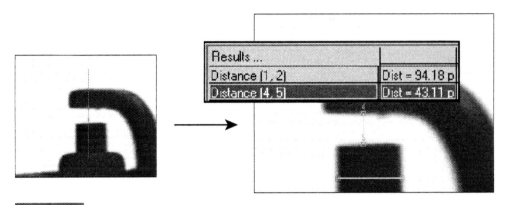

Figure 6-12.
Example of gauging analysis. (Reprinted with permission of National Instruments.)

in Figure 6-12, you can measure critical distances and compare with a tolerance; then, you can reject the flawed products. You can also find the position, angle, and sharpness of an edge, later calculating edge distance measurements with subpixel accuracy.

Computer Technologies for Machine Vision: Making Machine Vision Easier

National Instruments is introducing a new set of functions targeted at machine vision applications. The new machine vision functions are a set of high-level, task-based functions that allow you to easily and quickly design your machine vision system. The machine vision functions make many machine vision tasks very simple by incorporating commonly used utilities such as coordinate systems and nondestructive overlays. Furthermore, machine vision fundamentals, such as identification, inspection, and gauging, are presented in an easy-to-use, task-based manner. By incorporating many commonly used utilities and simplifying the interface to the algorithms, the machine vision functions give you extremely powerful tools to help solve your machine vision applications.

Machine Vision Functions

Because the machine vision functions are presented in an easy-to-use manner, you can often perform a task with only one function. Here are some of the most common functions:

- The edge locators can find edges using rectangles and circles as search areas. Instead of just finding an edge point, the edge locators find the best-fit line or circle present in the search area.

- Distance measurements allow you to easily perform horizontal and vertical clamping measurements. You define a rectangular search area and the software returns a distance in user-calibrated units.

- Feature location uses pattern matching technology to easily find features with one function.

- Counting and measuring objects allow you to easily count and measure bright or dark objects in a given search area.

- Intensity measurements provide light meters that compute the intensity of a region. You can use intensity measurements to detect the presence or absence of a feature or to perform cosmetic testing.

Machine Vision Application Development

The machine vision functions simplify tasks needed for all machine vision applications. Because the tasks are integrated in the machine vision functions, you only need to program the tasks to solve your application.

Many machine vision applications require you to define a region of interest. Now you can use only one function to interactively define a region of interest. In many applications, the part to be inspected moves within the image. To account for this movement, you must define a coordinate system and make all measurements relative to the coordinate system. These two steps are now incorporated into the machine vision functions.

When making measurements, it is important for the software to output the results in user-defined units and not in pixels. Use spatial calibration to output results in user units and correct for lens distortion. After performing an inspection task, users often want to see an image of the part with the results indicated on the image. Use nondestructive overlays, which are integrated in the machine vision functions, to indicate the result of the inspection task.

Application of Machine Vision Functions

The image of a floppy disk in Figure 6-13 illustrates the use of the machine vision functions, including locating edges, measuring distances, finding features, and counting objects. This example also demonstrates the use of relative coordinate systems and nondestructive overlays to indicate search areas and results.

Example 6-1 (VBI CD\Sections\Vision\Demos\IMAQLV50Demo\disk .llb) on the accompanying CD-ROM demonstrates how to solve a real-world application using the machine vision VIs, including interactive region of interest selection, inspection of a part, and display with nondestructive overlay.

Figure 6-13.
Example of machine vision functions. (Reprinted with permission of National Instruments.)

Color Analysis in Machine Vision

Today, monochrome images are typically used to solve a large majority of the computer vision applications. For some applications, however, the image data contained in a monochrome image may not be sufficient to solve the problem at hand. Using a color image acquisition board and camera can add an extra dimension of information.

You can use color inspection in pharmaceutical applications to inspect color-coded objects such as IV bags or tablets. Automotive applications, such as fuse box inspections, make use of color machine vision when you can distinguish objects based on color. The food industry uses color inspection to inspect fruit for ripeness and bruises. In all these applications, traditional approaches using monochrome image information can present problems. Color capabilities are suited for machine vision applications where color is the unique identifier for the device under inspection. Color in machine vision applications is a tool whose greatest possibilities are now just being realized. However, with new technologies, such as MMX, PCI, and the availability of low-cost machine vision software, color for machine vision is steadily increasing in appeal.

Color Pattern Matching

You can use color pattern matching to quickly locate known reference patterns, or fiducials, in a color image. With color pattern matching, you create a model or template that represents the object for which you are searching. Then, your machine vision application searches for the model in each acquired image and calculates a score for each match. The score relates how closely the model matches the pattern found. You should use color pattern matching to locate reference patterns that are fully described by the color and spatial information in the pattern. Color pattern matching can be faster than monochrome.

Gray-scale (monochrome) pattern matching is a well-established tool for alignment, gauging, and inspection applications. In all these application areas, you can often use color to simplify a monochrome problem by improving contrast or separation of the object from the background. The images shown in Figure 6-14 demonstrate the benefits of using color information when locating color-coded fuses. In the gray-scale image of the fuse box, the gray-scale pattern matching tool finds it difficult to clearly differentiate between fuse 20 and fuse 25 because of similar gray-scale intensities and the translucent nature of the fuses. In the color image of the fuse box, color helps separate the fuses and

Figure 6-14.
Example of gray-scale (monochrome) versus color pattern matching. (Reprinted with permission of National Instruments.)*

can distinguish between fuse 20 and 25. In such applications, color helps improve the accuracy and reliability of the pattern matching tool.

Color Pattern Matching Features

Color pattern matching should work robustly and reliably under various conditions. In automated machine vision applications, the visual appearance of the materials or components under inspection can change due to factors such as orientation of the part, scale changes, lighting changes, and so on. The color pattern matching tool should maintain its ability to locate the reference patterns despite these changes.

A color pattern matching tool should be able to locate the reference pattern in an image even if the pattern in the image is rotated and scaled. When a pattern is rotated in the image, the pattern matching tool detects the pattern in the image and returns the orientation of the pattern with respect to the template image. Also, color pattern matching can find multiple instances of a model in the image with subpixel accuracy.

The pattern matching tool finds the reference pattern in an image under conditions of uniform changes in the lighting across the image. Pattern matching can find patterns in images that have undergone some transformation due to blurring and addition of noise. Blurring usually occurs due to incorrect focus or depth of field changes.

---

*A color image of Figure 6-14 is provided on the accompaying CD-ROM (VBI CD-ROM/ sections/vision/Color Figures from chapter 6).

Color Location

You can use color location to quickly locate regions of specific colors in an image. With color location, you create a model or template that represents the colors for which you are searching. Then, your machine vision application searches for the model in each acquired image, calculating a score for each match. The score relates how close the model matches the colors found. You should use color location to locate objects that are fully described by the color in the object.

In many applications, an object can be fully described by its color, and spatial information is not required. Since color location uses only the color information and not spatial information, searching for an object takes less time. By using color location instead of color pattern matching, you can improve the speed of searching for an object. For example, you can count the number of color-coded tablets in an image using color location as shown in Figure 6-15.

The example `blister.llb` (VBI CD\Sections\Vision\Demos\IMA QLV50Demo\blister.llb) on the accompanying CD-ROM will demonstrate how to solve a real-world application using color pattern matching. Other practical examples of machine vision and virtual instrumentation are discussed in the following three examples. The first example, `lcd.llb` (VBI

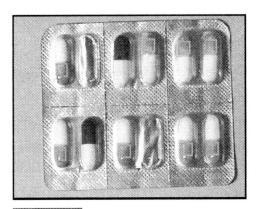 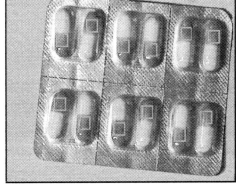

Figure 6-15.
Example of color location to count the number of color-coded tablets. (Reprinted with permission of National Instruments.)*

---

*Color image and interactive demo of Figure 6-15 is provided on the accompanying CD-ROM (VBI CD-ROM/sections/vision).

CD\Sections\Vision\Demos\ IMAQLV50Demo\lcd.llb), shows how to use IMAQ Vision's Seven-Segment display reader VIs to read values that are displayed on an LCD panel. Once the numeric value for each display has been determined, it can be shown in the form of a digital indicator as illustrated in Figure 6-16.

This next example shows how to use IMAQ Vision's meter reading VIs to read the position of a needle. These VIs are useful in many applications

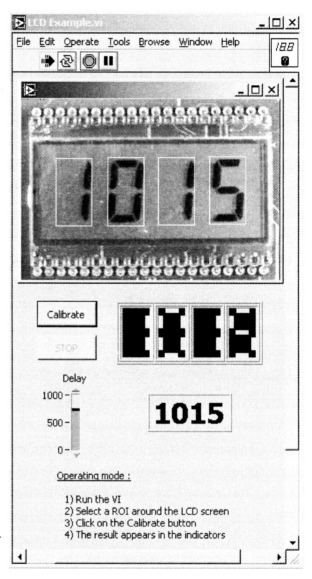

Figure 6-16.

Seven-Segment LCD example. (Reprinted with permission of National Instruments.)

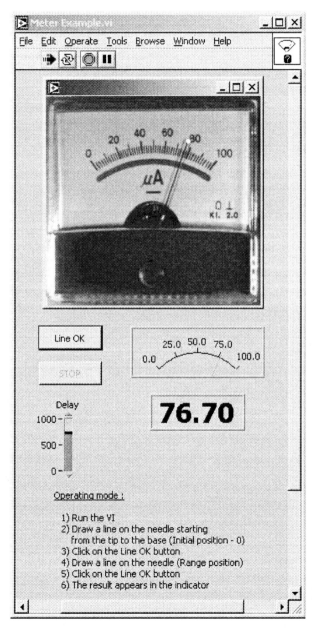

Figure 6-17.
Meter example. (Reprinted with permission of National Instruments.)

such as the calibration of speedometers and gauges in industries such as the automotive and medical industry. As indicated on the VI's front panel (see Figure 6-17), the user simply runs the VI then uses LabVIEW's Picture Control Toolkit functionality to calibrate the instrument. This is achieved by drawing a line on the needle (from the tip to the base) and clicking on the Line OK button to establish the initial position of zero. Once the initial position is established, the user performs a similar procedure at the maximum value to establish the range of the meter. At this point, the VI has been calibrated and any position of the actual meter will be translated into a corresponding value on the virtual meter on the VI's front panel. This example is also provided on the accompanying CD-ROM (VBI CD\Sections\Vision\Demos\ IMAQLV50Demo\ meter.llb).

The last example (VBI CD\Sections\Vision\Demos\IMAQLV50Demo\ barcode.llb) discussed in this section illustrates how to use IMAQ Vision's 1D Barcode reader VIs to read standard barcodes. Again, the user performs a simple calibration procedure by selecting a region of interest (ROI) that crosses the barcode area, and the VI will then automatically interpret standard barcodes into their numeric equivalent (see Figure 6-18).

Cameras and Interfacing

A decade ago, camera choices in the machine vision market were limited to simple interlaced cameras. These cameras, which followed the RS-170 and CCIR analog video standards, limited acquisition size and frame rate. Over the past several years, higher speed parallel digital cameras have become the cameras of choice for higher speed and higher performance applications. Basic RS-170 and CCIR monochrome cameras have remained the workhorses of the simple, low-speed, or extremely cost-sensitive machine vision applications.

Today, two new digital standards are emerging that have been designed to complement and possibly replace the existing standards. IEEE-1394, a PC-centric interface technology (also referred to as *firewire*), was designed to meet the needs of medium- to low-performance machine vision applications. Offering an easy-to-use interface and small camera connector, IEEE-1394 cameras give users a digital, feature-rich alternative to standard analog cameras—at a price.

Camera Link is a new interface standard codeveloped by many of the machine vision industry camera and frame grabber vendors. Using a high-speed serialized protocol, Camera Link promises to achieve parallel digital levels of performance with a simplified and smaller cabling scheme. Serial and timing communications are included in the specification, allowing camera companies to define and implement even more full-featured digital cameras.

Figure 6-18.
Barcode example. (Reprinted with permission of National Instruments.)

Interfacing with Digital Cameras

With all of the digital camera standards on the market, which should you choose? The answer to this difficult question depends largely on your particular application. Camera resolution and frame rate are often the two key factors in choosing the right camera. However, timing and triggering modes, as well as camera form factor, can also play an important role in making your decision.

As different standards appear on the market, however, other important criteria you should consider include camera availability and upgradeability. Will the interface standard that you choose be around in the future? As your application needs change, will you be able to easily switch to a different camera without changing your software?

Parallel Digital Cameras

Parallel digital cameras have been around for many years. They are the industry's highest performing cameras and are used globally to solve vision applications in a wide variety of fields. While there is no official standard interface for these cameras, the installed customer base has led to a host of support and interfaces from various manufacturers. Look for an EIA-644 (LVDS) interface to achieve the highest possible data rates. LVDS has been rated to run up to a 1.2 GHz clock rate, however, no camera or frame grabber on the market today supports this rate. This rating does, however, provide a foreseeable upgrade path to a higher performing camera in the future.

Because of the lack of a standard, cameras from different manufacturers have various connector and cabling schemes. As a result, upgrading or changing cameras can be a somewhat arduous task. Also, because the interface is a parallel digital interface, the connector and cable size are sometimes quite large.

With a parallel digital camera and a digital camera board (such as National Instruments 142x series IMAQ boards for PCI and PXI/CompactPCI), you can acquire images at thousands of frames per second with greater gray-scale resolution and with higher spatial resolution (up to 4,000 $\times$ 4,000 pixels).

Camera Link Cameras

As we mentioned earlier, a new standard for digital camera data transmission called Camera Link is being developed jointly by many of the key frame grabber and camera companies. This standard is based around a high-speed serial data transmission link called Channel Link that was developed by National Semiconductor. The standardization will improve the connectivity of digital cameras to digital frame grabbers while ensuring an extremely high level of performance. This standard combines much of the ease of use of an IEEE-1394 standard with the high-performance levels of the parallel digital standards. Although currently there are only a few Camera Link-compatible cameras on the market, many of the parallel digital camera companies have committed to camera development. While the industry acceptance of this standard is promising, currently there is no installed base of users. For customers planning future projects, Camera Link is a viable option to consider, but until the standard has proven itself on the market, there is some risk of long-term availability.

IEEE-1394 (Firewire) Cameras

The promise of IEEE-1394 has always been lower costs and simpler serial connections. Originally designed to connect multiple peripherals to a personal computer, IEEE-1394 has found a niche in the transmission of digital video data between cameras and PCs. After years of promises, there are only five cameras on the market today. These cameras are of reasonable quality and most have extensive feature sets that you can control over a serial link. All IEEE-1394 cameras on the market today follow the digital camera specification for IEEE-1394 cameras. While this specification ensures interoperability between the cameras, it limits the video content to either 8-bit monochrome or YUV color area scan, making IEEE-1394 cameras suitable only for specific applications. The future of IEEE-1394 is unclear, as it is extremely dependent on the PC industry as a whole. The lack of any backing by Intel and only partial backing by Microsoft has opened the door to two other serial standards, USB 2.0 and Serial ATA. IEEE-1394 may be a viable solution for customers with very simple and fixed requirements, but it is a far riskier alternative than choosing a parallel digital solution and much costlier than using an analog solution.

Interchangeable Cameras

NI-IMAQ driver software for image acquisition is designed to scale between many types of acquisition methods. You can start using a low-cost, RS-170 camera and image acquisition board and then upgrade to a faster, higher-resolution camera and board with minimal software changes. To start using a new camera, you configure the new board and camera in Measurement and Automation Explorer (MAX); then, you can immediately reuse your software. Because the NI-IMAQ driver software uses one set of function calls that work for a wide variety of cameras, there is no need to rewrite your software. This capability is extremely important in that it preserves your investment in the software development by providing a scalable solution that can take advantage of increasing bandwidth transmissions and new technologies.

Future Camera Technologies

What does the future hold for the camera industry? Faster camera interfaces are a top priority. Camera Link has been developed to meet this need.

Another developing technology is the use of CMOS (complementary metal oxide semiconductor) sensors. While CMOS technologies do not currently yield the same accuracy as traditional CCD processes, CMOS is improving daily. The promises of CMOS include much cheaper image sensors and tighter camera integration, with more of the camera electronics embedded on the same die as the image sensor. Additional benefits of CMOS sensors include higher possible frame rates, especially for the larger megapixel cameras and smaller camera bodies.

The rapid advancements in CMOS and CCD sensor technologies have contributed to corresponding advancements in a myriad of exciting medical imaging applications. These applications include advances in endoscopy, digital X-ray imaging, computerized tomography (CT), magnetic resonance imaging (MRI), and Picture Archiving and Communication Systems (PACS). In addition to these broad categories of imaging modalities, the development of specialized virtual instrument applications such as wound measurement, cell and particle counting, and eye-motion tracking will also be discussed.

Camera Advisor

Camera Advisor is a one-stop Web resource for engineers and scientists to select an imaging camera. The address for this application on the Internet is www.ni.com/cameras. Using this catalog of cameras, you can view features and specifications for more than 100 cameras. Camera Advisor explains how various cameras work with National Instruments hardware and software. When visiting this new section of the National Instruments Web site, you can compare different models and makes of cameras, such as line scan, area scan, progressive scan, and digital and analog cameras. You can also use Camera Advisor to compare technical details of various cameras.

Motion Systems

A motion system is generally made up of the following basic components: a controller, a drive, a motor, and a feedback device. Controllers generate trajectories, which the motor follows. Drives then take the signals sent by the controller and change them into current signals that will actually move the motor. Feedback devices are used to close the control loop in closed-loop systems. The parts of the system that National Instruments makes are the controllers, the software to run the controllers, and the drives.

There are many different kinds of controllers, drives, motors, and feedback devices. You choose each component based on the requirements for your application. For example, consider an application that requires high torque, high speed, and precise control. Since servomotors generally have higher torque at high speeds, a servomotor would be the most appropriate. After choosing the type of motor, you can then find a controller that controls servomotors. Since you want precise control, a PCI-7344 motion control board would be appropriate; it can control servomotors and is precise. After choosing the motor, you can also choose the type of drive to use. If the servomotor is a DC-brushed servomotor and requires less than 8 amps continuous current and 20 amps peak, you can use the nuDrive with an adaptor to connect to the PCI-7344. Since you are using a servomotor, you also need to consider the type of feedback device to use for closing the control loop. If you want to do position control, you can use an incremental encoder that mounts on the motor and easily interfaces with the PCI-7344 controller.

As previously discussed, you need to consider many aspects when building a motion control system. You may want to ask yourself some of the following questions:

- What type of motion is required for the system?
- What size motors will be necessary?
- What type of drive works your type of motor?
- What type of environment will the system be in?
- How accurate does the motion need to be?
- Will feedback be necessary and how accurate does that feedback need to be?

With National Instruments hardware, such as data acquisition (DAQ™), image acquisition (IMAQ), and motion control, you can use the Measurement and Automation Explorer (MAX) to quickly test connectivity, acquisition,

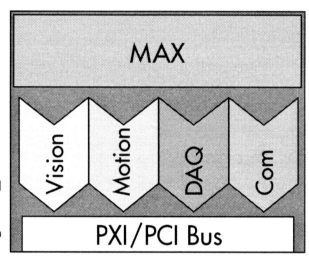

Figure 6-19.
Integrated Measurement and Automation Explorer (MAX) for machine vision, motion control, and data acquisition and analysis.

and control. In addition, you can rapidly display signal, image, and data results. You can use MAX to configure the features of your boards in a consistent manner. For data acquisition this could mean setting the input for differential signals, and for IMAQ this could mean setting the board for RS-170 camera input. The goal of MAX is to deliver a consistent interface for managing and configuring all of your measurement devices.

LabVIEW and National Instruments driver software give you a common software platform for numerous types of measurements. For example, the NI-DAQ and NI-IMAQ driver software application programming interfaces (API) are similar. If you know how to use NI-National Instruments DAQ driver software, you can use this knowledge with product lines such as IMAQ, motion control, and GPIB. In addition, NI-National Instruments driver software is scalable. The same function, IMAQ Snap, can acquire an image from a low-cost monochrome camera, a color camera, or a high-resolution digital camera. MAX simplifies configuration and testing and reduces code rewrite when you want to move between hardware products in the same product line (see Figure 6-19). In addition, your integrated measurement environment needs to have an open architecture. From LabVIEW you can call dynamic link libraries (DLLs) and interface ActiveX controls. Other programming benefits include logically named functions and consistent parameter names, attributes, examples, and error handling. With LabVIEW, you take advantage of the benefits of a rapid development environment and Windows connectivity. Plus, you can easily add the reliability of embedded real-time control.

You can build next generation machines with the ability to network to enterprise-wide systems. By using off-the-shelf PXI/CompactPCI and PCI hardware, you create a system that is easy to maintain and scale for future needs. LabVIEW and PC-based machine control systems reduce system downtime by providing detailed diagnostics. The LabVIEW flexible user interface can show the machine status on the PC screen and prompt your machine operator through repair steps. By building diagnostics into the LabVIEW application, machine operators can clear faults and repair equipment failure by following graphical directions displayed on screen. Overall, this intelligent and fast troubleshooting dramatically reduces downtime.

Example: Designing a Motion Control Project

Oftentimes, motion control is a core part of building a machine. More and more, machine builders are discovering the benefits of using off-the-shelf components and leveraging the PC architecture for the motion and numerical control of their machines. In a traditional machine, much of the design effort is spent on proprietary hardware. When building an open architecture machine using motion control, systems integrators are now forced to spend a tremendous amount of time and money developing custom application software. Thus, software productivity becomes an important factor of building the motion component of a machine. Using LabVIEW and an open architecture FlexMotion™ controller from NI, the user can create an application ten times faster than traditional methods. We guide you step by step through creating a simple motion project in a very short time. For this application, imagine you are interested in drilling a hole, roughly 5 inches in diameter, in a block of wood or metal. This would be one task of a machine that manufactures car parts, a subwoofer box, or furniture.

Figure 6-20 illustrates the components of a motion control system. Motors, feedback devices, and mechanical fixturing greatly vary from application to

Figure 6-20.
Motion components: Motion software, motion controller, motor drive and motors, feedback, and mechanical fixturing.

application. For example, if you are building a semiconductor wafer handler, your motors, feedback, and fixture are totally different than if you were building a biotech test tube sampler or a computer-controlled lathe. Selection and integration of these hardware components require your specific vertical market expertise.

Regardless of your system or machine, users can meet all their drive, motion controller, and software needs using off-the-shelf components from National Instruments. The drive passes current through your motor, modulating the signal as necessary for different sizes and types of stepper and servomotors. The motion controller is specialized hardware designed to intelligently command and output any trajectories. It handles features such as limit switches, breakpoints, encoder feedback, and more. You send high-level commands to your controller, such as Load Target Position and Start, with the motion software. It is also where you can create an interactive user interface.

To control any motor, nearly every setting you can think of is software configurable—stepper output mode, inhibit polarities, rising/falling edge/level position triggers, PID parameters, and more. Configuration for your mechanical system is often a one-time process using an easy-to-use, interactive interface, Measurement and Automation Explorer (MAX). As shown in Figure 6-21, there are hundreds of parameters you can optionally configure to optimize your system. Then, within your application, call the Initialize function and all the parameters you have configured are set.

In addition, you can run diagnostics and test individual axes of motion using this program. These diagnostics include a two-axis virtual stage. This is useful for our demonstration, because it illustrates real-world motors in motion. It traces the path as well. This virtual stage is shown in Figure 6-22.

Now let's return to our example application. Using the NI-Motion software and controller, a basic point-to-point move involves only three commands.

Note: Although you most likely do not have a motion controller board or motors, the concepts and processes of motion control will be demonstrated in this example.

- First, initialize the board. This configures all motion parameters to a known state and energizes the motors.

- Next, load the target position to which you want to move.

- Finally, command the board to Start Motion.

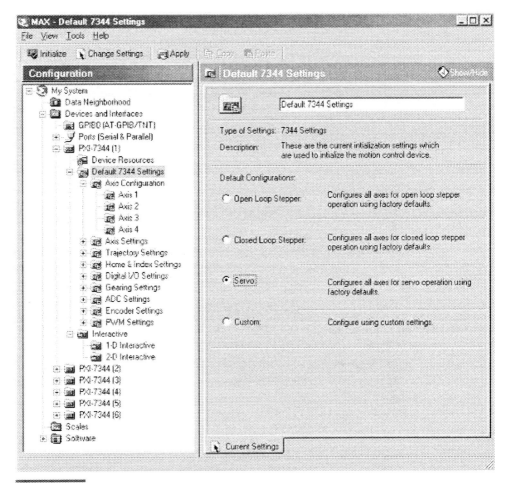

Figure 6-21.
Measurement and Automation Explorer (MAX) Configuration screen. (Reprinted with permission of National Instruments.)

Note that, after you run this program, LabVIEW completes immediately while the motor is still moving. This is because the motion board has a real-time trajectory generator and onboard intelligence to complete the commanded move.

The block diagram in Figure 6-23 illustrates how the motion board follows a trapezoidal velocity profile using your default velocity and acceleration parameters, arriving exactly at the commanded target position.

For our next step in the simple point-to-point move, we add a Velocity and Acceleration control. Setting each of these parameters requires one function

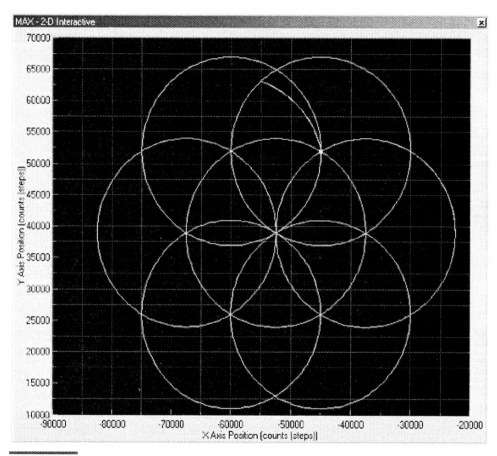

Figure 6-22.
Interactive 2-dimensional motion control graph. (Reprinted with permission of National Instruments.)

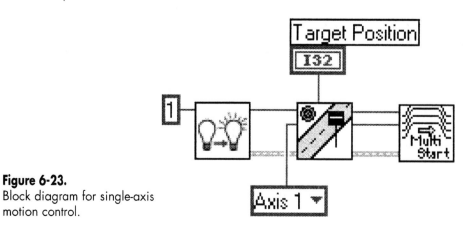

Figure 6-23.
Block diagram for single-axis
motion control.

call. Users can load velocity and acceleration using two different units, depending on their personal preference. With counts/s and counts/s^2 users can easily specify a low-level, precise value. RPM and RPS/s are higher level real-world units. Users can also benefit from moving at speeds below 1 count/s (because they are floating point parameters).

Once the move begins, we can plot the current position of the moving axis while the move is not complete. This requires two commands, Check Move Complete and Read Position. We use a LabVIEW while loop to poll our motion board. We can set the trapezoidal profile values, and monitor the move until it is completed.

Now let's introduce a second axis by simply copying and pasting the block diagram of one move. Also, we create separate copies of the position, velocity, and acceleration controls in order to set independent values. This block diagram is shown in Figure 6-24.

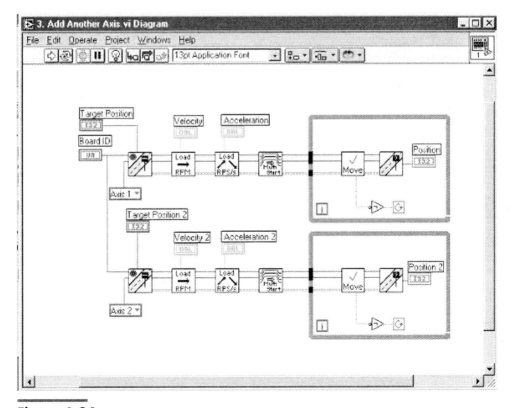

Figure 6-24.
Block diagram for adding a second axis for motion control.

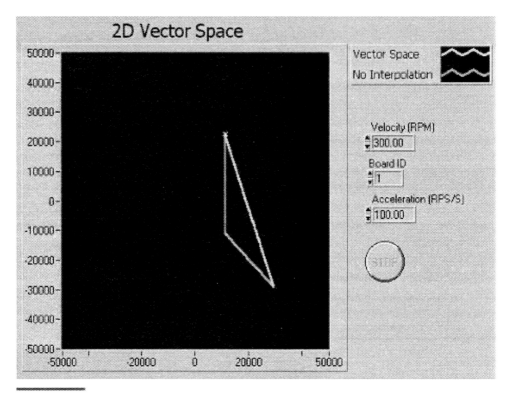

Figure 6-25.
2D vector space (no interpolation vs. linear interpolation).

If we set velocity and acceleration to be equal on both axes, but with different target positions, users see that the motion starts on both axes (roughly) at the same time. However, one axis completes its move before the other one does. If we were moving a two-axis stage or positioner, this would not be acceptable to perform vectored moves. We would have to resolve our desired vectored move into component vectors, in terms of velocity and acceleration. Even then, the move does not truly start and end at the same time, because the LabVIEW code is in a basic race condition and may be off by several milliseconds. Figure 6-25 illustrates the problem of no interpolation versus linear interpolation, also called using a vector space (VS). The National Instruments motion interface supports onboard calculations of vectored information, causing motion on two (or three) axes to start and stop at exactly the same time (within microseconds).

To use a vector space, first configure the vector space (X, Y, and Z axes). These can be in any order, and any axis can be disabled for a 2D VS instead of a 3D VS. The Help window for the `Configure Vector Space.flx` VI is illustrated in Figure 6-26.

Next, replace the individual Read Position functions with a Read Vector Space Position. Users do not have to change Load Velocity, Load Acceleration, and Start Motion; the program will intelligently operate on the vector instead of an individual axis.

Keeping in mind that our original goal was to move in a circular pattern, we now introduce the circular interpolation mode of the NI-Motion controller. This is similar to linear interpolation in that it operates on a vector space; however instead of loading an (x, y, z) coordinate, you specify three parameters of a circular arc.

In Figure 6-27, radius corresponds to label 6, start angle is 4, and travel angle is 2. Using these three parameters you can define any circular arc.

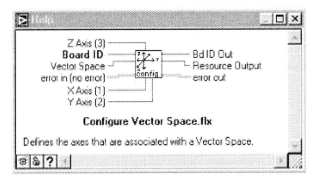

Figure 6-26.
Help window description for `Configure Vector Space.flx` VI.

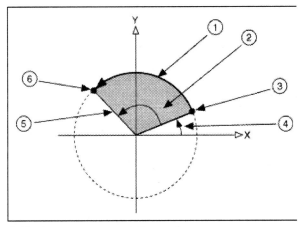

1. Circular arc
2. Travel angle
3. Starting position
4. Start angle
5. Radius
6. Ending position

Figure 6-27.
Defining a circular arc.

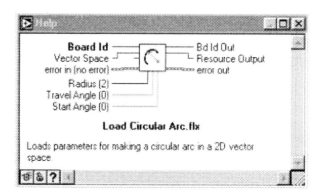

Figure 6-28.
Help window description for
Load Circular Arc.flx VI.

The help window for the `Load Circular Arc.flx` VI is illustrated in Figure 6-28. The start angle must be between 0–3608, but the travel angle can be outside of this range for multiple arcs or reverse movement.

Recall that we are drilling a circular hole. In order to prevent an uneven cut, we should avoid dropping the drill bit at the edge of the path. This may cause splintering. So, we would like the profile to follow two arcs as shown in Figure 6-29.

In the LabVIEW code, just add another Load Circular Arc command and another Start command. We also have replaced the while loop with a high-level function called Wait for Move Complete. This is a blocking function (shown in Figure 6-30) that suspends itself until the move is complete.

Now, you can use the two-axis position plot (virtual stage) to monitor your profile and ensure that the six parameters you have chosen for the two arcs are correct.

To further reduce the risk of splintering, it would be nice to move smoothly through the two arcs instead of stopping motion after the first arc and then starting a new motion. With blending, users can superimpose two trajectories between two independent moves, so that you can move from point A to B to C without stopping motion at B. The charts in Figures 6-31 and 6-32 illustrate a velocity profile with and without blending. In our code, we replace the Start Motion with a Blend Motion. Also, we replace the Wait for Move Complete with a Wait for Blend Complete.

In a few short steps, we can create a motion profile that would take significant hardware effort in a traditional, proprietary system. In a real machine, we could continue to increase the performance and functionality of this task. For instance, you could download all the code to the controller and have it execute onboard. Then, you could have it wait on a digital trigger. You could also have it stored in nonvolatile flash ROM and automatically run after

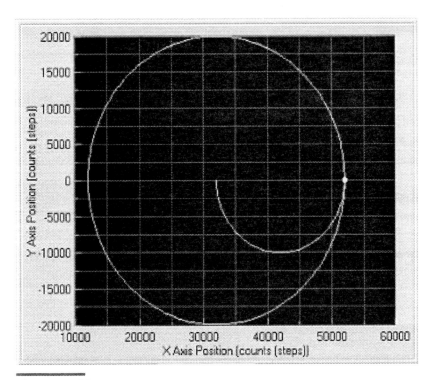

Figure 6-29.
Multiple arc profile.

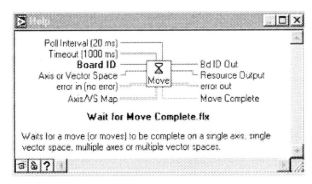

Figure 6-30.
Help window description for `Wait for Move Complete.flx` VI.

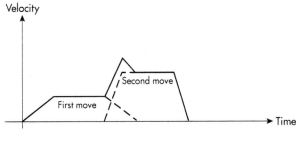

Figure 6-31.
Velocity profile with blending.

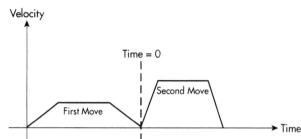

Figure 6-32.
Velocity profile without blending.

a power cycle. You could also run two of these tasks in parallel on a four-axis controller, using the onboard, real-time, multithreaded operating system. We will now examine several real-world examples of biomedical applications that use machine vision or motion control technologies.

A LabVIEW-Based Wound Management System

The Wound Management System is a simple, portable, computer-based approach to wound assessment that implements both digital and infrared imaging technology. The digital image is analyzed to determine wound area and color properties, while the infrared image will be analyzed to determine blood perfusion properties of the wound.

Purpose of System

The Wound Management System provides a simple and cost-effective method by which healthcare personnel can quantify the healing rate of a wound, assess the effectiveness of wound care treatment techniques, and document wound healing progress. The computer-based system improves the repeatability and effectiveness of both wound area measurement and analysis of wound coloration as compared with current practices.

Advantages of System

Wound measurements currently involve invasive and inconsistent measurement techniques utilizing manual measurement tools or plastic wound overlays and requiring estimation of wound size using geometric simplifications (e.g., considering only the length and width of a wound and using a square area approximation of wound size). Current practices also involve qualitative, subjective clinician assessment of wound colorization and inconsistent use of photographs to document patient progress.

Significance

Wound care currently encompasses a wide variety of patient types, from acute, simple lacerations seen in the emergency room to devastating wounds resulting from trauma and surgical interventions to chronic lower extremity ulcers and pressure sores. There is fragmentation of wound care personnel and facilities within the medical community. Presently, new treatment modalities are appearing at a very rapid rate. It is often beyond the ability of a single individual to comprehend and incorporate the new information into improved and cost-effective techniques for patient care.

Chronic wounds represent a major issue in healthcare today, particularly among the elderly. Nonhealing wounds can persist for years, causing pain to patients and placing them at risk for secondary infection or loss of limb. The financial burden of the care of these patients is phenomenal: Institutional care of nonhealing wounds is estimated to cost approximately $1,000 per day per patient. A recent study estimates that approximately eight million people in the United States presently suffer from nonhealing wounds, resulting in over $10 billion additional annual healthcare costs.

As a result, wound care has become a specialty in itself, and, over the last 15 years, an enormous amount of wound care products and services have emerged claiming to empower clinicians with the ability to effectively manage the healing process. In addition, several wound measurement products have also been introduced to the market. These products range from simple rulers and pliable transparent sheets (with bull's-eye target patterns) that cost less than $20 to high-tech computer-based measurement systems that cost more than $80,000.

Several factors recently led to making wound care a medical discipline by itself:

• Dedicated wound care programs are increasingly caring for Americans with chronic wounds and could offer the potential to produce enhanced outcomes at reduced costs for payers and consumers of healthcare services. Evidence of this trend can be found in a variety of recent reports and studies.

- A report by Frost and Sullivan indicates the U.S. wound management products market is now $1.74 billion and should grow to $2.57 billion by 2002. The cost of treating the chronic wound is estimated at $5 to $7 billion per fiscal year, and these wounds are increasing at a rate of 10% per year, according to the report.*

- The development of the American Academy of Wound Management (AAWM), a national, nonprofit certifying board, is another indication that wound care is coming into its own as an industry. Board certification is now available for physicians, nurses, therapists, researchers, and other healthcare professionals involved in wound care. Over 1,000 wound care professionals have requested applications for board certification through AAWM.

Presently, in most hospital settings, the daily clinical routine for wounds involves little or no quantitative documentation. Measurements are generally performed with the help of a ruler or a disposable transparency where the wound is drawn. In some cases, clinicians will periodically take pictures with a digital camera, but no formal protocol has been established, no lighting considerations are taken into account, and the purpose is more educational and qualitative than quantitative. Figure 6-33 illustrates an example of tools and techniques that are currently used to measure wounds and healing rates.

The LabVIEW-based Wound Management System eliminates subjectivity and approximations by using computer algorithms to (1) calculate the wound area based on the true, irregular wound outline; (2) determine the percentage of wound area that is red, yellow, and black (the colors most indicative of wound healing progress); and (3) evaluate or present information on wound vascularization (via infrared imaging). In addition, because the system is based on obtaining successive images of the wound as treatment progresses, clinicians have access to an image archive of patient progress. Clinicians can also add personal notes to the file for a patient to create an inclusive database of images, analysis, and clinical progress. The system has been designed to be portable and easily used by a typical clinician. The system will be self-contained and will utilize a fixture to address concerns such as lighting, distance from camera to patient, and photographic angle.

---

*For more information, see www.pslgroup.com/dg/daf6.htm.

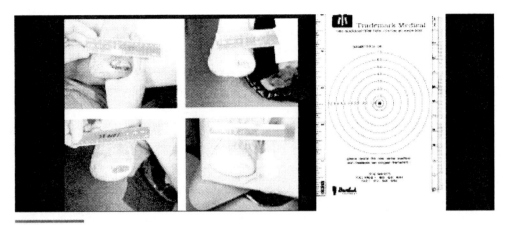

Figure 6-33.
Wound documentation as it is ideally performed today. (Source: Hartford Hospital, Hartford, CT, USA.)

Indicated Population

The applicable patient population includes both sufferers of acute wounds (lacerations, trauma, surgical wounds, etc.) and chronic wounds (arterial, venous, and diabetic ulcers, pressure sores, etc.). Application of the wound healing system will focus on the chronic wound population, which accounts for the most extensive costs to healthcare and would result in the most patient benefit. Initial proof-of-concept testing occurred at Hartford Hospital's foot clinic, which focuses on diabetic foot ulcers.

The following figures illustrate some of the image acquisition, analysis, and data management modules of the Wound Management System. The panel in Figure 6-34 illustrates how any patient (and his or her respective wound images) can be selected for historical and visual comparison. In addition, a particular image can be selected for advanced analyses as shown in Figure 6-35.

Figure 6-35 illustrates how a selected wound can be automatically analyzed in terms of its area and coloration. Computer algorithms are employed to identify the region of interest and perform specific analyses including wound contour, area, and coloration. Future analyses that are being developed include infrared imaging and volumetric analysis.

Figure 6-36 is an example of thermal imaging on a diabetic foot wound. This technique may be an effective method to monitor wound healing and revascularization as well as predict future wounds and ulcers.

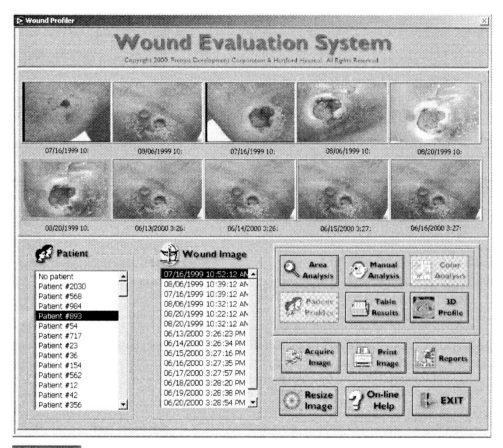

Figure 6-34.
Patient and Wound Selection panel.

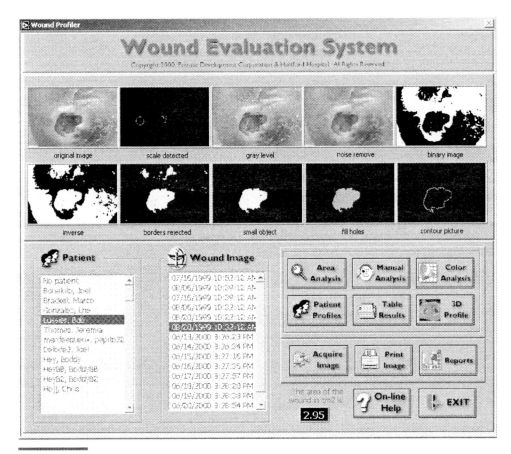

Figure 6-35.
Automated Wound Analysis panel.

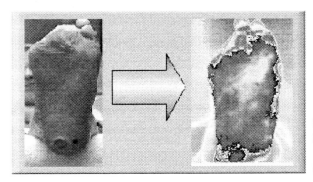

Figure 6-36.
Infrared/thermal imaging.
(Courtesy of ImageTherm
Engineering.)*

*A color image of Figure 6-36 is provided on the companion CD-ROM (VBI CD-ROM/ sections/vision/Color Figures from chapter 6).

Benefits of Wound Management Technology

The benefits of wound management technology can be divided into three primary categories: (1) benefits to the clinician, (2) benefits to the patient, and (3) benefits to the healthcare organization. Specific benefits of each category are described next:

1. Benefits to the Clinician

- Consistent and repeatable method for measuring wounds
- Objective and quantitative data for assessing wound care approach
- Access to historical data on wound healing rates based on treatment type and patient-specific variables
- Easily accessible image archive of patient progress

2. Benefits to the Patient

- Noninvasive wound measurement and assessment
- Reduced likelihood of nosocomial infection
- Enhanced client understanding of and participation in wound management (e.g., measured improvement and visual presentation of wound progression motivates continuation of proactive patient practices)

3. Benefits to Healthcare Organization

- Facilitates more rapid assessment of wound care effectiveness, thereby decreasing treatment time and related costs
- Decreases time required by clinician to document wound condition
- Allows remote sharing of information between sites for teaching, collaborating, or both

Independent Solution Articles

IMAQ and LabVIEW Automate the Study of Eye Motion

by Philippe Sauvan-Magnet, Graftek Imaging, Inc.

The Challenge: Automating the monitoring of human eye motion to aid physicians in researching internal ear diseases.

The Solution: Developing a flexible PC-based image processing system using IMAQ and DAQ boards, IMAQ Vision software, and LabVIEW.

Introduction

Graftek France, a National Instruments Alliance Program member, was retained by CCA Biodigital, a company specializing in medical research equipment for otolaryngologists (ear, nose and throat doctors), to develop a new generation of nystagmography research systems. Nystagmus is a particular motion of the eyes occurring in specific situations, such as watching the landscape race by while riding in a train.

In these situations, a low-speed motion (when the eyes are focused on an object) is followed by a high-speed motion (to locate the next object).

The new system was required to simulate this condition for the eyes with moving objects, such as a 6-foot-long horizontal LED bar, a light ball projecting a light beam on the wall, or a rotating chair. The patient can see the simulation equipment through a 45-degree semitransparent mirror, while the patient's eyes are being illuminated by infrared lights and monitored by two cameras. Both the infrared lights and the cameras are mounted on open goggles.

An Inexpensive Solution

The goal was to create an innovative solution without using expensive onboard DSP chips and to achieve an expected system life of 10 years. In the past, dedicated image processing boards incorporating DSP chips have been used. Because two such boards are required to track the two eyes simultaneously, a system is quite expensive. Plus, DSP programming is time-consuming and not at all flexible. A much less expensive virtual instrumentation solution is available thanks to IMAQ and the power of today's Pentium processors; image processing boards with onboard DSP chips are no longer necessary.

The Image Processing System

Our system consists of a Pentium computer running Windows 95, equipped with LabVIEW, IMAQ Vision, two IMAQ PCI-1408 image acquisition boards, and one PCI-DIO-96 digital

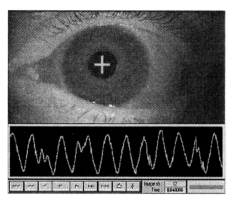

Figure 6-37.
LabVIEW and IMAQ vision track human eye movement.

I/O board. The two cameras monitoring the patient's eye movements provide two video signals that are routed to the two IMAQ boards.

We chose IMAQ boards and IMAQ Vision software because we believe that the new PCI technology, combined with the fast Pentium engine, offer enough power to avoid using DSP chips. An important benefit of using LabVIEW for such applications is the ability to integrate image acquisition and processing with data acquisition and control functions, such as controlling the LED bar with the PCI-DIO-96 board.

Using the LabVIEW graphical user interface (GUI), we created an easy-to-use operator interface for setting the parameters of the application, which are critical. They must be easy to adjust by a doctor concerned with the patient rather than the computer. Calibration, which includes configuring the board to adapt to the patient's eye lighting, differs from one patient to another.

We also interfaced the application to a database for recording the results in the patient's records. The LabVIEW data analysis library offered numerous possibilities for calculating the different coefficients (speed, acceleration, position, and so on) necessary for the diagnosis. The entire development of the application, including clinical tests, took five man-months compared with the man-years of development invested in DSP-based solutions.

The finished product can acquire 100 images/s and process it on the fly. We did this by using a special feature of the interlaced video, where each image is composed of two fields acquired with a 20 ms separation. We acquire 50 fields/s for each eye and the software must process 100 images/s with a resolution of 768 by 256 for each field. Using the LabVIEW CIN toolkit, we developed an optimized LabVIEW virtual instrument (VI) that performs the eye tracking by software—a real-time 100 images/s processing algorithm.

Conclusion

Using LabVIEW and IMAQ products, we were able to develop a research system in a very short time. By developing a LabVIEW and IMAQ system based on PC-based virtual instrumentation technology, instead of DSP technology, we reduced cost significantly.

For more information, contact Philippe Sauvan-Magnet, Graftek France,
Le Moulin de L'image —26270 Mirmande —France,
tel (33) 4 75 63 00 29, fax (33) 4 75 63 03 65,
e-mail sauvan@graftek.fr

Endothelial Evaluation of Corneal Transplants by Knowledge-based Digital Image Processing

by Norbert Dahmen, Professor, and Georg Toszkowski, Scientific Assistant, FH Niederrhein, University of Applied Sciences, Krefeld, Germany

The Challenge: Automating the endothelial cell density analysis of human corneal transplants for successful keratoplastics.

The Solution: Development of a facilitated and reliable diagnostic tool for endothelial cell density evaluation using a knowledge-based digital image processing approach.

Abstract

The objective determination of the endothelial cell density by inverse phase contrast microscopy after cell border swelling in hypotonic solution is of crucial importance for successful keratoplastics and thus for a long-lasting recovery of the patient's visual capabilities. Up to now, endothelial cell density had to be determined by manually counting the endothelial cells within a fixed frame on widely blurred and low-contrast Polaroid images. Since it is difficult to distinguish living from necrotic cells, the results are significantly subjective with a high error rate. Moreover, the complete evaluation procedure is rather expensive and very time consuming.

Based on LabVIEW, IMAQ Vision, and the Fuzzy-Logic Toolkit, a facilitated and reliable automatic diagnostic tool for endothelial cell density determination was developed, avoiding the disadvantages of manual cell counting. The diagnostic system uses scans from Polaroid images as well as digital images directly acquired by a CCD camera.

The Lions Cornea Bank, located at the Heinrich-Heine-University in Düsseldorf, Germany, is one of the first institutes capable of determining the endothelial cell density for corneal transplants automatically. The diagnostic system operates fast with a high reliability and reproducibility.

Problem

The endothelial cell density is an important quality characteristic of corneal transplants. Endothelial cells do not regenerate well. Loss of cells as a consequence of injury or metabolic disease, for example, is compensated by a spatial expansion of the surrounding cells. Corneas with low endothelial cell density are thus unsuited for transplantation.

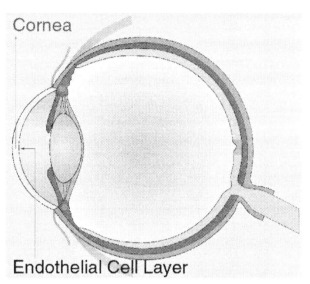

Cornea

Endothelial Cell Layer

Figure 6-38.
Cornea and endothelial cell layer.

Figure 6-39.
Corneal transplant taken from a donor.

Being cultivated for up to six weeks at a maximum, the corneal transplants are periodically inspected for quality control purposes. After a first generalized inspection of the whole endothelial cell layer at low magnification by phase contrast microscopy after cell border swelling in hypotonic solution, the most representative area of the cell layer is selected for acquiring a Polaroid image at a high optical magnification. An investigator can then determine the endothelial cell density by manually marking and counting the cells within a fixed frame put on the best perceptible area of the Polaroid image.

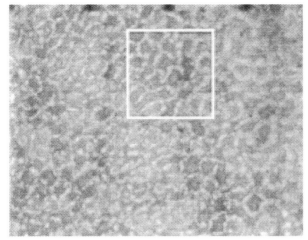

Figure 6-40.
Polaroid image with count frame.

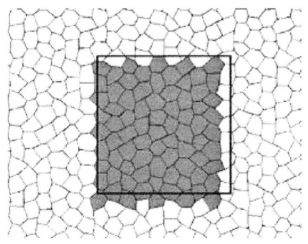

Figure 6-41.
Cell count scheme.

The number of cells within the count frame is multiplied with a magnification specific factor leading to the cell density given in cells per square millimeter. The whole procedure is very time consuming (about 15 minutes for a single transplant), and the results are rather subjective since cells within large areas of the endothelial cell layer are not exactly perceptible and thus hardly to be distinguished from necrotic (diseased) regions.

Solution

The problem for the implementation of a computer-based endothelial evaluation tool mainly consists in the optically difficult image acquisition scenario during phase contrast microscopy of the transplant's endothelial cell layer leading to a widely blurred and low-contrast image. Additionally, the cells form a tightly closed pattern that cannot easily be separated from the background. Due to geometric reasons—the transplant has a spherical shape—cell borders are not necessarily closed within the photographic plane. Looking at the gray-scale distribution shown in Figure 6-43, it is quite obvious that simple thresholding techniques cannot be applied to separate the cells from the background. The endothelial cell layer image is additionally characterized by an inhomogeneous brightness distribution.

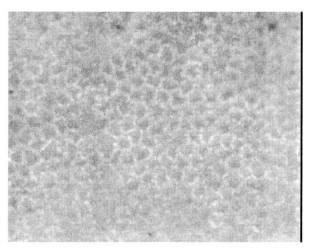

Figure 6-42.
Endothelial cell layer image.

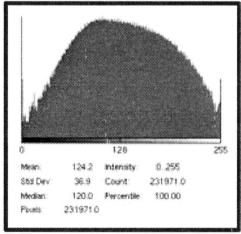

Figure 6-43.
Gray-scale distribution.

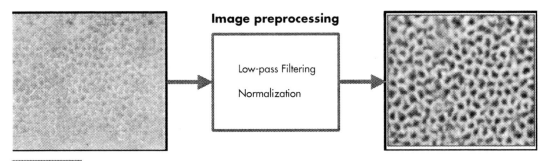

Figure 6-44.
Image preprocessing by low-pass filtering and normalization.

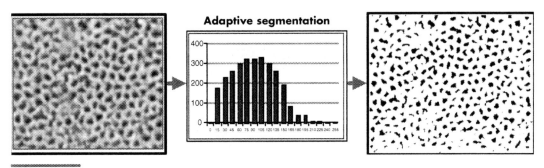

Figure 6-45.
Object segmentation.

Due to compatibility reasons with respect to the cornea transplant data accumulated at Lions Cornea Bank so far, they requested that the fixed frame method be implemented first. The diagnostic tool should accept scans from (existing) Polaroid images as well as images acquired directly by a CCD camera mounted on the inverse phase contrast microscope.

Before starting the object recognition step, the gray-scale image is first improved by conventional image preprocessing procedures like low-pass filtering and normalization. The result of this step is shown in Figure 6-44.

Subsequently to the image preprocessing step, the improved gray-scale image is transformed into a binary image using a specifically designed adaptive segmentation algorithm. This had become necessary since each endothelial cell layer image has its own unique threshold needed for optimal object separation. A typical result of this step is shown in Figure 6-45.

Based on the binary image obtained from the segmentation step, the object classification is carried out with special emphasis to the distinction of single objects (endothelial cells) and multiple objects (either endothelial cells without completely swelled borders or necrotic regions).

For the object classification, the following object characteristics have been taken into account: object area, compactness (object area related to the area of the object's surrounding rectangle), and circularity (object periphery related to the periphery of a circle with the object's area). A lot of transplant data has been statistically evaluated without finding any crisp thresholds to reliably distinguish between single and multiple objects regarding the object characteristics mentioned earlier.

To improve the automation degree of the diagnostic process with regard to the imprecise limits between single and multiple objects, a knowledge-based approach has been chosen for the object classification. For this the Fuzzy-Logic Toolkit, available from the LabVIEW PID Control Toolset, has been applied.

After object classification, the cell density is calculated and displayed on the front panel of the diagnostic system. As requested by the Lions Cornea Bank, multiple object regions

Area:	Area (multiple object)	>>	Area (single object)
Compactness:	Compactness (multiple object)	<<	Compactness (single object)
Circularity:	Circularity (multiple object)	>	Circularity (single object)

Figure 6-46.
Object characteristics for a Fuzzy-based classification.

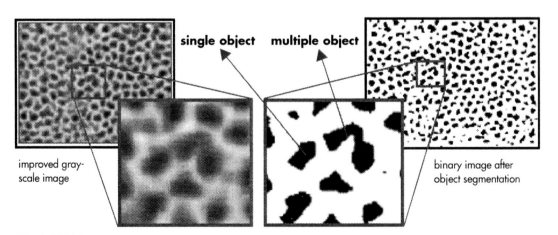

Figure 6-47.
Object classification problem.

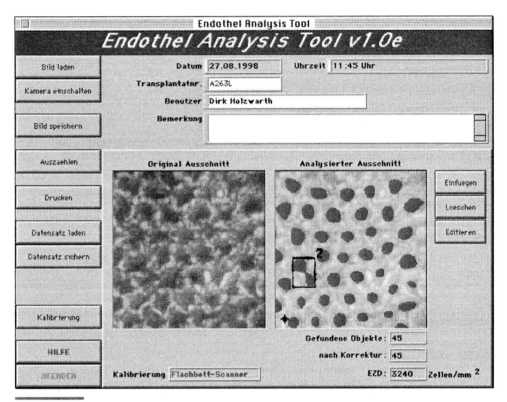

Figure 6-48.
Front panel of the endothelial analysis tool.

are marked as potentially necrotic regions and a suggestion is additionally calculated indicating the number of cells representing the multiple object. A medical expert can correct the diagnostic data in any case of misinterpretation.

The image processing steps described previously are integrated into a user-interactive diagnostic tool offering several functions such as loading and storing images including the transplant-specific data. A setup mode is available for calibration purposes with regard to the image acquisition method (scans from Polaroid images or CCD camera).

A print facility is also available as an on-line-help reference describing the several system functions. For the Lions Cornea Bank, a specific link to the transplant database system has been established. The tool has been successfully tested in the eye clinic at Heinrich-Heine-University in Düsseldorf, Germany. Figure 6-49 shows the test implementation at the Lions Cornea Bank during the introductory presentation.

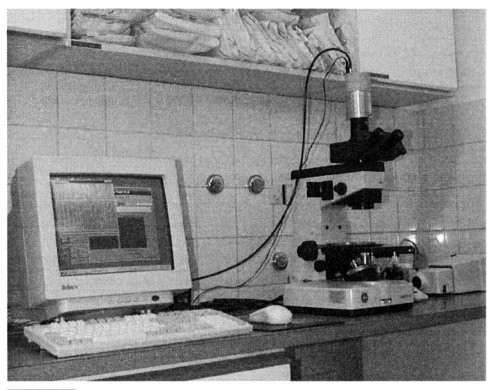

Figure 6-49.
Endothelial analysis tool with inverse phase contrast microscope and CCD camera.

Conclusion

The Lions Cornea Bank, located at the Heinrich-Heine-University in Düsseldorf, Germany, is one of the first institutes capable of determining the endothelial cell density for corneal transplants automatically. Using the new analysis tool within an extensive clinical test, intra- and inter-individual variability of endothelial cell count was statistically significantly low compared with expert evaluation by hand. Furthermore, significantly less time and money for evaluation and documentation had to be spent during the test by using the computer-based evaluation system. In the future, neuronal techniques will reduce further the necessity of expert corrections. Multiple fixed frames will be implemented in order to derive a quality profile for the whole endothelial cell layer including complete cell morphology.

References

Haberäcker, P. (1995). Praxis der digitalen bildverarbeitung und mustererkennung. Munich: Carl Hanser Verlag.

Jamal, R., & Krauss, P. (1998). LabVIEW—Das grundlagenbuch. Upper Saddle River, NJ: Prentice Hall.

Jamal, R., & Wenzel, L. (1995). The application of the visual programming language LabVIEW to large real-world applications. Proceedings of 1995 IEEE Symposium on Visual Languages (pp. 99–105).

Kahlert, J., & Frank, H. (1993). Fuzzy-Logik und Fuzzy-Control. Verlagsgesellschaft mbH, Braunschweig: Friedr. Vieweg & Sohn.

National Instruments. (1997). Fuzzy Logic for G toolkit reference manual. Austin, Texas: author.

National Instruments. (1999). IMAQ vision user manual. Austin, Texas: author.

Counting Particles or Cells Using IMAQ Vision

by John Hanks

Introduction

To count objects, you use a common image processing technique called particle analysis, often referred to as *blob analysis*. Blob analysis is the process of detecting and analyzing distinct two-dimensional shapes within a region of the image. Blob analysis can provide your application with information about the presence or absence, number, location, shape, area, perimeter, and orientation of blobs within an image. In machine vision applications, blob analysis is used for applications such as detecting the presence of flaws on silicon wafers, finding the orientation of an integrated circuit on a plug-in board, and locating objects in motion control applications when there is significant variance in part shape and orientation. Overall, National Instruments IMAQ Vision software can measure more than 49 different parameters of a blob. This application note introduces you to blob analysis and shows you how to use IMAQ Vision software to count components, particles, or cells in an image.

Blob Analysis and Counting Objects Overview

A simple definition of a blob is a group of connected pixels. In general, blobs are thought of as a group of contiguous pixels that have the same intensity. Image processing operates on these blobs to calculate the area or perimeter, or to count the number of distinguishable blobs. Before you can apply blob analysis you must preprocess the image by converting a gray-scale image (an image with 256 levels) to an image with only two gray scales—zeros and ones. The objective is to separate the important objects, blobs, from the unimportant information, background. A technique called thresholding appropriately separates the blobs from the background. The result of the thresholding process is a *binary* image, which is an image of pixel values of only ones and zeros. The blobs are represented by the connected pixels of ones, and the background is represented by the zeros. By binarizing the image into ones and zeros, the task of writing image processing algorithms for blob analysis is made easier. For example, to find the area of a blob, you simply need to count the pixels with values of one that are connected. Another benefit of binarizing the image for blob analysis is that the blob analysis calculations are fast. This application note will introduce blob analysis with a simple cell counting application. Realize that blob analysis is used not only in biomedical imaging applications, but also in industrial inspection applications for counting components and calculating the locations of objects. Each of the IMAQ Vision image processing steps is discussed—acquiring the image, preprocessing, and then blob analysis.

Here are the steps for counting objects using IMAQ Vision:

1. Acquiring the image
2. Histographing to identify the threshold values
3. Thresholding to create a binary image
4. Filtering to remove noise and particles on the border of the image
5. Particle (blob) analysis to count cells

Step 1—Acquiring the Image

You can use a National Instruments IMAQ 1408 board to acquire images from standard and nonstandard analog cameras and microscopes. You can easily configure the board for acquisition using a point-and-click user interface. The configuration software saves the settings for the board to a configuration file, which is then used by the NI-IMAQ driver software in the development environment to simplify acquisition of images.

Configuring Your Camera or Microscope

For most particle or cell counting applications it is easier to configure and acquire images from a camera that uses standard video. Many cameras use popular analog video standards that include the VSYNC and HSYNC timing information as well as an image size. The standard video formats are EIA RS-170 and CCIR for monochrome video, and National Television Systems Committee (NTSC) and Phase Alternate Line (PAL) for color video. For example, camcorders most often output color-composite video in NTSC or PAL formats, combining the luminance (brightness) and chrominance (color) components into a single analog signal. NTSC is used in the US and Asia, while PAL is popular in Europe. Overall, you can easily configure the IMAQ board for standard video using an interactive software utility that is shipped with the board, the NI-IMAQ Configuration Utility. You can use this utility to set up the IMAQ hardware to acquire gray-scale images from a color video signal (NTSC or PAL). In hardware, an antichrominance filter is applied to remove the color information. By selecting this filter, the incoming color video is translated to an 8-bit gray-scale signal.

Some devices, such as microscopes, may output the analog video signal and the synchronization signals, such as VSYNC and HSYNC, on separate lines. Using the IMAQ hardware and the IMAQ-2514 cabling option, you can input each of these separate signals (PCLK, HSYNC, VSYNC) and composite SYNC (CSYNC) inputs to the IMAQ-2514 cable assembly. Plus, the IMAQ-2514 provides connections to all video sources (Video 0, 1, 2, 3, single-ended or differential), the external digital I/O lines and triggers, and external sync lines.

Step 2—Histographing to Identify the Threshold Values

In the image shown in Figure 6-50, some cells have been chemically stained, or tagged, to make them easier to count. The stained cells have a higher contrast from the background and the other cells. This is a common technique used in cell counting applications. In factory automation or component counting applications, you can adjust the lighting to get the maximum contrast between the background and the particles you wish to count.

A histogram of an image gives the frequency (count) of the number of pixels per gray level value. It provides a general description of the appearance of an image and helps identify its various components, such as the background, blobs, and noise. Figure 6-51 is an 8-bit image with 256 gray scales. The black background is the spike at 0 and the white-stained cells are the gray-scale values above 110. For this application, the gray-scale values above 110 are the pixels we are interested in processing.

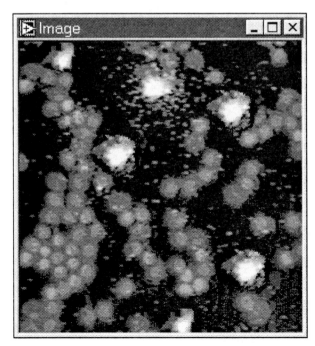

Figure 6-50.
Stained cells.

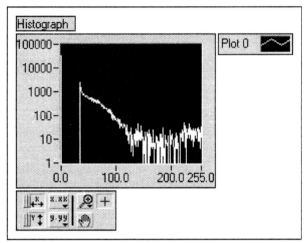

Figure 6-51.
Histograph of cell image.

The LabVIEW diagram in Figure 6-52 loads an image from file, then a histograph function is applied to the image. The histograph function is used to determine the upper and lower gray-scale threshold limits.

Step 3—Thresholding to Create a Binary Image

The diagram shown in Figure 6-53 has been modified to add the threshold function. The threshold function segments an image into two regions, an object region and a background region. In this process, all pixels that fall within the gray-scale interval defined as the threshold interval are given the value one. All other pixels in the image are set to zero. In this cell counting example, all pixels above gray-scale value 110 are set to a value of one, and all pixels below 110 are set to zero.

The result is a binary image—an image with zero and one values. Binary images contain only the important information, and often can be processed very rapidly. Generally, algorithms to process binary images are faster than algorithms for gray-scale images.

Several new functions are used in Figure 6-53. The **IMAQ Create** function creates an image buffer for the image. Plus, a border size input has been added to the **IMAQ Create** function, which adds 3 pixels around the perimeter of the image. This border is necessary because many image processing functions use several adjacent pixels to process the image. Adding pixels around the image ensures that calculations on the outermost pixels in the original image can be used in processing the image. The **IMAQ Threshold** function calculates the threshold, and the **IMAQ GetPalette** function sets up the display window to display a binary image. A window number constant has been added to the **IMAQ WindDraw** function to identify the raw and binary images. The range input controls the

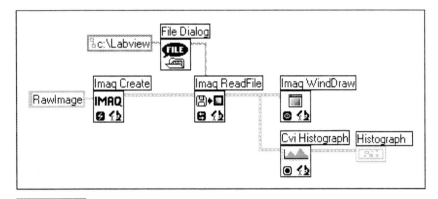

Figure 6-52.
LabVIEW diagram for loading the cell image from file, displaying the image, and calculating the histogram.

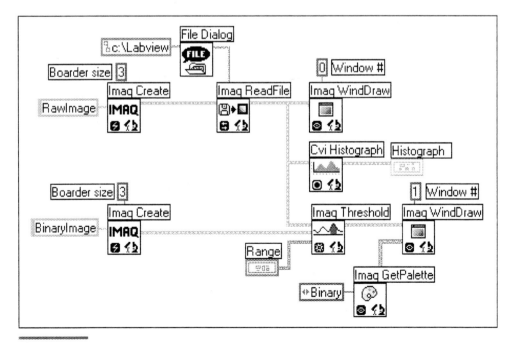

Figure 6-53.
LabVIEW diagram for calculating the threshold and displaying a binary image.

upper and lower threshold values. The result of this diagram is the binary image shown in Figure 6-54. The white pixels are "one" values and the black pixels are "zero" values.

Step 4—Filtering to Remove Noise and Particles on the Border of the Image

The section of code in Figure 6-55 operates on the binary image. The **IMAQ Remove-Particle** function filters or removes the particles below a certain pixel size. The **IMAQ RejectBorder** function removes the particles on the border of the image, which is a com-

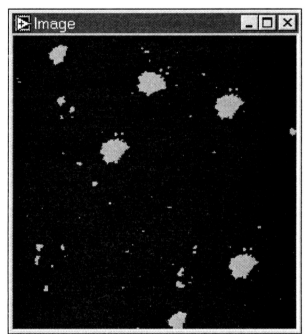

Figure 6-54.
Binary image of cells.

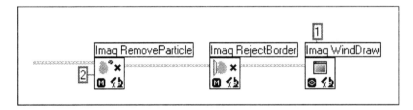

Figure 6-55.
LabVIEW diagram to filter noise and remove particles from the border of an image.

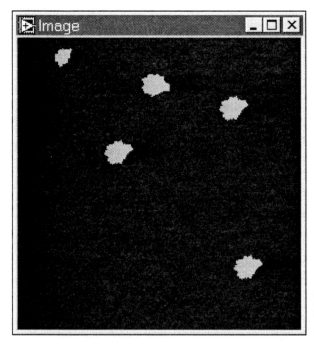

Figure 6-56.
Cells after thresholding
and filtering.

mon technique, because you cannot accurately determine the size of particles on the border of an image.

After the thresholding and filtering, as seen in Figure 6-56, only the cells or blobs of interest are left.

Step 5—Particle (Blob) Analysis to Count Cells

You can use the IMAQ Vision blob analysis function to analyze blobs in an image. You can count, label, and measure cells and objects. Calculate the area, perimeter, orientation, and location and 49 other parameters with blob analysis functions. Plus, to make counting easier, change the shape of blobs with morphology functions, and remove and filter blobs of certain size with spatial filters.

There are many blob calculations such as area, perimeter, moment of inertia, orientation, mean chord, width, height, ellipse axis, elongation factor, circularity factor, type factor, projection, location, bounding rectangle, and many more.

IMAQ Vision includes many different functions for manipulating particles or blobs. For example, if two cells are overlapping, you can use the erode function to remove the pixels from the perimeter of the particles until there are two distinct particles. Once the particles are separated, they can be counted. Use morphology functions to erode, dilate, fill holes, convex (fill holes on the edges), reject objects on the border, and separate blobs.

The processed binary image is then passed to the **IMAQ BasicParticle** function (shown in Figure 6-57). The Basic Reports output returns the area of each blob in pixels. There are five cells in Figure 6-56 with sizes of 189, 379, 374, 374, and 374 pixels. Plus, the **IMAQ BasicParticle** function returns the upper left and lower right pixel coordinates of a rectangle around each blob.

For more complex measurements, you can use the **IMAQ ComplexParticle** function (shown in Figure 6-58). This function calculates up to 49 different parameters for each blob. You can use each of these parameters in inspection and counting applications to uniquely quantify parts or particles. For example, you can use the perimeter, area, and center of mass calculations to quantify the difference between particles. Use more parameters for finer quantification.

Table 1. *IMAQ ComplexParticle Function Output.*

	Parameter	Description
0	Area (pixels)	Surface area of particle in pixels
1	Area (calibrated)	Surface area of particle in user units
2	Number of holes	Number of holes
3	Hole's Area	Surface area of the holes in user units
4	Total Area	Total surface area (holes and particles) in user units
5	Scanned Area	Surface area of the entire image in user units
6	Ratio: Area/Scanned Area %	Percentage of the surface area of a particle in relation to the scanned area
7	Ratio: Area/Total Area %	Percentage of the surface area of a particle in relation to the total area
8	Center of mass (X)	X coordinate of the center of gravity
9	Center of mass (Y)	Y coordinate of the center of gravity
10	Left column (X)	Left X coordinate of bounding rectangle
11	Upper row (Y)	Top Y coordinate of bounding rectangle
12	Right column (X)	Right-hand X coordinate of bounding rectangle
13	Lower row (Y)	Bottom Y coordinate of bounding rectangle
14	Width	Width of bounding rectangle in user units
15	Height	Height of bounding rectangle in user units
16	Longest segment length	Length of longest horizontal line segment
17	Longest segment left column (X)	Leftmost X coordinate of longest horizontal line segment
18	Longest segment row (Y)	Y coordinate of longest horizontal line segment
19	Perimeter	Length of outer contour of particle in user units
20	Hole's Perimeter	Perimeter of all holes in user units
21	SumX	Sum of the X-axis for each pixel of the particle
22	SumY	Sum of the Y-axis for each pixel of the particle
23	SumXX	(Sum of the X-axis)^2, for each pixel of the particle
24	SumYY	(Sum of the Y-axis)^2, for each pixel of the particle
25	SumXY	Sum of the X-axis and Y-axis for each pixel of the particle
26	Corrected projection X	Projection corrected in x
27	Corrected projection Y	Projection corrected in y
28	Moment of inertia Ixx	Inertia matrix coefficient in xx

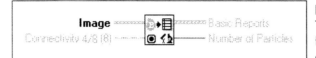

Figure 6-57.
The **IMAQ BasicParticle**
function calculates the area of
each blob.

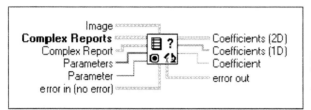

Figure 6-58.
The **IMAQ ComplexParticle**
function calculates up to 49
parameters for each blob.

Table 1. (continued)

	Parameter	Description
29	Moment of inertia Iyy	Inertia matrix coefficient in yy
30	Moment of inertia Ixy	Inertia matrix coefficient in xy
31	Mean chord X	Mean length of horizontal segments
32	Mean chord Y	Mean length of vertical segments
33	Max intercept	Length of longest segment
34	Mean intercept perpendicular	Mean length of the chords in an object perpendicular to its max intercept
35	Particle orientation	Direction of the longest segment
36	Equivalent ellipse minor axis	Total length of the axis of the ellipse having the same area as the particle and a major axis equal to half the max intercept
37	Ellipse major axis	Total length of major axis having the same area and perimeter as the particle in user units
38	Ellipse minor axis	Total length of minor axis having the same area and perimeter as the particle in user units
39	Ratio of equivalent ellipse axis	Fraction of major axis to minor axis
40	Rectangle big side	Length of the large side of a rectangle having the same area and perimeter as the particle in user units
41	Rectangle small side	Length of the small side of a rectangle having the same area and perimeter as the particle in user units
42	Ratio of equivalent rectangle sides	Ratio of large side to small side of a rectangle
43	Elongation factor	Max intercept/mean perpendicular intercept
44	Compactness factor	Particle area (length x width)
45	Heywood circularity factor	Particle perimeter/perimeter of circle having same area as particle
46	Type factor	A complex factor relating the surface area to the moment of inertia
47	Hydraulic radius	Particle area/particle perimeter
48	Waddel disk diameter	Diameter of the disk having the same area as the particle in user units
49	Diagonal	Diagonal of an equivalent rectangle in user units

Conclusion

With just a few IMAQ Vision functions you can easily count particles or cells and find their area. IMAQ Vision is flexible and fast for both laboratory and factory automation applications. For advanced inspection applications you can use multiple parameters to accurately quantify the difference between the particles or cells you are inspecting. Although this note presents graphical examples of IMAQ Vision functions for LabVIEW and BridgeVIEW, IMAQ Vision also contains the identical functions in C for LabWindows/CVI programmers and ActiveX Controls for ComponentWorks programmers.

Part IV
Medical Device Development Applications

Medical Device Testing

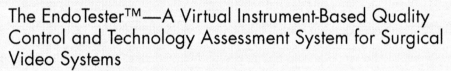

7

Over the past 15 years, virtual instrumentation in general, and LabVIEW™ in particular, have provided medical device manufacturers, biomedical engineers, researchers, and clinicians with an array of powerful software applications and integrated test systems to ensure the safety and efficacy of medical devices. Chapter 8, *LabVIEW in a Regulated Environment*, which discusses FDA certification, will detail how virtual instrumentation tools address the rigorous requirements for testing and validating medical devices.

It is noteworthy to recognize that many medical device manufacturers are in fact small companies (i.e., less than 300 employees). In today's dynamic, highly competitive marketplace, the ability to move quickly and bring new products to market directly impacts a company's profitability and very survival. This is particularly the case when a company must navigate through the highly structured and rigorous procedures of FDA approval.

This chapter will illustrate how LabVIEW-based applications have empowered biomedical engineers and medical device manufacturers alike to leverage the power of virtual instrumentation to quantitatively measure the quality, efficacy, and effectiveness of a wide array of medical devices. Case studies will range from quantifying endoscope performance to testing infusion pumps to defibrillators, ventilators, and implantable pacemakers. Due to the uniqueness of test setups for individual medical devices, this chapter will be largely devoted to the distinct developments of independent contributors.

The EndoTester™—A Virtual Instrument-Based Quality Control and Technology Assessment System for Surgical Video Systems

The use of endoscopic surgery is growing, in large part, because it is generally safer and less expensive than conventional surgery, and patients tend to require less time in a hospital after endoscopic surgery. Industry experts conservatively estimate that about four million minimally invasive procedures were performed in 1996. As endoscopic surgery becomes more common, there is an increasing need to accurately evaluate the performance characteristics of endoscopes and their peripheral components.

Introduction

The assessment of the optical performance of laparoscopes and video systems is often difficult in the clinical setting. The surgeon depends on a high-quality image to perform minimally invasive surgery, yet assurance of proper function of the equipment by biomedical engineering staff is not always straightforward. Many variables in both patient and equipment may result in a poor image. Equipment variables, which may degrade image quality, include problems with a rigid and flexible endoscope, either with optics or light transmission. The light cable is another source of uncertainty as a result of optical loss from damaged fibers. Malfunctions of the charge-coupled device (CCD) video camera are yet another source of poor image quality. Cleanliness of the equipment, especially lens surfaces on the endoscope (both proximal and distal ends) are particularly common problems. Patient factors make the objective assessment of image quality more difficult. Large operative fields and bleeding at the operative site are just two examples of patient factors that may affect image quality.

The evaluation of new video-endoscopic equipment is also difficult because of the lack of objective standards for performance. Equipment purchasers are forced to make an essentially subjective decision about image quality. A team of biomedical engineers, surgeons, and software engineers at Hartford Hospital and Premise Development Corporation have developed an instrument, the EndoTester™, with integrated software to quantify the optical properties of fiber-optic endoscopes. Figure 7-1 illustrates this application's Main Menu.

Materials and Methods

The EndoTester (see Figures 7-2 and 7-3) is a specialized optical bench used for the quantitative testing of the fiber-optic path and the lens system in rigid and flexible endoscopes. In addition to the specialized test station, the Endo-Tester requires

- A high-intensity variable light source,
- A flexible fiber-optic cable,
- A CCD video camera processor,

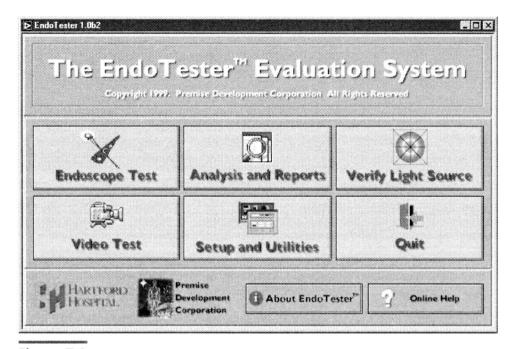

Figure 7-1.
The EndoTester Main Menu.

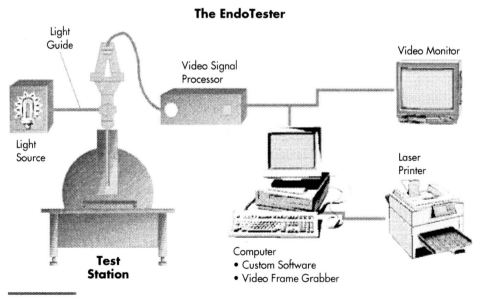

Figure 7-2.
The EndoTester system diagram.

- An optional video monitor (recommended),
- A standard PC and a video capture board, and
- A laser printer (600 dpi or higher recommended).

A standard PC containing either an 8-bit gray-scale or a 24-bit color video digitizing board (MRT Videoport, MRT Micro, Inc, Boca Raton, FL or IMAQ-PCI-1408, National Instruments, Austin, TX) and custom software allows for easy acquisition and analysis of the endoscope's optical properties. Collectively, the optical test bench and specialized software allow the user to perform a series of tests on the endoscope and its peripheral devices. Specifically, these tests include

- Relative Light Loss,
- Reflective Symmetry,
- Percent of Lighted (Good) Fibers,
- Geometric Distortion, and
- Modulation Transfer Function (MTF).

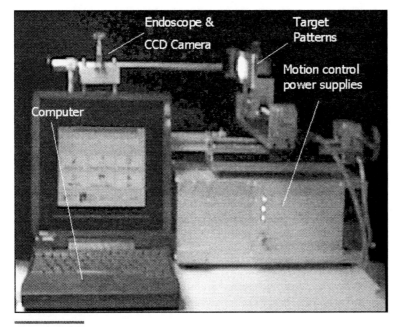

Figure 7-3.
EndoTester basic test fixture.

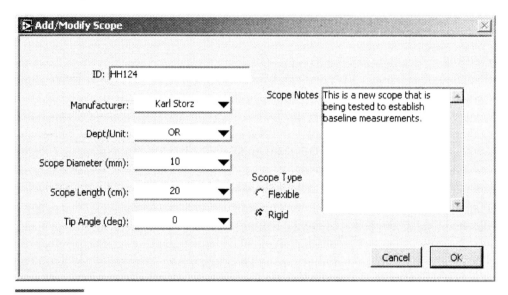

Figure 7-4.
Endoscope Information Profile screen.

Each series of tests is associated with a specific endoscope to allow for trending and easy comparison of successive measurements. Specific information about each endoscope (i.e., manufacturer, diameter, length, tip angle, department/unit, control number, and operator), the reason for the test (i.e., quality control, pre/post repair, etc.), and any problems associated with the scope are also documented through the electronic record. In addition, all the quantitative measurements from each test are automatically appended to the electronic record. Figures 7-4 and 7-5 illustrate the Information Profile and Problem Entry screens of the EndoTester. Figure 7-6 illustrates the Test Sequencer, in which the various tests can be either manually or automatically selected.

Endoscope Tests

Relative Light Loss in Optic Fibers

The *relative optical light loss* measurements quantify the degree of light loss from the light source to the distal tip of the endoscope. The relative light loss will increase with fiber-optic damage. Changes in the light source intensity or the condition of the fibers in the fiber-optic cable are normalized out of the

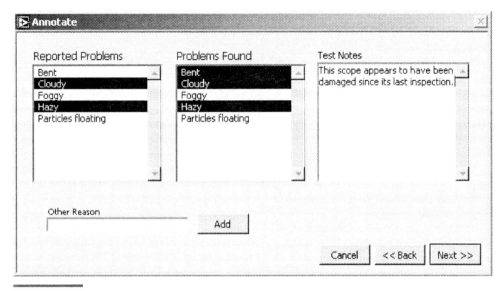

Figure 7-5.
Problem Entry screen.

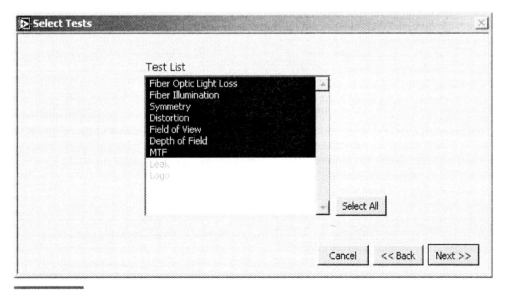

Figure 7-6.
EndoTester Test Selection screen.

relative loss calculation since the relative light loss is determined directly from the endoscope's light output and light input.

Having a simple means to quantitatively measure performance variation with respect to time is desirable for all endoscopes but appears to be particularly valuable for the evaluation of disposable endoscopes. These units may be projected to have a rated life of 30 uses, but measurements of performance change under actual operating and sterilizing conditions at a given institution are proof positive.

The relative light loss for the optic fibers of the endoscope under test is calculated by the following equation:

$$\text{Relative Light Loss} = 10 \log \left(\frac{\text{Light Out}}{\text{Light In}} \right)$$

The Light In value is measured by configuring the output connection end of the fiber-optic cable to a test fixture that holds the cable a fixed 2-inch distance from a photometer (Edmund Scientific, Model INS-DX100). The Light Out value is measured by the photometer illuminated by the endoscope under test.

Reflective Symmetry

Reflective symmetry is a measure of light amplitude in the endoscope's field of view. This value is important in that it quantifies the effective distribution of light. By employing five magnitude comparators, it is possible to transform the continuous illumination pattern into five annular rings of decreasing grayscales. The pattern of gray rings produced by this test should be nearly centered on the image. The pattern is circular for zero-degree endoscope tip angles and sometimes is elliptical for angled endoscope tips. A histogram graphs the number of pixels from each comparator output (light intensity). For each ring that is displayed in the filtered image, the user can see graphically how many pixels exist for each band of intensity. In order to pass this test, the Percent Bright Area is required to be greater than or equal to 50% of the maximum brightness. The following equation defines how Percent Bright Area is calculated. Gray 1 and Gray 2 refer to the two brightest (innermost) rings.

$$\text{Percent Bright Area} = \left(\frac{\text{Gray 1} + \text{Gray 2}}{\sum \text{pixels}} \right) \times 100$$

In addition to calculating the Percent Bright Area of the field of view, the boresight error can also be measured. The *boresight error* is defined as the dif-

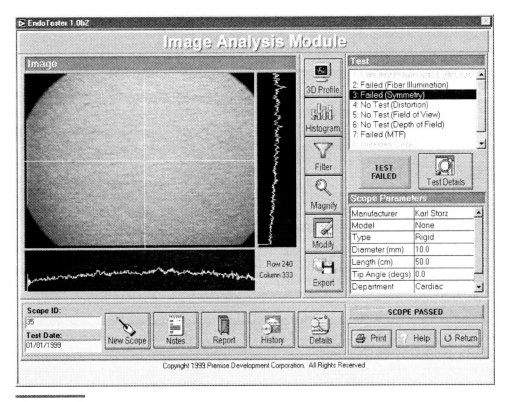

Figure 7-7.
Reflective symmetry test.

ference between the center of the field of view and the center of illumination provided by the endoscope. A grid of 0.2-inch × 0.2-inch squares, whose center square is solid black, is positioned in the center of the field of view. Combining this grid with the ring pattern shows the geometrical distance from the center of illumination to the center of the field of view. By counting the 0.2-inch square grid, a quantitative measure of the boresight error is determined for the endoscope at a 2-inch tip distance. Figure 7-7 shows the Image Analysis Module of the reflective symmetry test.

Lighted Fibers

A close-in reflection (less than 0.25-inch separation) of the distal end of the endoscope from a polished mirror or Lucite surface is captured by the CCD

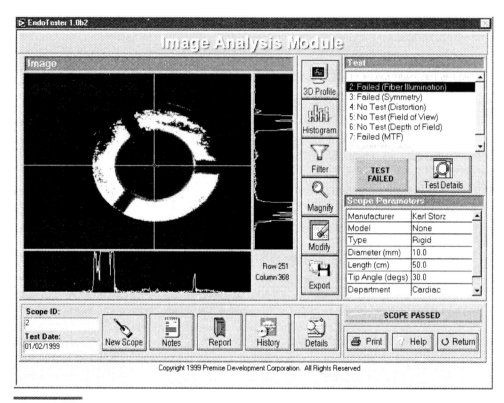

Figure 7-8.
Endoscope fiber illumination test.

camera. This provides a record of the pattern of lighted optical fibers for the endoscope under test. The number of lighted pixels will depend on the endoscope's dimensions, the distal end geometry, and the number of failed optical fibers. New fiber damage to an endoscope will be apparent by comparison of the lighted fiber pictures (and histogram profiles) from successive tests. Statistical data is also available to calculate the percentage of working fibers in a given endoscope. Figure 7-8 shows the Image Analysis Module of an endoscope *fiber illumination test.*

In addition to the 2D profile of lighted fibers, this pattern (and all other image patterns) can also be displayed in the form of a 3D contour plot as shown in Figure 7-9. This interactive graph may be viewed from a variety of viewpoints in that the user can vary the elevation, rotation, size, and perspective controls.

Figure 7-9.
Endoscope fiber illumination (3D contour profile).

Geometric Distortion

The *geometric distortion test* is used to quantify the optical distortion of the rod-lens system. The EndoTester's video frame grabber captures the image of a square grid pattern. Distortion is defined as the change of magnification at points around the field of view with respect to the maximum magnification occurring at the center of the field of view. By measuring the diagonal length of the central square of the pattern, with respect to the diagonal length of any other square, the geometric distortion can be determined. The percent distortion is calculated by the following equation:

$$\text{Percent Distortion} = \left[\left(\frac{\text{Other Square Diagonal Length}}{\text{Central Square Diagonal Length}} \right) - 1 \right] \times 100$$

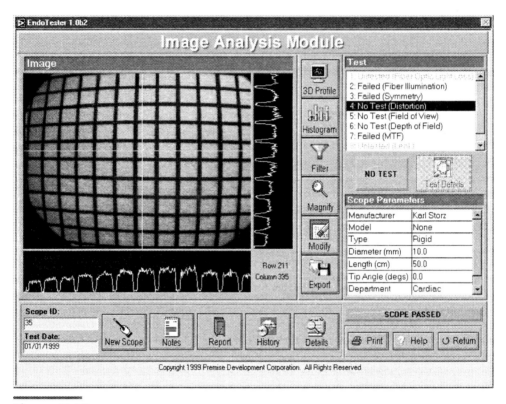

Figure 7-10.
Distortion Analysis panel.

Figure 7-10 illustrates the screen that is used to measure the geometric distortion of the endoscope under test.

Modulation Transfer Function

Aperture response is a universal criterion for specifying picture definition and other aspects of imaging system performance. It can be used for film images, camera lenses, television camera imagers, receiver picture tubes, and the human eye. The aperture response is measured as a contrast ratio by square wave (contrast transfer function [CTF]) or sine wave patterns (modulation transfer function [MTF]).

The square wave response data can be converted to equivalent sine wave data by mathematical manipulation, however, the sine wave method is the most direct approach. Variable density film targets are now available as sine transmission targets, hence, a direct measurement of MTF can be performed. In fact, even a single space frequency measurement can provide a good index of performance for an optical lens system, such as an endoscope. This is the approach used in the EndoTester. The endoscope is tested with a one cycle per millimeter sinusoidal transmission target (Sine Patterns, Penfield, NY).

The MTF of the lens system is measured at a spatial frequency of six cycles per degree of apparent field of view. Measurements at this frequency are considered to be an accurate indication of good optical instrument performance, when the MTF is high. Thus a single spatial frequency on the test target provides a good quantitative index of the local spot performance of the endoscope's lens system. Typically, the center region of a lens has the best optical performance, and the edges of the lens have less visual sharpness. Therefore, the MTF test measures performance at the left and right edges, the top and bottom edges, as well as the center of the lens. Studies have shown that a user's perception of image quality can be correlated to high values of modulation transfer function.

The MTF test uses a film target whose transmittance of white light varies sinusoidally across the X axis, and is constant with respect to the Y axis. The fiber-optic cable is removed from the endoscope under test and is used to illuminate the sine target. The endoscope views the transmitted sinusoidal light pattern. The frame grabber captures the sinusoidal light variation along the horizontal sweep axis of the television image of the target. Modulation is defined as:

$$\text{Modulation} = \frac{(T_{max} - T_{min})}{(T_{max} + T_{min})} \quad \text{where } T \text{ is transmitted light intensity.}$$

When this measurement is made over a range of spatial frequencies, a plot of the curve of the modulation transfer ratio as a function of spatial frequency is called the modulation transfer function (MTF). The MTF of a system is equal to the product of the MTFs of each component of the system. Thus, the MTF of the signal captured by the frame grabber is the product of the endoscope's MTF and the TV system's MTF. Figure 7-11 illustrates the screen that is used to measure the modulation transfer function of the endoscope under test.

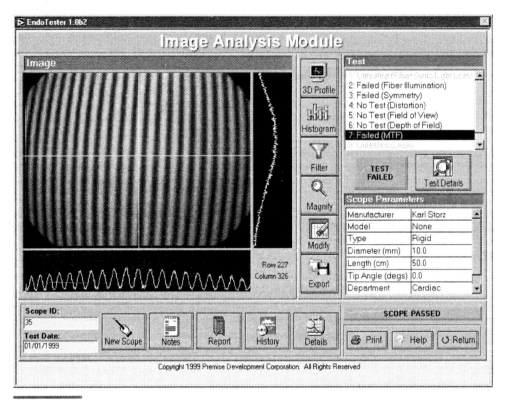

Figure 7-11.
Modulation transfer function (MTF) panel.

Results

New and used 0-, 30-, and 70-degree endoscopes from several manufacturers were evaluated with the test system. The system was able to quantify light loss related to fiber damage and define reflective symmetry and geometric distortion. Numerical evaluations of endoscopic lens acuity were made. These are values of the modulation transfer function at one space frequency (6 cycles/apparent degree), which has been found to be representative of a lens system's performance.

As was shown in Figure 7-2, the EndoTester works in conjunction with a light source, a fiber-optic light cable, and a CCD video camera. Normally, data is taken with the same set of these auxiliary components so that variations of the endoscopes under test can be measured. Variations in light source intensity and light cable losses are normalized out with this technique. However it is apparent that a relative efficiency test of other light sources and light cables can be performed by substitution and comparison of actual brightness values. In a similar manner, a relative evaluation of another CCD camera and signal processor can be made by substitution. Note that a good quality endoscope should be used as the fixed part of a substitution test.

Discussion

An easy-to-use optical evaluation system for endoscopes has been developed. This system allows objective measurement of endoscopic performance prior to equipment purchase and in routine clinical use as part of a program of prospective maintenance. Measuring parameters of scope performance can facilitate equipment purchase. Vendor claims of instrument capabilities can be validated as a part of the negotiation process.

The adoption of disposable endoscopes raises another potential use for the EndoTester. Disposable scopes are estimated to have a life of 20–30 procedures. However, there is no easy way to determine exactly when a scope should be thrown away. The EndoTester could be used to define this endpoint.

The greatest potential for this system is as part of a program of preventive maintenance. Currently, in most operating rooms, endoscopes are removed from service and sent for repair when they fail in clinical use. This causes operative delay with attendant risk to the patient and an increase in cost to the institution. The problem is difficult because an endoscope may be adequate in one procedure but fail in the next, which is more exacting because of clinical variables such as large patient size or bleeding. Objective assessment of endoscope function with the EndoTester may eliminate some of these problems.

Equally as important, an endoscope evaluation system will also allow institutions to ensure value from providers of repair services. The need for repair can be better defined and the adequacy of the repair verified when service is completed. This ability becomes especially important as the explosive growth of minimally invasive surgery has resulted in the creation of a significant market for endoscope repairs and service. Endoscope repair costs vary widely throughout the industry with costs ranging from $500 to $1,500 or more per repair. Inappropriate or incomplete repairs can result in extending surgical time by requiring the surgeon to switch scopes (in some cases several times) during a surgical procedure. Given these applications, we believe that the EndoTester can play an important role in reducing unnecessary costs, while at the same time improving the quality of the endoscopic equipment and the outcome of its utilization.

It is the sincere hope of the authors that this technology will help to provide accurate, affordable, and easy-to-acquire data on endoscope performance characteristics, which clearly are to the benefit of the healthcare provider, the ethical service providers, manufacturers of quality products, the payers, and, of course, the patient.

FluidSense Innovative IV Pump Testing

Introduction

FluidSense Corporation, located in Newburyport, Massachusetts, is an innovative leader of the IV therapy process. FluidSense extends efforts to continually improve the IV therapy process, reduce IV medication errors, and enhance overall patient satisfaction. FluidSense's on-site Infusion Service's Program incorporates process management techniques to bring more efficiency and effectiveness to hospital infusion therapy. The company assures that a clean, tested, fully powered, ready-to-use infusion pump is available to the caregiver when needed.

As a part of their ongoing efforts, FluidSense set out to create a complete, customized system to test and calibrate the FS-01 FluidSense IV pump (see Figure 7-12). The intent was to build a sophisticated LabVIEW™-based test system that interacts via infrared (IR) with a dual-processor embedded controller device.

The Test System

Once the FS-01 infusion pump is returned to the FluidSense on-site office for service, it may contain several days of continuous drug delivery and patient information. Technicians may retrieve this data and recalibrate and test the infusion pump to ensure proper performance before it is returned to the patient floor for reuse.

LabVIEW was used to develop the System Tester application to download the data, calibrate, and test the pump. The requirements were that this application be simple and easy to use for the FluidSense field service technicians. The total process must be efficient.

Figure 7-12.
FluidSense FS-01 infusion pump
system test setup. (Reprinted
with permission of FluidSense
Corporation.)

The test system comprises a computer laptop running the System Tester application and an external IR interface device that is connected to the laptop's serial port. The infusion pump is put into reconditioning mode and one of the pump's three IR ports is placed in front of the IR interface device. The FluidSense service technician then needs only to click Run Test and the software will begin the automated process of downloading data and testing the infusion pump.

LabVIEW's serial communication virtual instruments are an integral part of the System Tester application. These VIs were used to create an IR-compatible serial port interface driver that allows easy communication from the PC to the infusion pump. We use many components of this driver throughout the entire testing process to send commands and receive responses from the infusion pump.

The initial stage of the process includes receiving basic information about the pump such as the serial number and embedded software information. The data collected during the last IV therapy session is downloaded and stored into a set of files that we use for later analysis. If the pump is in a faulty state, the application will detect this faulty state and prompt the service technician to fill in questionnaires that describe the physical condition and behavior of the pump.

In addition to infusion data collection, the System Tester software tests several key components of the FS-01 pump. The software issues a series of commands to sequence through the pump's failsafe mechanisms and status indicators. The FluidSense service technicians play an integral role in this part of the testing process. They are asked to confirm that the status LED, audio sound, user input devices, and failsafe mechanism are all properly functioning before proceeding. A second portion of the test includes a more in-depth analysis of the FS-01 pump and its internal electromechanics. One test includes assessing an optical sensor that the pump uses to coordinate motion and position with other components of the internal hardware. The unique waveform associated with this sensor's readings is displayed to the FluidSense service technician using LabVIEW waveform VIs.

Specific measurements are made to determine the infusion pump's ability to measure fluid delivery, to detect and quantify air in the cassette chamber, and to confirm integrity of the cassette. For example, the Power Test measures the energy used per fill stroke (the infusion pump delivers fluid by filling and emptying a cassette). With a primed cassette inserted into the infusion pump, the System Tester application commands the pump to perform a fill stroke. The data is acquired. We used a combination of the waveform analysis and statistics VIs to create an algorithm that took this data and measured the total energy consumed during one fill stroke and compared it with our pass/fail criterion. This pass/fail criterion is read into memory using LabVIEW's configuration file utilities.

Training Emulator

Since the FluidSense technicians play such an important role in the testing process, we needed to create an application that would be used to train the staff on how to properly test an infusion pump. This LabVIEW application, called Training Emulator (see Figure 7-13), allows a test administrator to create simulated test sessions where System Tester would fail a certain test or two and the service-technician-in-training would be measured on his or her response to that failure. The application required a scaled-down version of System Tester. This version would not require data downloaded from an infusion pump since we would pass or fail the test based solely on the simula-

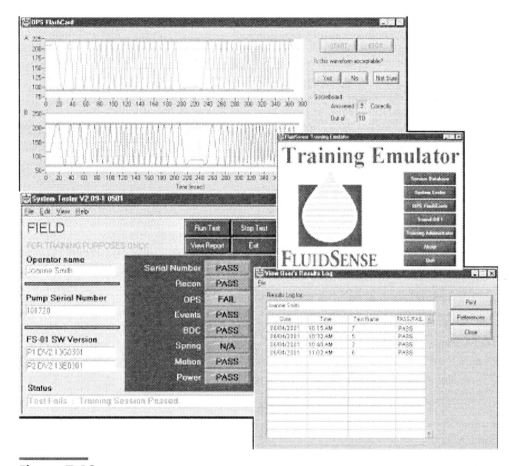

Figure 7-13.
FluidSense Training Emulator.

tion environment. Additional utilities were required to fully support the Training Emulator as a standalone application. Utilities for user and test management were created to allow the administrator to create tests and monitor a technician's performance. Additional teaching tools were added to the Training Emulator application, including the OPS Flashcards Module, which helps teach the technician how to distinguish between acceptable and unacceptable waveforms produced by the infusion pump, and a Sound Off! Module, which allows the user to play sound files for each of the audible sounds that an infusion pump might make before, during, or after an infusion.

Summary

The System Tester application is used extensively to qualify a pump throughout its life cycle. Data collected from System Tester allow us to improve and validate the infusion pump's design and function both in manufacturing and at the hospital. Its simple interface allows for easy processing with minimal operator intervention. The Training Emulator allows us to quickly teach and certify our on-site technicians on the testing process because we are able to put together an application that is easy to use and configure. LabVIEW has been an invaluable resource as we continue to develop and modify our data collection and testing process without sacrificing time or money. With LabVIEW, we are able to demonstrate that it is easy to introduce, implement, test, and deploy a new concept in a comparatively small amount of time without sacrificing usability or reliability.

Independent Solution Articles

LabVIEW and DAQ Board Accelerate Development of Medical Diagnostic Instruments

by David G. Edwards, President, FemtoTek, Inc.

The Challenge: Efficiently developing a comprehensive PC-based system for testing blood coagulation timing instruments.

The Solution: Using an AT-MIO-16X board to collect the data along with LabVIEW for its integrated acquisition, analysis, and presentation capabilities.

FemtoTek, Inc. has developed a data acquisition (DAQ) application for Medical Laboratory Automation, Inc. (MLA) to use in developing its medical diagnostic equipment. The application, MLA-DAQ, performs simultaneous signal capture and data analysis. MLA-DAQ is used extensively in the final product qualification and field trial analysis of MLA's latest diagnostic instrument, the Electra 1600C.

Introduction

When Medical Laboratory Automation, Inc. decided to accelerate the development of its medical diagnostic equipment, the company turned to National Instruments for DAQ boards and LabVIEW software and to Alliance Program member FemtoTek, Inc. for application software development. MLA had previously used PC-based hardware and single-function software packages.

MLA, an industry leader in the design and manufacture of blood coagulation timers, features diagnostic equipment designed for fast and accurate processing of medical samples. The equipment combines physical measurements with advanced mathematical analysis routines. To provide effective development and checkout tools for their latest instrument, the Electra 1600C, MLA needed to integrate the functions of data acquisition, user interface and data display, advanced mathematical analysis, and data storage and retrieval into one PC-based application. FemtoTek developed this application, MLA-DAQ, using National Instruments DAQ boards and LabVIEW. Functions that previously required data transfer between separate data acquisition, spreadsheet, and mathematics software packages are now integrated into a single LabVIEW program.

The Coagulation Analyzer

Blood coagulation timers are used for both analysis and screening of clotting disorders. A number of methods exist for measuring the time it takes blood to coagulate. MLA equipment measures changes in optical density of the sample. Automatic measurement in the Electra 1600C starts when an automatic pipette system places test samples of plasma in disposable cuvettes. A linear belt transport mechanism slowly steps the samples through a series

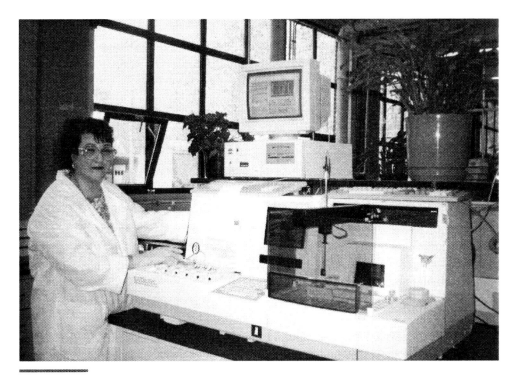

Figure 7-14.
Electra 1600C Coagulation Analyzer with MLA-DAQ Data Acquisition System.

of heated locations and finally to the optical photometers. Heating brings the samples to a controlled body temperature. Before the samples reach the optical detectors, clotting is initiated by adding a start reagent from one of six reagent pumps. In the Electra 1600C, cuvettes move in batches of four, one immediately after the other. The operator measures four samples at the same time using four separate optical channels.

The time a sample spends at any station, such as the optical detectors, varies from 40 to 200 seconds. Moving a new batch of samples into the optical detector station takes between 0.5 and 10 seconds, depending on the demands on the various pumps in the reagent delivery system. The instrument analyzes the optical detector signal to give the coagulation time result.

Connecting to the electrical signals of the Electra 1600C coagulation analyzer requires a DAQ system that can measure eight analog inputs for the optical detector and second derivative; measure four digital inputs for the trigger signals; and generate four digital outputs to drive the derivative clamp signals. The hardware for MLA-DAQ is a National Instruments AT-MIO-16X board installed in a 486-DX2/66 MHz personal computer with 8 MB of RAM. The AT-MIO-16X board has the necessary 16-bit analog accuracy as well as sufficient digital I/O lines, while the 486 PC provides enough processing power to run the combined MLA-DAQ data acquisition and analysis tasks.

Eliminating the Processing Bottleneck

The MLA-DAQ application measures the analog optical detector signal and its analog second derivative. A key requirement for the MLA-DAQ application is to keep up with the sequence of samples in the coagulation analyzer—recording and analyzing the signals for each sample as it moves through the analyzer. Accurate analysis of the signals demands that the operator record the full length of the signals, leaving little time between samples for data analysis and storage. FemtoTek removed the processing bottleneck by using the multi-processing inherent in LabVIEW; MLA-DAQ captures current signals while it is analyzing the previous signals.

FemtoTek designed MLA-DAQ so that the different functions that run the data acquisition, data analysis, signal display, and RS-232 connection are independent LabVIEW loops. These functions can run independently and asynchronously yet still exchange data, using first-in-first-out (FIFO) buffers, which remove the need to program interlock features into the independently running loops.

The TTL trigger signals monitored by MLA-DAQ mark the start and end of the optical detector trace for each sample. The starting triggers and the optical detector signals on the four channels are offset in time, because the start reagent is added to each sample at a different time. MLA-DAQ places the captured signal data in the data arrays normalized to the trigger start time for the respective channel so that the same analysis routines process the data for all channels.

An important part of MLA-DAQ is the mathematical analysis algorithms implemented by MLA in their coagulation analyzer. The extensive library of analysis routines in LabVIEW simplified the transfer of the algorithms to MLA-DAQ from the mathematical packages where they were developed. Integrating the analysis algorithms into the LabVIEW program is a higher performance solution than linking MLA-DAQ to a mathematical package through dynamic data exchange (DDE) or file transfer.

Using the System

To keep the user interface as simple as possible and still meet the needs of the users for complete information, FemtoTek consolidated the main user interface on one front panel. For many operators, the most important feature on the front panel is the graph showing the signal output of the optical detectors in the instrument. To achieve a compact but flexible user interface, FemtoTek made extensive use of the powerful local variable and attribute node features of LabVIEW.

FemtoTek has designed MLA-DAQ to monitor a coagulation analyzer during long periods of use. The operator can record many thousands of samples. An automatic save feature generates a data file for each set of four samples. The operator can save the corresponding assay type and sample ID number to an information file with a matching filename. In addition to facilities for capturing and saving the optical density data, FemtoTek has provided operators with a zoom function to analyze specific portions of the signal. The operator can also print out complete signal or zoom portions for later analysis. The operator can even reload old data files into the system for comparison, analysis, and printout.

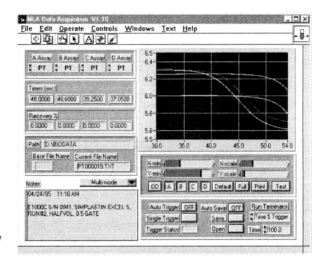

Figure 7-15.
Main front panel of LabVIEW
program.

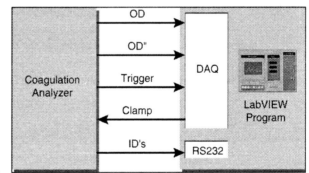

Figure 7-16.
MLA-DAQ system diagram.

Summary

MLA-DAQ is a complex application. However, thanks to a clear definition from Medical Laboratory Automation and FemtoTek's experience developing PC-based systems, the development time for MLA-DAQ was short and the resulting product went efficiently into use. Originally intended for internal R&D use, the MLA-DAQ application has been dubbed an unqualified success and is now used extensively in the final product qualification and field trial analysis of MLA's latest instrument. Several installations of the package are in use at MLA, with further installations planned for the laboratories of reagent suppliers and key end-users of the Electra 1600C in the U.S. and Europe.

For more information, contact FemtoTek at 560 Fellowship Road, Mt. Laurel, NJ 08054, tel (609) 235-4435, fax (609) 722-0153.

Measuring Medical Pump Accuracy with LabVIEW

by Amy S. Pomaybo, Advanced Development Engineering, Medrad, Inc.

The Challenge: Measuring the very low flow rates of medical pumps accurately and analyzing and displaying the data efficiently.

The Solution: Building a LabVIEW-based automated test system.

Introduction

Medical pumps are an indispensable tool to medical personnel because of their convenience, low price, and ability to deliver fluids accurately and safely. These pumps are used in a wide range of clinical situations, such as surgical procedures, treatment of pain, and diagnostic procedures. Typically, medical pumps use a disposable syringe, a connector tube, and a catheter to deliver small quantities of fluids such as anesthetics, analgesics, and diagnostic agents over long periods of time.

Because medical pumps are used to deliver potent drugs and to obtain diagnostic information, pump accuracy, as specified by the manufacturer, is critical. Until recently, pump manufacturers specified the pump accuracy as just a percentage, with no indication as to whether this was the drive, volume, or flow rate accuracy. As a way of eliminating this confusion, a test standard was proposed by the International Electrotechnical Commission (Draft IEC-601-2-24). In this standard, performance of the pump is characterized by (1) measuring flow rate over time from the start of the infusion (start-up curves), and (2) determining the percentage variation in flow rate during various time periods (trumpet curves). With this data, users can ensure correct dosage and optimum patient safety.

Because most medical pumps operate at low flow rates (sometimes on the order of milliliters per hour), applying the test methods outlined in the standard is time consuming and computationally intensive. To address this problem, we chose LabVIEW for developing a computerized test system that could simplify data collection, perform advanced data analysis, and graphically display data.

The Test System

The test system is made up of several components—an electronic scale, a Pentium PC, LabVIEW software, and an interface cable. According to the test standard, fluid flow is defined as the weight of the fluid delivered, in grams, over a defined period of time. We placed a beaker on the scale and used it to collect the fluid delivered by the pump. To acquire data from the scale, we built a custom cable to connect the RS-232 serial port on the scale to the serial port on the PC. LabVIEW was ideally suited to operate the test system because of its flexibility and programming ease. The LabVIEW built-in serial port virtual instruments (VIs) provided a great starting point for performing bidirectional communication with the scale

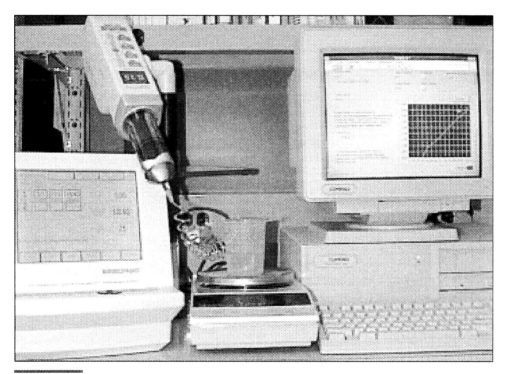

Figure 7-17.
The medical test system pump.

via RS-232. The LabVIEW acquisition VI periodically queries the scale for information, so you can start a test and leave it running with minimal operator intervention.

Once we collected data, we developed a separate LabVIEW VI to perform the start-up and trumpet curve analysis. In this analysis VI, the weight (grams) as measured with the scale was converted to flow rate (ml/minute) by dividing by the fluid density and test duration. We used a LabVIEW waveform graphic on the front panel to view the flow rate over time. This curve displays flow rate continuously from the start of the infusion, so users can visually observe flow rate uniformity and any delay in delivery due to mechanical compliance.

Unlike the continuous start-up curve, the trumpet curve displays data averaged over particular time periods or "observation windows." Trumpet curves are named for their distinctive shape, converging to the right with time on the x-axis and flow rate accuracy on the y-axis. The trumpet curve defines, for a programmed flow rate, the maximum and minimum percentage variation from the expected flow rate relative to the observation window. Over short observation windows, fluctuations in flow rate have a greater effect on accuracy as represented by the bell of the trumpet. As the observation window increases, short-term fluctuations have little effect on accuracy as represented by the flatter part of the curve.

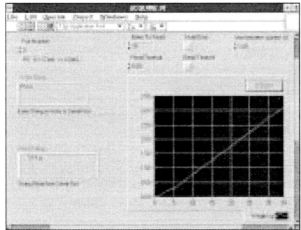

Figure 7-18.
The LabVIEW acquisition VI
front panel.

To implement the trumpet algorithm, we specified the observation windows and pro-grammed flow rate on the LabVIEW front panel. Flow rate variations were determined as a percentage by subtracting the programmed flow rate from the measured flow rate, then dividing this difference by the programmed flow rate. Within each observation window, the maximum and minimum variations were determined, and graphed relative to the observation window. With this type of analysis, you can correlate the half-life of a drug to be administered with the observation window interval in the graph and decide if the pump is suitable for that application. The built-in LabVIEW array functions greatly simplified the start-up and trumpet analysis, making it easy to "wire" the data directly into the array functions without any programming.

Conclusion

This LabVIEW-based test system has been an effective development and checkout tool for our R&D laboratory. This type of testing would have been a difficult task without LabVIEW, which simplified data collection and analysis, decreased testing time, and reduced the amount of operator interruption. LabVIEW also eliminated the need for separate data acquisition, spreadsheet, and mathematical software packages. With its numerous features, LabVIEW made implementing this new test and measurement method easy.

This test system was originally intended for internal R&D use. However because it has been such a valuable tool, several other departments are currently evaluating it. The demands on the test system are constantly changing, but with LabVIEW, we can easily modify the system.

For more information, contact Amy S. Pomaybo, Medrad, Inc., One Medrad Drive, Indianola, PA 15051, tel (412) 767-2400 ext. 4061, fax (412) 767-8899, e-mail apomaybo@medrad.com

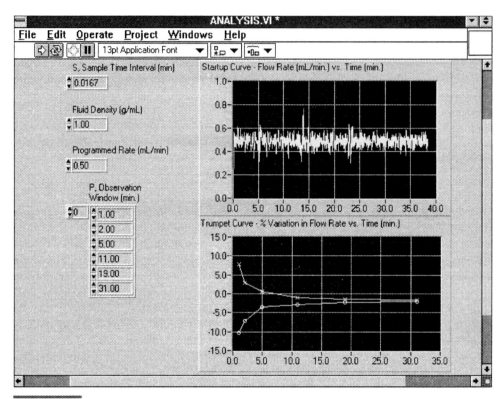

Figure 7-19.
The LabVIEW analysis VI shows a sample start-up curve and a trumpet curve.

Reducing Test Time of Implantable Pacemaker Hybrid Circuits with LabVIEW

by Jacopo Costella, Supervisor/Test Engineer, MEDICO SpA

The Challenge: Reducing the test time of implantable pacemaker hybrid circuits without having to purchase expensive automated test equipment (ATE).

The Solution: Implementing the necessary design for testability (DFT) structures on the hybrid and building a PC-based ATE system circuit using National Instruments data acquisition (DAQ) and GPIB boards controlled by LabVIEW.

Introduction

Traditionally, we have tested pacemakers by simulating the operation and observing the circuit response under some specific conditions, such as changes in battery status and heart behavior. With the growing complexity of these circuits, this testing approach began to require some 40 to 45 minutes, when it once took only 10 to 15 minutes for a complete test.

Automating the Tests

To remove this bottleneck in the production flow, we needed to automate testing of the circuit, despite its small size and the consequent poor availability of test points. To make automated testing possible, our first step was to insert a scan chain inside the digital controller IC of the pacemaker to add some test points on the hybrid. With these changes, we can now control and observe the whole circuit.

Our next step was to assemble the necessary test equipment. To control the scan chain structure inside the IC, we built a small microprocessor-based scan controller board, which exchanges data and commands through a 16-bit parallel port to the 8255 PIO of a National Instruments AT-MIO-16DE-10 DAQ board. With this new interface, we drastically increased the speed of the hardware and software debugging of the scan controller. By working on a single PC, we could run the C remote debugger for the scan controller software and LabVIEW for data and command exchange through the PIO port in multitasking mode.

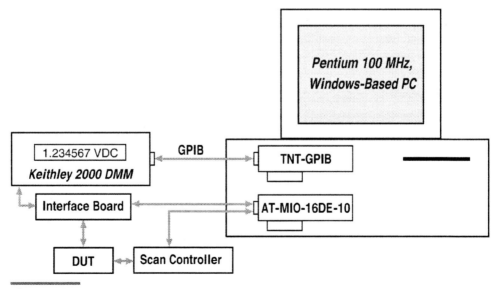

Figure 7-20.
Block diagram of the pacemaker hybrid ATE system.

We developed a complete LabVIEW command set for the scan controller to be used as a plug-and-play component in the final test program.

We also needed to develop the test algorithms. After the implementation of the DFT structures, we partitioned the circuit into a set of well-defined, fully controllable, and observable functional blocks. We can test each one separately with a special LabVIEW program written specifically for that functional block. We can also debug and optimize each functional block program before including it in the main test program. The extensive measurement resources built into the DAQ board—16 analog inputs, eight programmable function inputs for triggering, two analog outputs, and so on—drastically reduce the requirements for custom test interface circuitry. We needed only a set of analog buffers, logic-level shifters, and multiplexers for arbitrary signal and DC voltage generation through one of the two DACs.

We use a digital multimeter (DMM) controlled over the IEEE 488 bus by LabVIEW for parametric measurements on discrete components such as diodes, resistors, zeners, and transistors. The LabVIEW instrument driver for the DMM greatly simplifies setup of this process.

Using the DAQ board, we can do the following:

- Take timing measurements—gated/nongated frequency, period, and pulsewidth—with the DAQ-STC™ ASIC
- Test transient response to an arbitrary stimulus of filters and amplifiers by sampling analog signals that we have suitably triggered and/or synchronized to the inputs
- Test analog comparators for several reference voltage values—we find the 12-bit DAQ board accurate for such characterization
- Verify the frequency response and CMRR of filters and amplifiers
- Make analog time-constant estimates through analog signal sampling and post-processing
- Measure DC voltage

The Resulting System

Figure 7-20 shows a simplified block diagram of the resulting ATE. We structured the main test program to run the functional block virtual instruments (VIs) in sequence; we can separately enable or disable any of them. Figure 7-21 shows the main front panel. For each functional block VI, we can edit a set of pass or fail limits right from the main front panel. Each element of the set contains a minimum value, a maximum value, and a comment string that we typically use to describe the test and to specify the measurement unit. For each functional block, the screen displays a pass/fail indicator; we can also request more detailed results. With the functional block VI enabling or disabling capability, we can easily rerun a subset of the tests for rework or single block characterization, as necessary.

We can easily implement data management procedures, such as operator and serial number input, report printing, and data collecting for statistics, through the variety of LabVIEW library functions available.

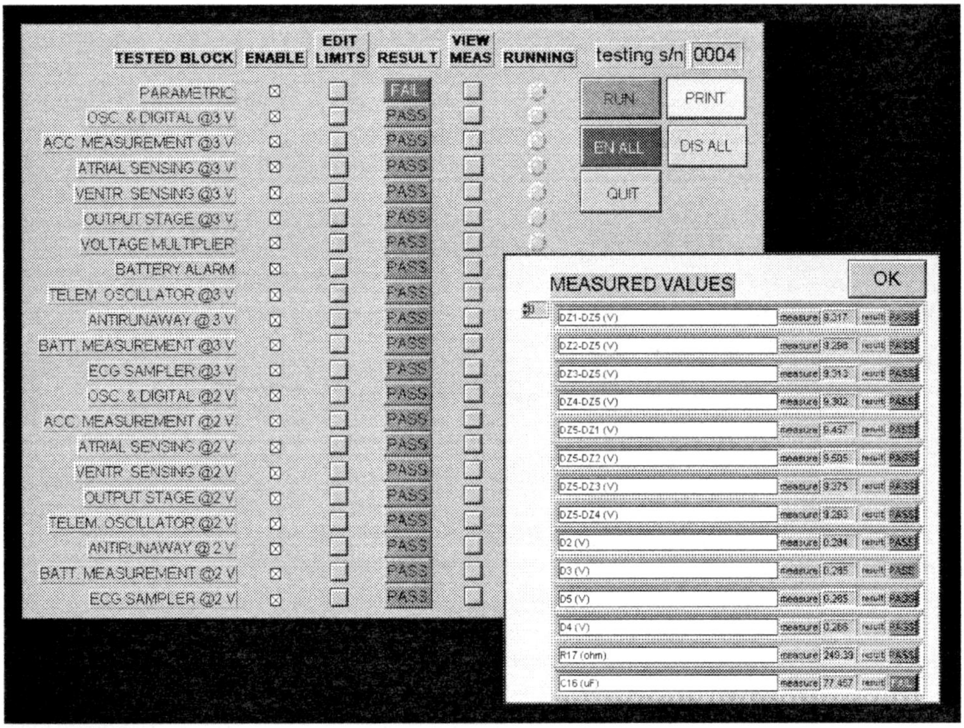

Figure 7-21.
Main front panel.

Conclusions

Usually a new test approach requires a long development time and a high ATE cost. With National Instruments products, we quickly developed an ATE system to comprehensively perform fault detection and isolation, test time, and so on, within a very short time and at low cost. The resulting equipment maintains a high level of flexibility and expandability for both hardware and software thanks to the variety of National Instruments DAQ board resources and their ease of use with LabVIEW.

For more information, contact Jacopo Costella,
MEDICO SpA, Via Pitagora 15, 35030
Rubano (PD) ITALY,
tel 39 049 8976755,
fax 39 049 8976788,
e-mail medico.red@interbusiness.it

Hybrid Evaluation and Development Using LabVIEW for Windows

by Allen J. Hunsaker, Medtronic Micro-Rel

In an environment of hybrid and IC technology, with both analog and digital signals present, the test, evaluation, and development of an electronic product require extensive and complex hardware and software solutions. In the past, these solutions involved programming in C, Basic, or other high-level languages, at a cost of hundreds of programming hours.

During product development and evaluation, programmers often work on-the-fly, sometimes spending hours figuring out how to continue where they were last programming. The cryptic nature of most programming, even with remarks, makes on-the-fly programming very difficult and time consuming.

Programming graphically with LabVIEW for Windows effectively eliminated this difficulty when testing pacemaker sensing capabilities, and a multitude of other pacemaker tests, using our engineering station. I found that graphical programming works much like creating a flow chart—when your flow chart is complete, your program is complete. The graphical images and icons of LabVIEW are so easy to understand that programmers can quickly pick up where they left off in their last programming session. On-the-fly programming becomes manageable, and programmers accomplish more work than when using a traditional, text-based language.

The Test System and LabVIEW

Our test system has a device under test (DUT) fixture; a rack of test equipment that includes digital voltmeters, counter timers, waveform generators, power supplies, current sources, oscilloscopes, and a proprietary programmer/decoder; a 486DX-50 PC running the Win-

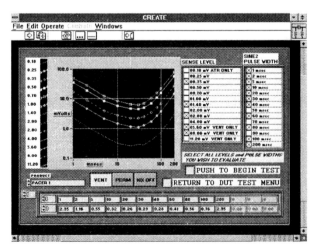

Figure 7-22.
The LabVIEW front panel for the pacemaker testing engineering station shows the results of a test of five different sense levels, each at 11 different pulse widths.

dows version of LabVIEW, equipped with a GPIB-PCIIA and a 48-line parallel interface board; and a custom interface box that includes signal conditioning, buffering, triggering, timing and delay circuits, relay switching, calibration, and loads.

We use LabVIEW to control GPIB instruments and the parallel interface, and to direct tests to the DUT. Through the main LabVIEW menu, users access five underlying submenus—DUT Test Menu, ULT (universal linear tester) Direct Control Menu, Equipment Direct Control Menu, Program DUT Menu, and Engineering Menu. These menus access individual tests and controls of the system. When developing software for our engineering station, I wanted to have pushbutton, user-friendly software. Windows made interesting graphics much easier to produce, but also required a significant learning curve to make custom graphical user interfaces (GUIs). LabVIEW for Windows integrated these features for me, and actually made my task of creating GUIs fun. With LabVIEW, changing instruments is as simple as swapping instrument drivers.

Testing Sensing Capabilities

Evaluating the sensing capability of a pacemaker was a difficult chore prior to using Lab-VIEW. Heart activity is simulated by applying sine^2 pulses of a specific amplitude and width to the sense input of the pacemaker. The pacemaker is designed to respond to only a narrow range of pulse amplitudes and widths. The pacemaker is subjected to these different amplitudes and widths to determine the amplitude at which the Pacemaker Sense Input responds to each pulse width. These data are plotted on a log-log chart to approximate the response curve of the sense circuit.

This evaluation was done at 6–8 pulse widths. It took 5–10 minutes per pulse width to set up and flip amplitude switches while observing pacemaker output for sensing activity. To complicate matters, the pacemaker would be programmed to several different sensitivity levels. This process took three to four days.

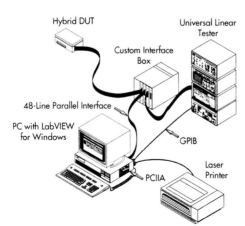

Figure 7-23.
The Medtronic Micro-Rel hybrid test and development system.

Using LabVIEW, I wrote a program to accomplish this same sense testing in just 12 hours. In addition to reduced programming development time, using LabVIEW netted about three times more data. Running the LabVIEW program took four hours, including printing reports. The best part is, all I did was start it—the testing was accomplished without operator intervention.

LabVIEW Use Expands

Several other departments, including Calibration Lab and IC Engineering and Development, are evaluating LabVIEW. Because each department uses different computer platforms, LabVIEW program portability will be a key feature. As for myself, I will continue to use LabVIEW in every way possible!

LabVIEW and SCXI Perform Automatic Test on Ventilators

by Lena Monvall, Frontec Datakonsulter Stockholm AB

Members of the Life Support Division (LSD) at Siemens-Elema AB were looking for a production test system to perform an automated test on their ventilators. Their need for a flexible, easy-to-use, and easily maintained test application led us, consultants to Siemens-Elema, to select the LabVIEW graphical environment for their application. We developed the test system using LabVIEW for Windows along with several SCXI modules for data acquisition and signal conditioning.

The ventilator SV300 consists of three modules; a pneumatic module, an electronic module, and a panel module. Breathing parameters are set by controls at the panel. The objective of the test system is to validate the electronic module of the ventilator in the production line by running a number of test cases. To perform a complete test of this module, other modules of the ventilator, such as the panel module and pneumatic module, were simulated. This requirement was difficult to meet, particularly the simulation of the panel module, because of the rapid updating procedure of the displays. Another requirement was that LSD should be able to maintain the test system and occasionally make changes to the source code. LSD also wanted to use the system to generate test reports with all test results logged.

The Test System

To meet LSD's test system needs, we used SCXI modules for data acquisition and signal conditioning and LabVIEW for program development. SCXI made it possible to meet the requirement of having an easily maintained, low-cost, and low-noise hardware environment. The extensive selection of LabVIEW Library virtual instruments (VIs) gave us great hardware integration capability and reduced development time.

An IBM-compatible Siemens Nixdorf computer running LabVIEW under Windows controls the system. The computer is connected to a SCXI-1200 module via the Centronics par-

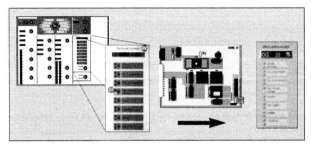

Figure 7-24.
The Spy, a custom SCXI module, retrieves information from the panel LEDs and displays. Using a software driver we developed, we integrated this information into LabVIEW.

allel port. The SCXI-1200 is situated in a SCXI-1001 rack that houses one SCXI-1163, 32-channel DIO; five SCXI-1124, 6-channel DACs; and one SCXI-1100, 32-channel MUX. The SCXI-1200 module removes the need for additional plug-in DAQ boards to control the SCXI system. The SCXI rack is connected to the ventilator via a self-developed interface unit that consists of a cross-connection board and a signal conditioning board.

The whole test is actually a sequence of several test cases. For performing the sequence, we chose the LabVIEW Test Executive to create automated test systems. Specific features we needed included test sequencing based on pass/fail status and dependencies, identification of unit under test, and logging of test results. The Test Executive even provided us with the facility of ASCII test report generation to a file. By a simple modification of the Test Executive, we added a printing feature to the test system so we can print test reports.

The Spy

To meet the requirement of simulating the panel, we developed our own SCXI module called The Spy. The Spy retrieves information on several LEDs and displays by "spying" on the data bus of the ventilator. An onboard program running on a Hitachi H8/325 is used to filter out the panel data. All interaction with the SCXI bus is made by a programming array logic (PAL) chip that we also developed. A software driver for The Spy was developed and implemented in C as a dynamic link library (DLL). We also wrote a code interface node (CIN) so we can integrate this driver into our LabVIEW code, making access to the DLL totally transparent for the LabVIEW programmer.

Benefits Pay Off

By using National Instruments hardware and software, we produced a PC-based test system with the same look and feel as a larger and more expensive test system. The high-quality GUI of LabVIEW, its ease of use for developing programs, and its capability for integrating DAQ hardware greatly reduced our development time. This environment helped us develop a flexible, easy-to-use, and powerful test application.

For more information, contact Lena Monvall at FDS, Gårdsvägen 7, 171 52 Solna, Sweden, tel +46 8 470 20 00, fax +46 8 470 21 99, e-mail Lena.Monvall@sth.frontec.se.

LabVIEW
in a Regulated
Environment

8

Key Characteristics and Terminology
Development Architecture

Automation Tools

Code Analysis (Static Tests)

Software Metrics

Metrics Visualization with LabVIEW

Applying Metrics Visualization

Inspecting Documentation Completeness

Requirements Traceability

Verification and Validation Test

Unit Testing

Automating Unit Testing with OverVIEW

Simulation

Full Project Testing and Regression Testing

Documenting the Project

Theory of Operation

Generating Validation Reports

Project Release Report

Verification after Release—Controlling and Maintaining the Software

Summary

LabVIEW™ users are demanding the software to perform ever-expanding roles in automating complex, mission-critical functions in industries such as defense, aerospace engineering, biomedicine, and pharmaceutical. These industries are pushing the latest design and development methodologies to their limits in order to deliver advanced products with superior quality on time and under budget.

One of the most challenging of these roles has been in the biomedical industry, exemplified in the automation of production testing of medical devices. Because these applications make pass and fail decisions on devices that make life and death decisions on people, accuracy and reliability are crucial. That is why the U.S. Food and Drug Administration (FDA) requires, by law, that the processes and methodologies be in place to ensure the released applications will not fail their intended tasks. In order to effectively compete, biomedical companies are constantly striving to automate as much of this process as possible.

An in-depth theoretical discussion on FDA certification is out of scope for this chapter and is the subject of many books. Rather, this chapter will outline some of the technical aspects of this regulated development environment directly pertinent to LabVIEW developers in this and similarly challenging industries. When reading this chapter, keep in mind that mistakenly failing good product affects only production yield, but mistakenly passing faulty product out to the field is an escape and presents a risk to the user and your organization's image. Preventing the latter is why this environment must strictly adhere to proven software engineering methodologies and go through rigorous validation and verification (V&V) practices.

In this chapter, we present an array of common issues and challenges, focusing specifically on how they relate to utilizing LabVIEW within this industry. As the foundation, we walk through an example process in detail to examine some of the challenges for each topic. Our goal, however, is to highlight the importance of how tools can assist with making the overall process more efficient and overcome some of the development challenges. Therefore, with each topic we look at some examples within the LabVIEW environment and discuss how one toolset, OverVIEW™,* can be implemented to address the various development components.

Key Characteristics and Terminology

The phrase *software validation* covers the entire software development life cycle, from planning through development and deployment to maintenance. It is an umbrella phrase covering both software verification and validation tests. We commonly hear the terms *validation* and *verification* used interchangeably, so the following serves to provide the subtle distinction.

Verification: Software verification is the performance of tests and analyses to determine whether products in each phase of software development are correct, complete, and consistent with respect to products in the previous phase.

Validation: Software validation is the performance of tests and analyses to determine whether the fully integrated computer software operates correctly, completely, and consistently with system specifications and requirements.

Auditability/Traceability: The primary component running through all aspects of a validation process is traceability. In the face of an internal or FDA

---

*OverVIEW™ is a registered trademark of TimeSlice, Inc. More information about Over-VIEW and the FDA regulatory process can be found at http://www.tslice.com.

audit, we must be able to prove (i.e., document) how the software that has been created meets the intended requirements.

Biomedical Product Testing

It is important to provide an understanding of the typical environment to which this chapter pertains. The arena for testing biomedical devices is interesting because it blends two important, yet opposing forces: *time to market* and *regulatory control*. It is safe to say that these forces are present with most organizations and industries and there is a balance achieved. However, the time-to-market force will typically dominate within organizations not held to the same level of regulatory control as those within the biomedical device industry. If your Palm Pilot™ fails to operate, you are inconvenienced, you might miss a couple meetings, and you're out a few hundred dollars. If an implantable medical device fails to operate, a life may be at risk. The industry is highly competitive and manufacturers attempt to avoid having their products viewed as a commodity. In other words, they want their products to stand out, be noticed, and not be common or simple or interchangeable. This means the complexity continues to increase while the time-window to release the product continues to decrease. The resultant impact to the test engineering community is simply to do more with less (see Figure 8-1).

Another dimension, which is typical, relates to the configuration of the production test systems. As with most industries, test management tries to maximize the usability of production test hardware. This typically results in having a high degree of product mix tested on the same platform, which means multiple test software applications are executed from day to day. We will ex-

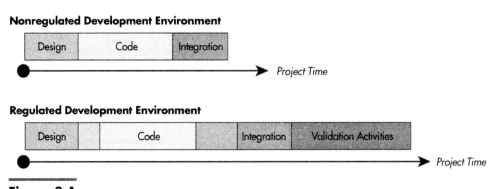

Figure 8-1.
Impact of validation and verification processes on project schedule.

amine the important issues that arise related to software if multiple test applications are executed on the same platform. We will also examine the importance of a well-defined architecture from both the hardware and software perspectives.

LabVIEW Scalability

Part of the reason LabVIEW has been implemented as a development platform within this industry is that it can shave valuable time off the development schedule—if implemented and utilized correctly. While LabVIEW addresses the productivity related to developing individual code modules, there are issues that arise with larger applications related to the aspect of controlling and managing the development process.

As applications grow in size and complexity, the tasks associated with development begin to change. Where a rapid prototype development methodology can suit a small one-developer project rather well, larger projects and mission-critical applications require an upfront strategy for system requirements, architectural design, integration, testing, documentation, quality, and so on.

As a project grows in size, the test effort will grow disproportionately in order to maintain a comparable quality level (see Figure 8-2). This can be

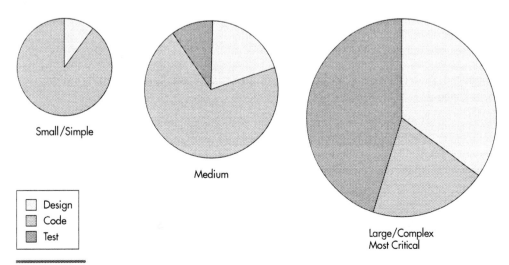

Figure 8-2.
Relevance of test as size and criticality of the application increase.

attributed to the number of interactions between different parts of the program. These unintended interactions can be minimized through proper attention to modularizing your design.

It is this scalability of LabVIEW, specifically for mission-critical biomedical applications, that we explore through the rest of the chapter. Some of the aspects and process components of how these challenges are addressed are shown here:

- Multiple developers per project
- Accurate project estimation
- Configuration management
- Documentation and coding standards
- Re-usable software
- Modular system architecture (drivers, test executive, utilities, etc.)
- Verification and validation activities

The Keys to a Successful Validation Program

The challenges that this environment presents, and specifically the methods for dealing with these challenges, are not all that different from good software engineering practices followed for other large complex applications that have a business need for quality. The most critical trappings in testing are typically nontechnical. They are consequences of the process and include the overall composition of the test team and whether the company follows well-integrated processes for formal requirements handling and change management.

The key to this entire discussion is that, while the techniques and process components are typically just good practice, we must have proof or verification that these good practices are actually being addressed. Another important consideration is that when implemented as a complete system, not only are the quality goals achieved, but they are achieved with a reduction in overall development cost and timing.

The remainder of this chapter presents three essential dimensions to a validation program: architecture, process, and automation tools. We briefly outline important components within each of these areas and highlight how they relate to implementation with LabVIEW.

Development Architecture

Prior to initiating development on a project, it is necessary to establish a well-defined software or development architecture. We examine this with respect to some general philosophies that are applicable for most software development activities and show how aspects of the architecture relate to the validation environment with LabVIEW.

Guiding Principles

While it is feasible to only discuss software architecture as a programming feature for how each developer should create LabVIEW code, we also look at it from the perspective of establishing an architecture for the department. This is similar in a sense to a process that would be mandatory for all test engineers to follow. Some guiding principles that should be addressed when configuring an architecture for LabVIEW development include the following. These terms are not necessarily mutually exclusive as there is some overlap in their relationship. However, it is appropriate to provide a simple definition or example for each.

- **Testability:** This guides development to assure that individual code modules can be easily tested in a unit testing environment, as well as the application easily being tested from the system level.

- **Modularity:** This keeps the system flexible so it can be enhanced or modified quickly and effectively. An example of modularity would be the utilization of a well-defined plug-in architecture in a test executive so that all tests have a common structure. Another would be the separation of functions into well-defined categories that can be adjusted independently.

- **Reusability:** This provides the means to share software code between multiple projects or multiple revisions of the same project. For this to be effective, the code must be well defined and must be constructed to execute in a variety of perspectives.

These guiding principles are no different than what would be found in any well-designed software architecture. They are highlighted here to examine the critical impact that they have in regard to validation testing. The key in developing and maintaining a software application is to release as quickly as possible without sacrificing quality and maintain a traceable link to requirements.

Testability

We will focus more on the aspects of testing later in the chapter, but for now we highlight how it applies to these strategies. Quite simply, if these guidelines are an accepted part of the architecture and developer mindset, then this will dramatically reduce the time for validation testing.

For instance, let's examine a test VI that has to measure some key pace pulse characteristics from a cardiac pacing device. Typically there would be several requirements defining the purpose of this test that need to be verified. If the VI is constructed without testability in mind, it will most likely delay the testing process and potentially sacrifice the verification of all requirements being satisfied.

It is easy to understand why software modules are not typically designed for testing. Testing is routinely thought of after code development is complete. Invariably, the resultant software is wedged into the test with hopes that the coverage is ample. The same engineering principles that are considered good practice for product development are useful for software. Most organizations subscribe to the tenets of concurrent engineering or design for X and try to get the various departments together at the outset of a product launch. These same principles should be in place with software development in that the testing should be a parallel effort that is undertaken at the outset of development (see Figure 8-3).

Modularity

The aspect of modularity has dramatic impact for the overall system as well as each test VI that a developer creates. Specifically, the overall system architecture should employ components such as instrument drivers, a test executive with a design for plug-in tests, and an independent reporting module. This will be explored in detail later.

As modularity relates to the design of each test VI, let's again explore the impact it has on testing and requirements traceability. If the test VI is constructed as a monolithic VI or even as somewhat modular without a well-

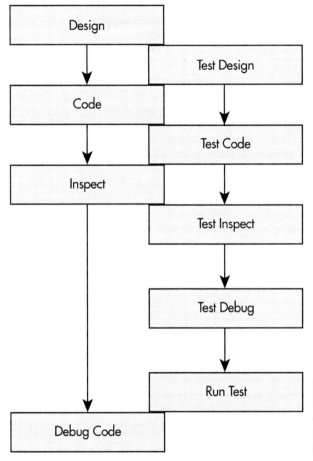

Figure 8-3.
Designing for test means the test case design and development should start with actual code development.

designed hierarchy, it will sacrifice test coverage and make the verification of requirements difficult. As we will show later, one advantage of unit testing is the amount of test coverage that can be applied by focusing multiple test cases on each unit. Modularity is the enabler of this process because the modules are the units that are tested. A well-designed, modular hierarchy allows for easier and more complete test coverage. It also leads to the other strategic guideline, which is reusability. Modules should be properly named for reuse. The VI's name should be its action—not its context or internal logic. If a module name that fits cannot be easily found, this module was probably not designed in such a way as to be reusable.

Software Recycling

There is little doubt that recycling previously generated software is important to shrinking development cycles and managing projects. Why reinvent when something is already available to be plugged into the application? Properly categorizing and organizing code for reuse is one important component. Another is assuring that the code is robust and truly modular so that it can be applied in various applications. Adhering to guidelines and utilizing inspection or testing tools makes code widely reusable.

Consistent Baseline and Directory Structure

The principles just discussed help shape the structure of how the software modules are developed. Another component related to this is directory structure and naming conventions of the software modules. The importance of these attributes allows control during development and makes for easier integration testing.

Part of this solution is having a well-defined system platform that is consistent, traceable, and easily recreated. This baseline would typically comprise the specific format for the computer, the operating system, applications such as LabVIEW, drivers, and so on. The baseline configuration must be a traceable component such that the environment in which a given test was run can easily be recreated. Let's say a device that was manufactured several years ago gets returned for analysis. It is vital to recreate the same test environment with the operating system, version of LabVIEW, and specific test application that was utilized at the time. The test applications might have a directory structure similar to that shown in Figure 8-4.

While this approach is consistent for all test applications, it creates a lot of redundancy. As mentioned earlier, it is important to have an architecture that is modular from the system perspective. This means the following components should be in place within the test engineering department:

- Consistent test platform using same measurement hardware for all tests
- Consistent fixturing and signal conditioning that is independent of test application
- A test executive with a plug-in architecture for adding and sequencing tests

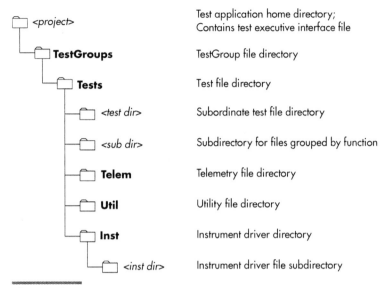

Figure 8-4.
Example of specified directory structure for test applications.

- A validated set of instrument drivers for the common measurement hardware

This approach isolates the task of the test engineer to what it should be, creating and validating test software—not system software. This provides the opportunity to modify the structure listed above and push elements such as instrument drivers, analysis functions, and utilities into a common area on the system that all test apps can share. Using a common architecture has the following advantages:

- Minimizes redundancy of code on the tester
- Minimizes the amount of code the test engineer is responsible for in project management
- Encourages and enforces modularity and software reuse
- Eliminates minor, uncontrolled changes to common software

The advantages of this common architecture must be balanced against the maintenance and proper naming conventions if the common software is updated. In other words, it is possible that the common software may need enhancements. It is vital to remain backward compatible and not adversely affect test applications that don't want or need the upgrade.

The policy for how common software is upgraded may vary. Some departments may enforce a policy that only one version of common software will exist on a tester. The burden is on the validation of the common software to remain backward compatible but also require a system-level retest from the test applications when a change is made. Another approach is to allow multiple versions of the common software where necessary. This means the different versions need to be segregated into uniquely named directory structures so they both can exist on the same test platform. While this eliminates the need for system-level retesting, it requires careful attention to assure that the proper version of the common software is being utilized by the test applications.

The Development Process

At the core of the validation program is the development process utilized to create test software. Just as the architecture previously described is like the landscape, the process is the navigational path describing how we get from start to finish. The primary theme that glues the process together as it relates to validation is traceability. In short, there must be a traceable path that essentially proves or verifies that the software code being executed for a given test application is based upon a certain set of requirements and all steps in between.

The process must be as efficient as possible within the constraints of the regulatory environment in order not to hinder the business objectives of innovation and new product development. An example process might look like that depicted in Figure 8-5. In general, this would be similar to a good software development process or design process within any industry. The key as it relates to traceability and control are the checkpoints that serve as the validation along the path. Note that several of the key stages in the process result in an associated document. Therefore, the process is essentially a combination of activities and resulting documents. The documents serve as the verification that a step in the process was achieved. They also serve as the traceable links throughout the process and as gates between adjacent stages.

For instance, an example process would not allow software development to begin until the requirements for the software are complete and reviewed. Fur-

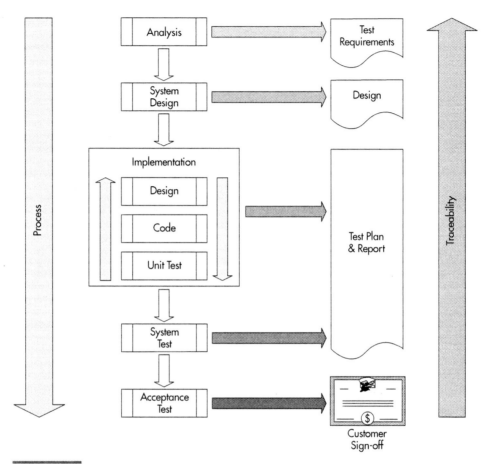

Figure 8-5.
Example software development process with validation checkpoints.

thermore, the actual software validation test could not proceed until the validation plan is complete and reviewed. This all relates back to the mindset of "say what you will do and do what you say," which dictates the development activities.

As it pertains to LabVIEW it is important to consider how the configuration management tool can capture files of various types, such as VIs, text files, and dynamic link libraries. It is also important to consider the implementation of file structure. For instance, the LabVIEW library (LLB) is a convenient container for grouping VIs within a project. However, since most configuration management utilities view the LLB as one entry or member in the project, information is lost regarding the details of individual VIs. This is important because part of a validation process and associated documentation relates to inspecting, testing, or modifying individual VIs. There needs to be a mechanism within the configuration management tool to track this progress. If the VI is buried in an LLB, this critical tracing is lost.

Project Management

A common technique for developing large software applications is to divide the functional modules between developers, which typically decreases the overall development time. This is done by incorporating a team of in-house developers, outside developers, or a combination of both. In this situation it is helpful to have guidelines or coding standards to create a consistent design and coding methodology. The need for configuration management is essential to prevent individual VIs from inadvertently being developed by more than one party at the same time and to manage the various revisions that may arise after initial release.

From the management perspective there is the need to properly allocate the resources involved with the project and manage the timeline. Neither of these can be achieved without accurate status information. It is challenging at best to allocate resources or assign tasks if there is little understanding of where the project stands. Another way of saying this is that there is no way to accurately estimate when you will reach your destination if you don't know where you are. Therefore, the project status is vital to allowing this to be achieved. Various categories of metrics and inspection criteria are needed to allow the proper level of project information to be ascertained.

Configuration Management

One of the most critical components in a validation environment is the configuration management of the software. This is the focal point for tracing source code to requirements and testing and the final implementation. It is also the necessary means with which to develop software in a shared environment. We could even say that if you do not currently have a configuration management strategy and tools currently implemented, stop reading, put down the book, and return when it has been implemented.

The configuration management allows code to be archived and retrieved with an identifier such as a label or a revision tag. This information is usually included as the traceable link within the associated documents in the process. For instance, it is useful to create a full project listing as part of the implementation or software release report. This typically includes all the VIs and associated files that comprise the project. The revision as given in the configuration management utility is shown adjacent to each file in the report.

Coding Standards

At the center of the development process is the actual construction of the software modules. The blueprint for this phase should be a set of guidelines specific to creating software within LabVIEW. There are several published items that provide a basis for a LabVIEW coding standard. We believe there can be no single, catch-all coding standard that addresses the specific needs of each organization. Therefore, the guidelines should be assembled as a combination of de facto industry standards, best practices in software development, LabVIEW-focused style, and methods that relate to the organization.

The importance of the LabVIEW coding standard cannot be overstated. It is the basis for the core stage of the software development process and is the traceable document that can be referenced to ensure a stable and consistent process is being followed. The coding standard is the rulebook or the guidelines for creating LabVIEW applications. The benefit of what it provides is a well-defined process for creation that dictates good programming practice and techniques related to the validation process. The enforcement of these guidelines is the self-discipline of each programmer, but more importantly the peer review process discussed next.

Since the coding standard is most effective when it is specific to the needs of the organization, it will most likely require an investment in its initial creation and ongoing maintenance. Careful thought should be taken in the creation of the document in order to maintain a high degree of quality and useful information, without drowning the user in unnecessary and inefficient procedures.

Figure 8-6 provides one potential technique for documenting the coding standards. The items within the coding standard should be categorized as Required and Recommended, similar to a requirements matrix, in order to maintain consistency and control without crippling the creative process. Typically,

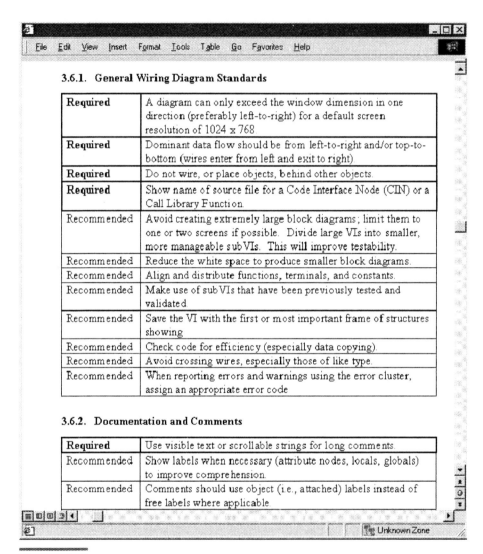

3.6.1. General Wiring Diagram Standards

Required	A diagram can only exceed the window dimension in one direction (preferably left-to-right) for a default screen resolution of 1024 x 768.
Required	Dominant data flow should be from left-to-right and/or top-to-bottom (wires enter from left and exit to right).
Required	Do not wire, or place objects, behind other objects.
Required	Show name of source file for a Code Interface Node (CIN) or a Call Library Function.
Recommended	Avoid creating extremely large block diagrams; limit them to one or two screens if possible. Divide large VIs into smaller, more manageable subVIs. This will improve testability.
Recommended	Reduce the white space to produce smaller block diagrams.
Recommended	Align and distribute functions, terminals, and constants.
Recommended	Make use of subVIs that have been previously tested and validated.
Recommended	Save the VI with the first or most important frame of structures showing.
Recommended	Check code for efficiency (especially data copying).
Recommended	Avoid crossing wires, especially those of like type.
Recommended	When reporting errors and warnings using the error cluster, assign an appropriate error code

3.6.2. Documentation and Comments

Required	Use visible text or scrollable strings for long comments.
Recommended	Show labels when necessary (attribute nodes, locals, globals) to improve comprehension.
Recommended	Comments should use object (i.e., attached) labels instead of free labels where applicable.

Figure 8-6.
Example section from a coding standard emphasizing levels of importance.

the items marked as Required in the coding standard are the items most heavily scrutinized during the peer review process. Items that are not compliant with the Required components of the coding standard should be justified with a traceable justification.

Automation Tools

To this point we have provided the backdrop by describing the critical challenge of balancing the time-to-market forces with those of regulatory control. Additionally, we have set the stage for development by describing a process and system architecture that might be typical within this environment. As we navigate through the process, it is essential to examine how tools are used to automate the tasks and make the process more efficient. With this in mind we will show how the important categories outlined in the rest of the chapter can be addressed with OverVIEW, a utility specifically created for the project control and validation attributes related to LabVIEW. A typical OverVIEW panel is depicted in Figure 8-7.

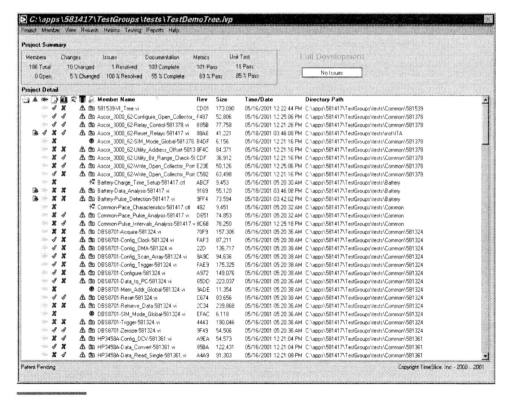

Figure 8-7.
OverVIEW provides project status, auto-inspection, and test automation to LabVIEW.

Code Analysis (Static Tests)

There are two basic types of module testing, static and dynamic. The FDA glossary defines *static testing* as the analysis of a program that is performed without executing the program—a code review or peer review. *Dynamic testing* is the analysis that is performed with executing the program, which is discussed later.

The FDA's guidance statements include, "Due to the complexity of software, dynamic analysis alone may be insufficient to show that the software is correct, fully functional, and free of avoidable defects. Therefore, static approaches are used to offset this crucial limitation." Two things are easily detected with peer reviews: documentation problems, which are maintenance issues, and code errors where subtle problems may remain in code undetected by other means. These problems are typically found quickly and the solutions are determined easily. Code problems typically found with a peer review or code analysis include:

- Invented code (not specified in the design)
- Missing code (code did not implement all of the design)
- Erroneous engineering units
- Erroneous input/output
- Semaphore race conditions (the code does a set flag—do action; should be do action—set flag)
- Interrupt race conditions
- Simple logic or math errors
- Difficult-to-maintain code (does not comply with coding standards)

Studies have shown that most errors and issues related to software are found during the peer review process. While it can be an involved process to pull resources into an inspection, there is good value in finding issues when it is cheaper to address them earlier in the development process. Therefore, it is important to take advantage of code inspection and analysis tools to speed up the inspection process and make valuable peer reviews as efficient and effective as possible. Let's examine a couple of areas where LabVIEW code inspection can be applied to achieve this efficiency.

Software Metrics

Software metrics are the quantitative measures of a given piece of software and are the cornerstone to automating aspects of the coding standard to speed up the peer review process. Simply stated, metrics are the various static and dynamic attributes of the software module.

Analyzing software metrics is an essential process step for predicting quality in test system software. Metrics analysis provides a quantitative assessment of how well the code is structured, which directly relates to the capability to maintain and enhance the application. Additionally, there are well-used visualization techniques to rapidly analyze code modules throughout the development process. These techniques provide the basis to reengineer code where necessary and avoid costly rework and maintenance efforts. Where complexity cannot be reduced through redesign, the metrics analysis indicates where concentrated inspections, debugging, and testing should be directed.

Applying visualization to software metrics is a valuable technique for rapid analysis and decision making. A popular method of visualizing software metrics on a given code module is the Kiviat diagram (see Figure 8-8). The key to analyzing complexity with the Kiviat diagram is pattern recognition and the use of proper limits. The metrics should be grouped by similar function on the diagram. For instance, all the metrics relating to memory usage should be adjacent to each other so this category is easily identifiable.

Inspecting individual code modules with the Kiviat diagram is useful for an in-depth analysis. For moderately sized and large applications, it becomes unreasonable to inspect each code module independently so it is important to visualize software metrics from the project perspective. Next, we will examine both project and component level methods for metrics visualization.

Metrics Visualization with LabVIEW

While software metrics visualization is well defined for conventional, text-based languages like C++, it is a new methodology for graphical languages such as LabVIEW. In the text-based software paradigm, the lines-of-code parameter is a very useful metric. Within LabVIEW, this metric has no meaning. This does not prevent a useful methodology of metrics visualization being applied to the graphical language. The same benefits that have proven to be successful for the text-based languages can be realized in LabVIEW as long

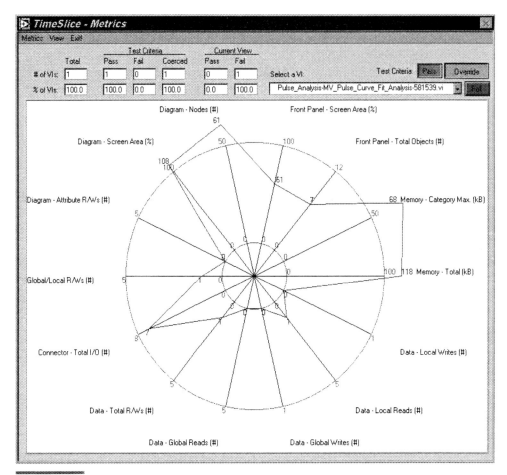

Figure 8-8.
Kiviat diagram for metrics visualization.

as we create and define the appropriate metrics and categories. An appropriate set of metrics might include those listed in Table 8-1.

Applying Metrics Visualization

Our example focuses on a simple analysis routine used as an elemental component within a test application. This analysis routine finds y for a given x value in a polynomial equation ($y = a_0 + a_1 x + a_2 x^2 + a_3 x^3 + \ldots$). This is one

Table 8-1. *Graphical language metrics.*

Category	Individual Metrics
Memory Usage	Total Memory Data Space Memory Code Memory Diagram Memory Front-Panel Memory
Front Panel	Number of Inputs Number of Outputs Total Objects Width of Panel Height of Panel Percentage of Screen Area
Icon Connector	Total Number of I/O Number of Inputs Number of Outputs
Wiring Diagram	Number of Structures Number of Nodes Number of Diagrams Depth Level Width of Panel Height of Panel Percentage of Screen Area Number of Attribute Reads Number of Attribute Writes
Data Coupling	Total Number of Read/Writes Number of Global Reads Number of Global Writes Number of Local Reads Number of Local Writes
External Calls	Total Number of Calls Number of DLL Calls Number of CIN Calls

of over a hundred VIs that comprise a project and serves as a useful analysis routine called by many higher level VIs within the test application.

As mentioned, it is challenging to inspect each VI individually using the Kiviat diagram, so we need a first-pass inspection at the project level. One method is to define an additional category of metrics that is used as test criteria. This set of test criteria is scanned automatically for all modules in the

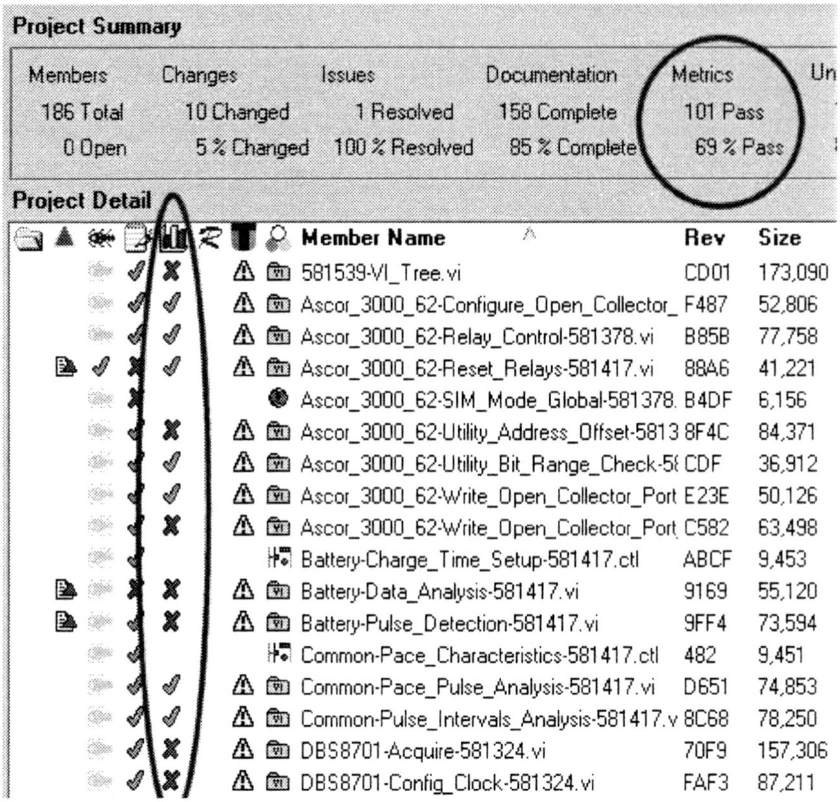

Project Summary

Members	Changes	Issues	Documentation	Metrics	Un
186 Total	10 Changed	1 Resolved	158 Complete	101 Pass	
0 Open	5 % Changed	100 % Resolved	85 % Complete	69 % Pass	

Project Detail

							Member Name ∧	Rev	Size
		✓	✗	⚠			581539-VI_Tree.vi	CD01	173,090
		✓	✓	⚠			Ascor_3000_62-Configure_Open_Collector_	F487	52,806
		✓	✓	⚠			Ascor_3000_62-Relay_Control-581378.vi	B85B	77,758
	✓	✗	✓	⚠			Ascor_3000_62-Reset_Relays-581417.vi	88A6	41,221
		✗		●			Ascor_3000_62-SIM_Mode_Global-581378.	B4DF	6,156
		✓	✗	⚠			Ascor_3000_62-Utility_Address_Offset-5813	8F4C	84,371
		✓	✓	⚠			Ascor_3000_62-Utility_Bit_Range_Check-5€	CDF	36,912
		✓	✓	⚠			Ascor_3000_62-Write_Open_Collector_Port	E23E	50,126
		✓	✗	⚠			Ascor_3000_62-Write_Open_Collector_Port	C582	63,498
		✓					Battery-Charge_Time_Setup-581417.ctl	ABCF	9,453
		✗	✗	⚠			Battery-Data_Analysis-581417.vi	9169	55,120
		✗	✗	⚠			Battery-Pulse_Detection-581417.vi	9FF4	73,594
		✓					Common-Pace_Characteristics-581417.ctl	482	9,451
		✓	✓	⚠			Common-Pace_Pulse_Analysis-581417.vi	D651	74,853
		✓	✓	⚠			Common-Pulse_Intervals_Analysis-581417.v	8C68	78,250
		✓	✗	⚠			DBS8701-Acquire-581324.vi	70F9	157,306
		✓	✗	⚠			DBS8701-Config_Clock-581324.vi	FAF3	87,211

Figure 8-9.
Automated metrics inspection from OverVIEW.

project, and an overall pass or fail status is provided for each code module, as depicted in Figure 8-9. This at-a-glance information can be used as a rough inspection that filters out which modules need further inspection with the Kiviat diagram.

The original design of our example code is shown in Figure 8-10. With this example we have determined at the project level that the original design of the polynomial VI showed a FAIL on the OverVIEW project inspection within the Metrics category. This prompted further inspection of the polynomial-analysis routine using the Kiviat diagram shown in Figure 8-11. It showed that both memory and wiring diagram components were exceeding the acceptable limits.

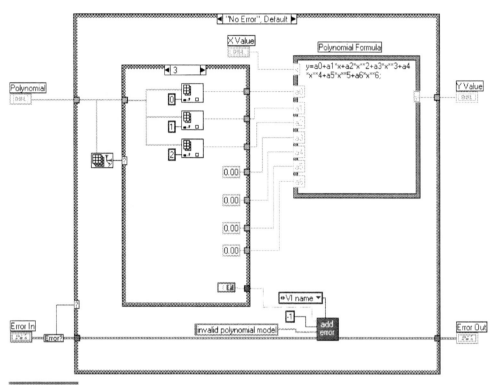

Figure 8-10.
Block diagram of original code design.

The analysis prompted a code redesign, shown in Figure 8-12, to make these areas more efficient. The resulting function met its objective by reducing the values of the metrics that had previously failed, as shown in the Kiviat diagram in Figure 8-13.

To understand why there was a benefit with the revised code, we first must analyze the approach taken to produce the intended algorithm. The design of the function is simple. An array of coefficients and a value for x are passed into the function representing a polynomial equation ($y = a_0 + a_1x + a_2x^2 + a_3x^3 + \ldots$). If only two coefficients are passed in, then it is a second-order equation of type ($y = a_0 + a_1x$). In this regard, the function adapts to the order by the number of coefficients.

In the original algorithm, this adaptation was done explicitly for each possible order between 0 and 7. The incoming array of coefficients first was sized

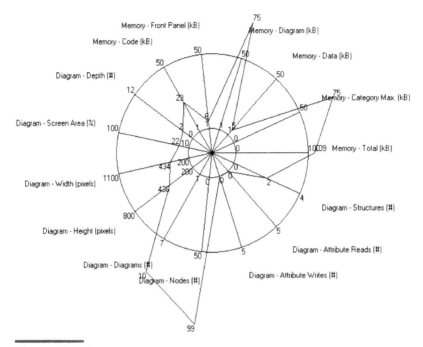

Figure 8-11.
Kiviat diagram of original code design.

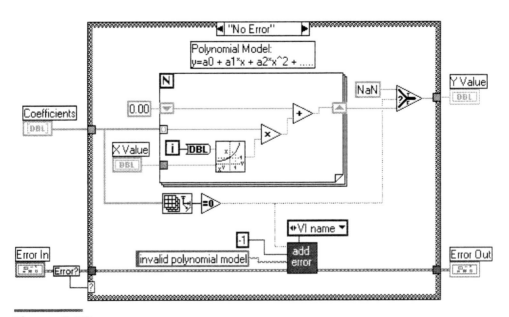

Figure 8-12.
Block diagram of revised design.

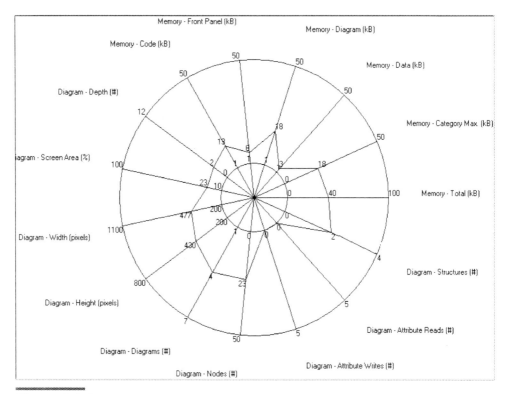

Figure 8-13.
Kiviat diagram of revised code design.

to determine the order. A unique case existed for each possible order between 0 and 7, and the logic flow then was exercised in one of these unique cases to extract the coefficients from the array. The resulting coefficients were passed out of the case to the mathematical equation that determined the Y value. Since there were eight unique cases [0 . . 7] and each case provided the resulting coefficients, the number of nodes was exceedingly high (99), and the number of diagrams also was high (10).

The number of nodes and diagrams (isolated sections of code) in LabVIEW are the best representatives of lines of code or function points in a text-based language. These text-based metrics typically are the gauge for determining the bug count within a function or within the overall application. It has been proven there are fewer bugs with fewer lines of code and fewer function points. As a result, we can safely say that the original algorithm in our example was much more likely to produce errors (60% to 75%) than the revised code.

It is important to note that the limits used in the metrics analysis can be somewhat arbitrary. This further substantiates the importance of the coding standard. The coding standard can be utilized to define the following:

- Which metrics will be analyzed
- Which metrics will be placed in the test criteria category for automatic inspection
- The limits for each metric value

We can see how a code inspection of the metrics complexity using Over-VIEW can drastically benefit the development and validation process for a few reasons. First, the automated inspection quickly and easily points out where there might be an issue within a large project. This can lead to a re-design, if appropriate, earlier in the process, before formal validation begins. Secondly, this metrics inspection can lead to better unit test designs for the VIs under test, and it can help predict the amount of time for the testing phase of the project.

Unit testing is a very comprehensive method of exercising an individual function. For the original algorithm in our example, we would need to make at least eight separate tests to properly cover the explicit cases that processed the coefficients. In the revised code, we could be confident with two or three tests since the algorithm was scalable and did not have explicit cases to handle the coefficients. Note that the metric called diagrams can serve as a predictor of testing effort, as it is directly proportional to the proper number of test cases. In our example, the value was ten for the original algorithm and four for the revised code.

Inspecting Documentation Completeness

We have discussed the vitality of a traceable link through all stages of the development process. This means that it should be required within the coding standards that each VI be sufficiently documented. The VI Description should adequately describe the functionality of the VI, and the input and output controls should be adequately documented.

This requirement can be very time consuming to enforce. Typically, there will be gaps where a subset of VIs within the project are out of compliance and do not have any traceable documentation. The process of opening each VI and looking at the help window for the appropriate descriptions is tedious at best.

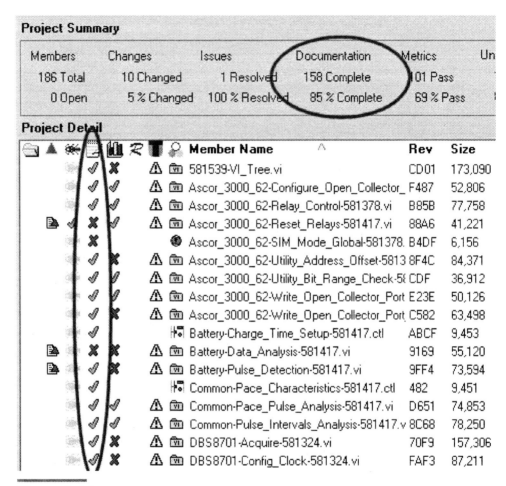

Figure 8-14.
OverVIEW Documentation status.

In the same fashion that the metrics complexity was analyzed automatically for the entire project, OverVIEW can quickly examine and modify the gaps that might exist in the documentation for all the VIs within an application (see Figure 8-14). This can literally shave days off of the process of inspecting and modifying descriptions.

The pass or fail status can be set to examine only the VI description or both the VI description and all the controls and indicators on each VI. It is one thing to inspect and point out where there are gaps, it is another to make the change. Therefore, it is important to have this same type of top-level approach on the

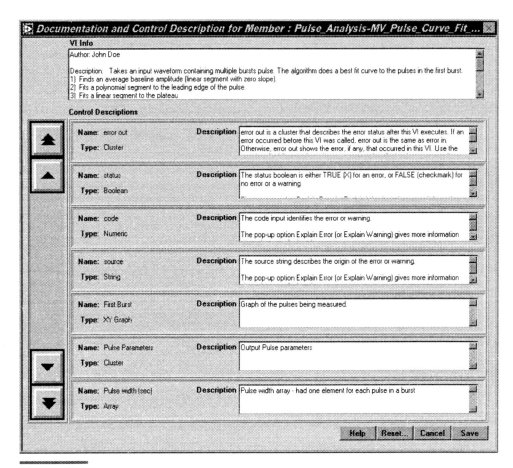

Figure 8-15.
Documentation Editor window from OverVIEW.

editing process as well. By opening an editing window (see Figure 8-15) we can see which components may be missing the appropriate documentation and make the change. This reduces the number of steps compared to performing this task on an individual VI basis through the properties window.

Requirements Traceability

The requirements test matrix is the traceable link to ensure that there is a test for each requirement. Each uniquely identified software requirement must be traceable to a specific validation test. With the requirements traceability matrix, the FDA expects to find either a software requirements traceability matrix or, alternatively, a numbering scheme that is devised to allow traceability.

To make sure that all components and assembly are being tested, use a test coverage matrix. This matrix should have all major features listed along the right column. The column headers should have all pertinent test levels and process steps. A check mark can be placed in spaces that test the feature of that row at that test level or process step. A more elaborate key of symbols can be used to show how that feature is verified.

With the test coverage matrix complete, compiling the requirements for each of the test levels is straightforward. Scan down the column for each test level. For each row that has a check mark, write down that requirement. A suite of tests needs to be decided upon to test each of these requirements individually or in groups. The specifications for each of the tests need to be traceable back to the applicable design documents. This is so that when the design eventually evolves, changes in the design documentation will trigger changes in the test documentation, which will result in appropriate changes in the corresponding tests.

Once the production tests are developed, it is time for validation of these requirements. Remember "Say what you do — do what you say," a mantra for testing for the FDA. You must have a signed-off and archived plan before starting to execute that plan. At the end of the V&V process, you will have a test plan and the report that documents the results of running that plan.

Verification and Validation Test

The objective is to follow a well-defined process and put as many effective inspections in place so issues can be addressed through inspection and good practice. However, there is still a need to properly test the software when the project is held to regulatory control. It is difficult to imagine the software that makes pass and fail decisions on implantable devices not being tested. With this in mind we outline a few of the various test levels and highlight how unit testing folds into the architecture of modularity and testability described earlier.

Unit Testing

We mentioned earlier how static analyses (peer reviews) were utilized, and we described the types of errors that are typically found with that process.

Dynamic analysis is concerned with demonstrating the software's run-time behavior in response to selected inputs and conditions. The code can be executed in several ways. It can be run on the actual target hardware or on a simulator. However run, test drivers and stubs are created to control the known inputs to the process and to capture outputs. This is most prevalent in unit testing, which can address both structural and functional testing.

Software structural (white-box) validation testing is based on the structure of the codes and is meant to challenge the decisions made by the program with test cases based on the structure and logic. It exercises the program's data structures and its control and procedural logic. Software functional (black-box) testing is based on the function (inputs/outputs) of the codes. These tests use both the product specifications and software requirements for the test development and acceptance criteria.

LabVIEW is a good environment for performing unit testing, primarily because any VI of any complexity level can be examined and executed independently. This allows a test shell to be configured that any type of VI can be plugged into for test execution. The shell should be able to perform the following functions:

- Quickly plug in and wire up the VI under test.
- Allow for editing and rearranging of multiple test cases on the same VI under test.
- Execute all the test cases.
- Determine pass and fail for all test cases.
- Generate a report detailing test summary and test case details.
- Provide a method of running all unit tests for the entire project.
- Provide a method for regression analysis.
- Provide an automatic test case generator for common boundary test conditions.

Unit testing can impact the validation process in several dimensions. First, it presents a high degree of granularity on the entire test application, which means the test coverage is much higher than without unit testing. Second, it can impact the validation process, helping to significantly reduce the time leading up to project release. Additionally, it enhances the reusability of code by adding a level of verification.

We discussed earlier the benefit of having a modular design for the software application because it increases the flexibility and requirements trace-

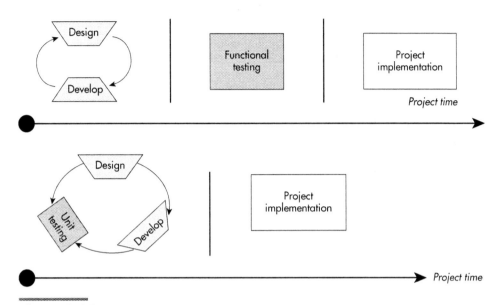

Figure 8-16.
Unit testing integrated into the development process.

ability. This is realized when unit testing is applied to these modules. By isolating each VI and focusing on its intended functionality with a comprehensive test, the overall test coverage of the application increases exponentially.

With a test strategy focused heavily on unit testing, the validation process can be modified to shrink the development time. A traditional approach is to create all the VIs for the application, develop a system test plan, and execute the test plan. The inefficiencies in waiting until all software is complete before formal testing begins plus the invariable reality that there will be corrections means this approach is not optimal. With unit testing the formal testing can be executed in parallel during the development phase (see Figure 8-16). As each VI is created and reviewed, the unit test can be applied. This incremental approach means that a simple functional validation test is all that is required at the end.

Many other benefits can be realized once a unit testing strategy has been implemented. The benefits with greatest impact involve reusability and regression testing. Most software projects require maintenance and upgrades. Also, it is common for new projects to be derivatives of previously released projects. In these cases it is most beneficial to utilize code that has already been written (and verified).

The last thing we want to do when upgrading a project is delay the release by retracing unnecessary steps. This is where regression testing can be beneficial. Assume we have adopted a strategy where there is a unit test for every VI in the project. Let's say that the project upgrade requires 3 VIs out of 200 to be modified. A regression analysis would show the hierarchical links between these modified VIs (see Figure 8-17). From a testing perspective it is

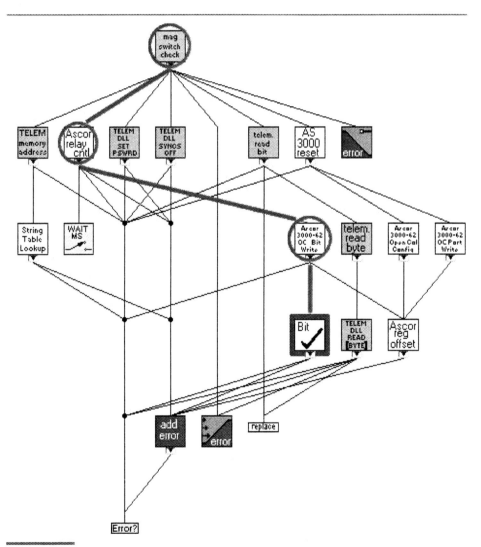

Figure 8-17.
Regression analysis on VI hierarchy.

important to verify that all affected code is compliant with requirements. Therefore, let's say that the modified VIs also affect 7 additional VIs that are dependent upon their functionality. Because we have already created unit tests for the VIs, the process is time-efficient and straightforward since we only have to run ten unit tests and a simple functional system-level test as part of the testing.

It is important to have tools that make this process efficient. If the unit test cases are preserved, then the act of executing these is very efficient. Also, the regression analysis becomes the traceable link to verify that the proper software was addressed during validation testing.

Code reuse is very important to streamlining the development of new and modified software applications. We have already mentioned how a system architecture with shared libraries for utilities and drivers is an important part of a code reuse strategy. Another dimension relates to sharing or reusing code at a higher level from project to project. With a unit test that travels in conjunction with the code module to be reused, the efficiency and confidence increase dramatically.

Remember, we are dealing with an environment that has an extra step added to everything. In other words, in most environments the aspect of using VIs that are already written is a great benefit. In the biomedical arena the benefit is cut in half if it is just the code that is shared. With an associated unit test the full benefit of code reuse is realized because there is verification and traceable proof that the VI performs to meet requirements and it minimizes the testing effort for the developer adopting the reused code.

Automating Unit Testing with OverVIEW

A fundamental component of the OverVIEW toolset, the unit test shell (UTool) is a utility for automating testing on LabVIEW modules. Because unit testing against functional requirements requires knowledge of how the VI operates, the process of arriving at test cases and evaluation criteria is an important component to the test design.

There are a couple of key aspects in UTool that leverage the effort put into the valuable process of arriving at test cases. First, multiple test cases can be created for the same VI under test. This allows a full range of known data to be passed through the VI under test. Second, the test cases persist within the UTool shell. This allows for easy and convenient execution of the test cases and efficient regression testing.

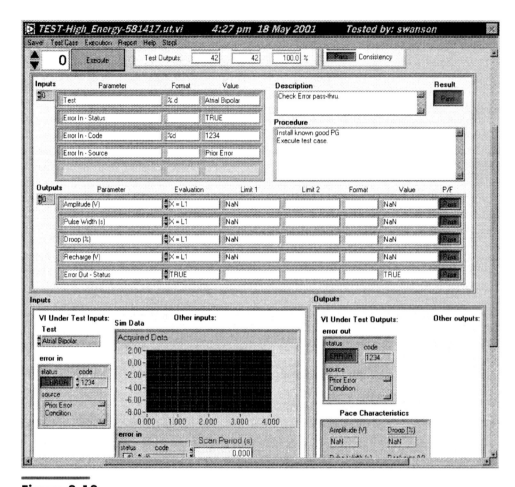

Figure 8-18.

Unit test shell within OverVIEW provides a persistent test shell with multiple test cases to be applied to the same VI under test.

LabVIEW's native data types and the fact that effective testing requires user knowledge mean that there is some manual intervention with configuring and editing a unit test. Once this is established, however, the rest is automated. The limits can be configured with a flexible range of evaluation criteria. Each test case can have its own set of inputs and outputs. And since reporting is the proof of compliance this tool provides a flexible and convenient automatic report generator. Figure 8-18 illustrates the UTool shell in a testing situation. Figure 8-19 shows the same unit test but with different criteria being applied.

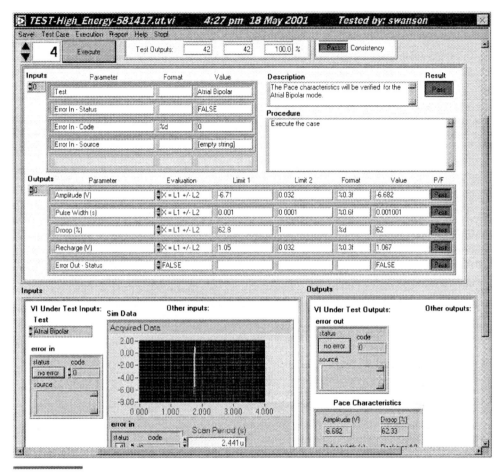

Figure 8-19.
The same unit test as in Figure 8-16 with different criteria being applied.

Simulation

Using the actual, real-world sensors and environment is best if possible and practical. However, if the real-world inputs cannot be used (e.g., because the inputs cannot be controlled for known values or it would put the testers at risk), the sensors or the environment can be simulated. What are the advantages of using simulations? Some benefits include being able to construct tests with explicitly known raw data, nominal input data, anomaly insertion, or known range limit tests. Again, if thorough validation testing can be achieved

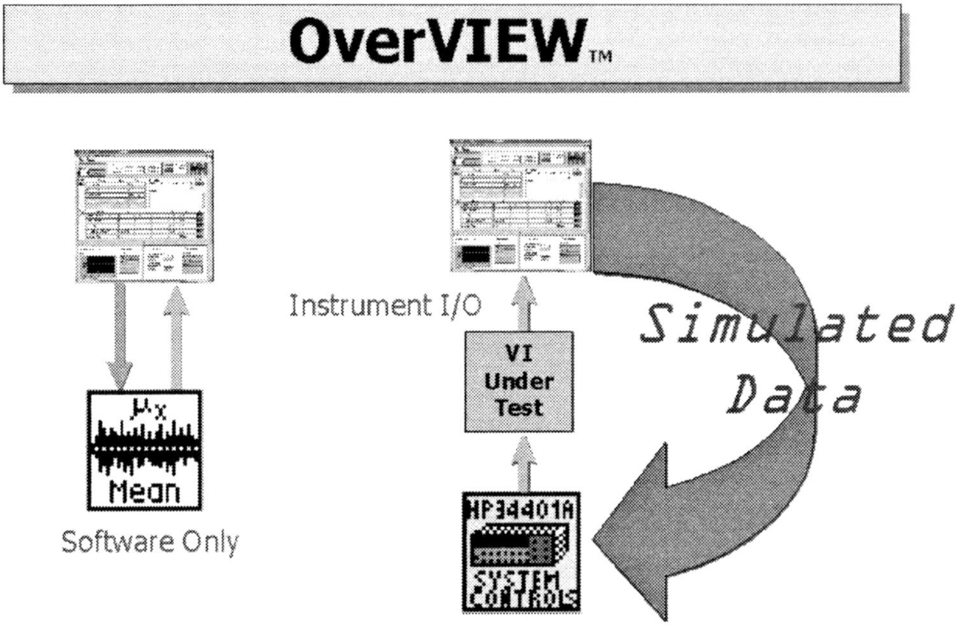

Figure 8-20.
Simulation data flow.

with the actual processor and environment, they should be used. Often, the optimum method is a combination of real and simulated environments.

Simulation is a proven technique to test the boundaries of a system. Typically, test and automation engineers test their systems by capturing product data with measurement instrumentation. It is challenging at best to gather enough products at the limits of specification to truly exercise the full range of the software. For low-volume and expensive products, engineers are lucky to obtain a few samples to test their system—and these are usually in the middle of the specification range. The resulting problem is that the software project is not tested to its full range of capabilities until it is too late—after it is released. The solution is to simulate the product data (see Figure 8-20).

We have discussed the structure of the UTool in that it provides a persistent means to pass known inputs and compare with expected outputs. This can also be mapped to enhance the simulation process. A LabVIEW project focused on testing product is going to have a combination of different code types. Some are going to be analysis or utility type VIs that we classify as software only. Others are going to have some relationship to the test equipment and therefore we classify them as hardware dependent. It is this class of hardware-dependent VIs on which the focus of unit testing and simulation is applied.

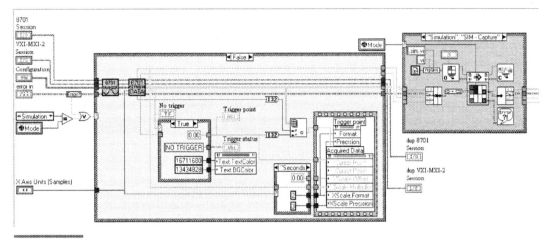

Figure 8-21.
Modified instrument driver with simulation in place.

The approach we focus upon is to apply simulation to the instrument drivers that are being called from VIs under test. In other words, it is the higher level test functions where the most critical unit testing can be applied. Assuming these VIs are dependent upon instrument drivers, then the objective is to properly simulate the instrument driver functionality and control the data that flows through the instrument drivers. In order to be effective, the simulation needs to leverage existing instrument drivers without requiring a significant modification. The process shown here is designed to quickly modify almost any instrument driver and allow it to work normally or in simulation mode (see Figure 8-21).

Full Project Testing and Regression Testing

With the simulation utilities folded into the Utool, there are new avenues for more comprehensive testing. Since OverVIEW provides a project-level perspective, it can automatically execute all the appropriate unit tests for some or all of the VIs within the project. This means that testing can occur almost daily if appropriate rather than waiting until the development phase is complete.

In addition to using this scheme to test all the VIs in the project, it can also be utilized for regression testing. Since the test cases persist within each unit test shell, a quick regression analysis would highlight which tests need to be rerun whenever there is a change to a lower level VI. This provides high confidence quickly as to the impact of a modification to a couple of VIs within the project.

The benefits to simulation and how it is applied with unit testing are numerous. First, it makes the testing process more efficient since it can occur early and repeatedly. Second, it speeds up development because testing is no longer dependent upon test system or hardware availability. This eliminates useless downtime waiting to execute on a tester. Finally, it is a convenient means to properly stress the entire software application by forcing known data sets that would not be easily available from actual products.

Documenting the Project

We have briefly discussed various components of documentation and their importance to the project. Typically, there are documents covering the following items in the test development process:

- Test Requirements
- Design or Theory of Operation
- Peer Review Reports
- Source Code Documentation
- Validation Plan and Results
- Project Release Report

All of these reports can be created using an accepted word processor or report-generation tool. The key is finding ways to automate this important, yet time-consuming step in the process. We will highlight several facets of how OverVIEW can automate portions of the documentation and reporting process in the following sections.

Theory of Operation

An attractive feature of LabVIEW in general is that the visual syntax of the source code can be easily interpreted if structured properly. This translates to a convenient method of describing the operation of an application. This further substantiates the need for a solid software architecture, a modular approach to programming, and a LabVIEW coding standard. When these components are in place, the VIs in the project can dictate the report describing the theory of operation. Simply printing the block diagram, the VI description, and the input/output descriptions goes a long way toward completing a very lengthy report (see Figure 8-22). The higher level portions of the report can then be filled in manually.

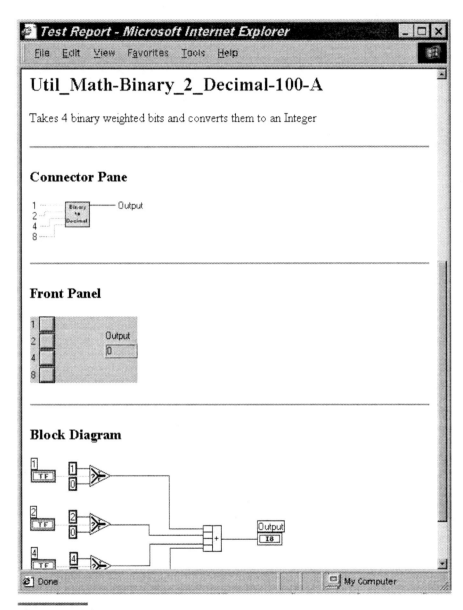

Figure 8-22.
Example block diagram with description.

Generating Validation Reports

It is essential to generate a validation test plan before the actual validation process begins. This plan outlines the specific description of what is being tested and how the test commences. Appropriate fields within the UTool are used to properly document the description and procedure for each test case on a given VI under test. The reporting module then allows a comprehensive plan to be generated for each VI under test. This outlines the reason and process of how to test, as well as the inputs and evaluation criteria, for each test case on a given VI. Figure 8-23 illustrates a validation plan for testing a battery.

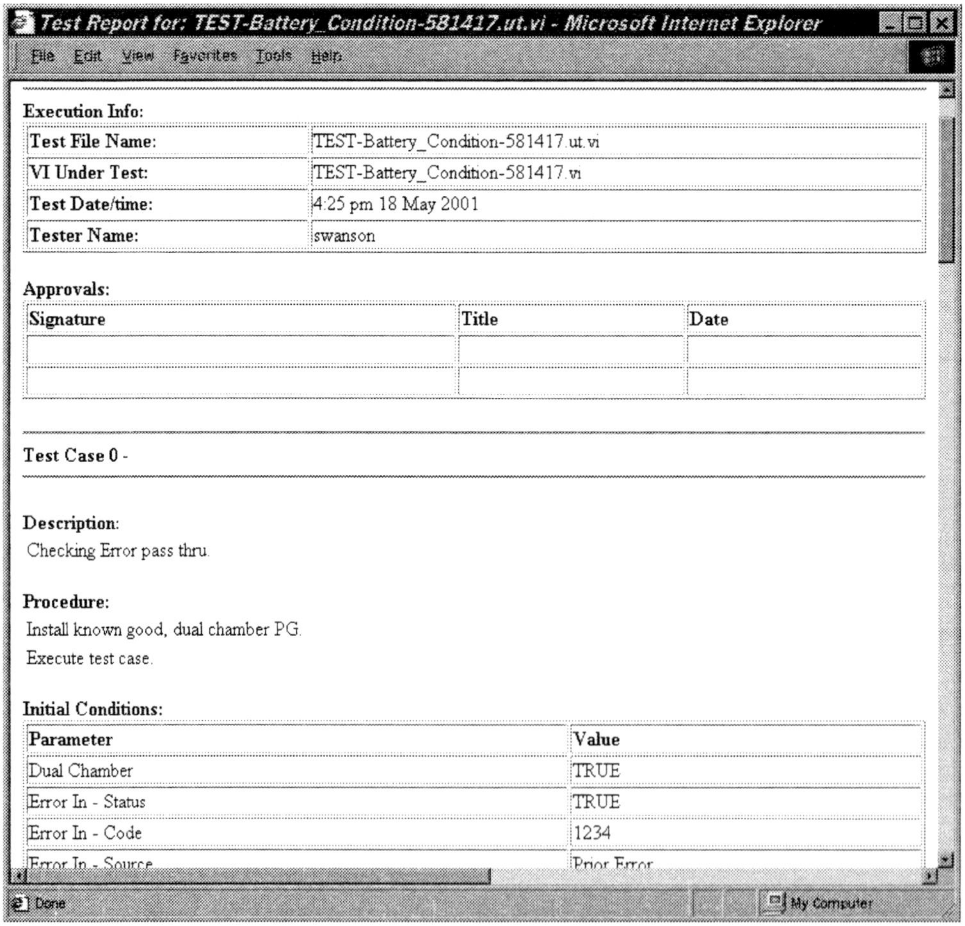
Test Report for: TEST-Battery_Condition-581417.ut.vi - Microsoft Internet Explorer

File Edit View Favorites Tools Help

Execution Info:

Test File Name:	TEST-Battery_Condition-581417.ut.vi
VI Under Test:	TEST-Battery_Condition-581417.vi
Test Date/time:	4:25 pm 18 May 2001
Tester Name:	swanson

Approvals:

Signature	Title	Date

Test Case 0 -

Description:
Checking Error pass thru.

Procedure:
Install known good, dual chamber PG.
Execute test case.

Initial Conditions:

Parameter	Value
Dual Chamber	TRUE
Error In - Status	TRUE
Error In - Code	1234
Error In - Source	Prior Error

Done My Computer

Figure 8-23.
Example validation plan.

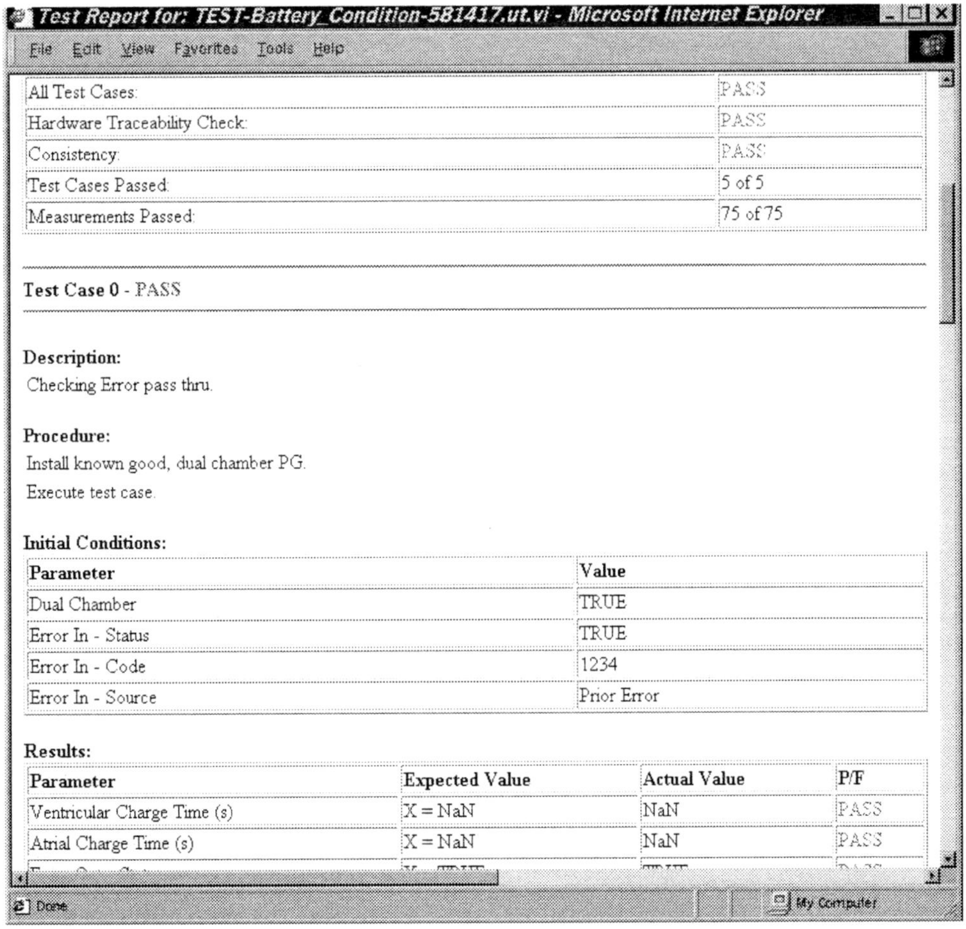

All Test Cases:	PASS
Hardware Traceability Check:	PASS
Consistency:	PASS
Test Cases Passed:	5 of 5
Measurements Passed:	75 of 75

Test Case 0 - PASS

Description:
Checking Error pass thru.

Procedure:
Install known good, dual chamber PG.
Execute test case.

Initial Conditions:

Parameter	Value
Dual Chamber	TRUE
Error In - Status	TRUE
Error In - Code	1234
Error In - Source	Prior Error

Results:

Parameter	Expected Value	Actual Value	P/F
Ventricular Charge Time (s)	X = NaN	NaN	PASS
Atrial Charge Time (s)	X = NaN	NaN	PASS

Figure 8-24.
Example validation report with results.

Once testing is complete, the reporting module can be executed again with a different focus, which includes the results. This outlines the same information as in the validation plan but also includes the specific results and the pass/fail criteria for each submeasurement for each test case (see Figure 8-24). This report becomes the final piece for a traceable and comprehensive validation report.

Project Release Report

Since so many of the important traceable components are located within the LabVIEW source code or the descriptions, it is vital from the project perspective to have an accurate understanding of the status of all members in the project. Just as the unit test and validation report is the proof that each VI meets requirements, so the project release report is the proof or validity that the project is at a stage ready for release. OverVIEW provides a convenient method for creating a project release report by gathering and reporting status on all project members. In addition to the project status report, a set of individual reports for each VI within the project can also be created showing the status and source code description. Since the report is created as an HTML document, it is easy to provide hyperlinks between the top-level project status report and the individual reports for each VI in the project (see Figure 8-25).

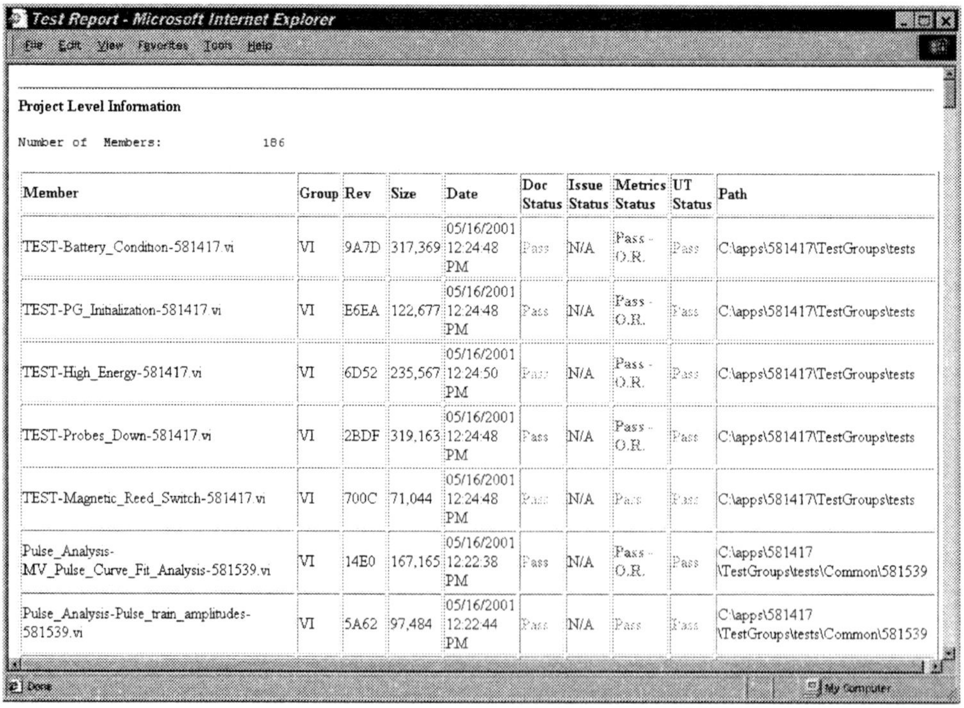
Test Report - Microsoft Internet Explorer

Project Level Information

Number of Members: 186

Member	Group	Rev	Size	Date	Doc Status	Issue Status	Metrics Status	UT Status	Path
TEST-Battery_Condition-581417.vi	VI	9A7D	317,369	05/16/2001 12:24:48 PM	Pass	N/A	Pass - O.R.	Pass	C:\apps\581417\TestGroups\tests
TEST-PG_Initialization-581417.vi	VI	E6EA	122,677	05/16/2001 12:24:48 PM	Pass	N/A	Pass - O.R.	Pass	C:\apps\581417\TestGroups\tests
TEST-High_Energy-581417.vi	VI	6D52	235,567	05/16/2001 12:24:50 PM	Pass	N/A	Pass - O.R.	Pass	C:\apps\581417\TestGroups\tests
TEST-Probes_Down-581417.vi	VI	2BDF	319,163	05/16/2001 12:24:48 PM	Pass	N/A	Pass - O.R.	Pass	C:\apps\581417\TestGroups\tests
TEST-Magnetic_Reed_Switch-581417.vi	VI	700C	71,044	05/16/2001 12:24:48 PM	Pass	N/A	Pass	Pass	C:\apps\581417\TestGroups\tests
Pulse_Analysis-MV_Pulse_Curve_Fit_Analysis-581539.vi	VI	14E0	167,165	05/16/2001 12:22:38 PM	Pass	N/A	Pass - O.R.	Pass	C:\apps\581417\TestGroups\tests\Common\581539
Pulse_Analysis-Pulse_train_amplitudes-581539.vi	VI	5A62	97,484	05/16/2001 12:22:44 PM	Pass	N/A	Pass	Pass	C:\apps\581417\TestGroups\tests\Common\581539

Figure 8-25.
Project status report.

Verification after Release— Controlling and Maintaining the Software

Up to this point we have discussed the issues during development and testing. It is essential to consider the implications of validation and verification even after the software is released. Since the primary consideration of this chapter is to focus on LabVIEW and validation, the most frequent usage of this is for production testing.

We have already highlighted some of the common aspects of a production testing environment in the biomedical arena. With test stations being utilized for multiple products, there are many released test applications installed at once on any given test station. This introduces some considerations on how to maintain control and traceability of the test software during implementation.

One method is to create standalone executables using the LabVIEW Application Builder for each and every test application. This has the benefit that all the software being utilized within a given test application is bundled and compiled together. However, there are some serious drawbacks, such that it requires additional compilation steps, which potentially can complicate the integration testing phase of the validation process. More importantly, it prevents some valuable strategies from being employed. For example, one of LabVIEW's greatest strengths is the ability to open any VI and execute it. This ability is valuable for debugging and troubleshooting. Also, a modular and reusable architecture involves code that is shared between applications on the same test platform. This shrinks development time and minimizes duplication.

If the LabVIEW modules are installed without compiling as an executable, then we need other tools to control the software when being executed. We have discussed how OverVIEW captures and maintains project information. This project information can be utilized as a control mechanism when executed with LaunchPad™, which is the run-time code verification engine for OverVIEW.

LaunchPad is intended to be utilized with a test shell such as a test executive or other vehicle that loads test applications for execution. It serves as a filter to warn and prevent testing if the following issues arise:

- A VI is in memory with an incorrect link.
- A VI or other file has been modified since release.

One of the issues LabVIEW users have faced is the fact that a calling VI will attach to a sub-VI by name only, even if it comes from a different location than expected from the calling VI. This can be a huge problem for production testers where there are multiple test applications installed. There are numerous reasons as to why a VI from a previously executed test could be left dangling in memory. If this occurs and another test application starts executing that uses the same names for sub-VIs, there is a risk of improper execution. Another common issue is that it is possible for VIs and other files to be modified once installed on the tester. This may occur during a debug operation and could be intentional or not. LaunchPad provides a Cyclical Redundancy Check (CRC) to assure that every file in the project is in the same state as when it was released.

Summary

What we have hopefully conveyed is that there are many dimensions to a successful validation process. In general, the key is to complete the project as quickly as possible without sacrificing the quality and necessary components of validation. Therefore, it is essential to incorporate architectural designs for the department as well as the individual software. It is vital to have a well-defined process that serves as the backbone for development. Finally, the complexities involved in managing a software project under regulatory control require automation tools.

It has been shown that LabVIEW is a very good software development tool as it dramatically improves development time, it is complementary to peer reviews, and it encourages a modular design. A complementary tool specifically designed to assist with the validation of LabVIEW software is OverVIEW. The project organization, automated inspection, metrics characterization, automated testing, and validation report generation all help to cut the overall development time of a project under regulatory control.

A significant investment needs to be made in order to facilitate these various components. This is no different than the investment that goes into a new facility or production line. The extra time and cost spent putting the proper design in place will reap tremendous benefits in terms of quality, cost of development, and time to market once this system is operational.

Independent Solution Articles

Operating a Commercial Medical Device with Certified, LabVIEW™- Based Software

by Jon B. Olansen, Ph.D., KaDa Research, Inc.

The Challenge: To develop a commercial user-friendly software application that would operate the newly developed KaDance 2000™ hand injury analysis tool while meeting all requirements necessary for FDA release to market.

The Solution: LabVIEW™ was used to develop the appropriate DAQ and control algorithms and user interface, while OverVIEW™ provided the necessary analysis, testing, and reporting capabilities. This combination effectively streamlined the software development process and enabled the validation and verification processes to be completed efficiently.

Introduction

When our hands are injured, we all hold objects in such a way as to minimize our discomfort. We do not squeeze any harder than necessary to maintain our grip on a drinking glass or doorknob as we try to continue our daily lives. Most of us are aware that we are "gripping" differently than we usually do. We may also notice that we are not as strong as we were, or that our grip tires quickly (we fatigue), or that our coordination has decreased, or that we drop things easily. We tend not to recognize more subtle differences in the behavior of our hands, which are only noticeable when viewed in the microsecond time frames available to advanced sensors.

By observing how an individual completes several repetitive tasks placed before them, we may observe many parameters, which may prove useful in understanding the nature of any injuries present in the hands under study. By studying these parameters, physicians will be able to make objective evaluations concerning the nature of any injury and document the efficacy of any treatment they have decided upon.

Purpose

The KaDance™ 2000 system is designed to aid in the identification of fine motor performance issues, which may be present in either or both hands of the subject. This is accomplished by measuring how the thumb (digit 1), index finger (digit 2), and little finger (digit 5) maintain a grip throughout several repetitive exercises. These measurements are basically the applied forces exerted by the digits measured as a function of time. Measurements may be made as rapidly as once every 2 *ms* for all three digits simultaneously so that a good understanding of the health of the hand's neuromuscular systems may be studied.

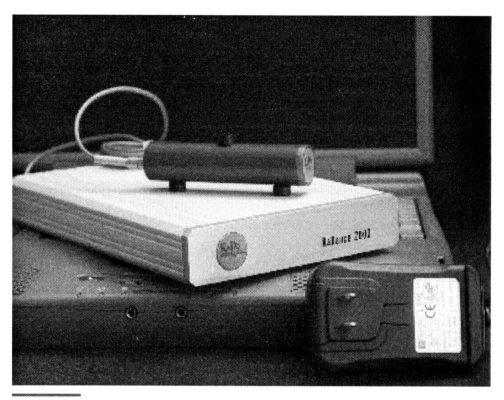

Figure 8-26.
The KaDance 2000™ initial development unit prepared for certification by the U.S. Food and Drug Administration.

Application

The software application developed for the KaDance 2000™ required a simple, user-friendly interface that would enable test conductors to perform the following functions:

- Enter appropriate user (test conductor) and subject information
- Control testing by producing a sequence of tasks for the subject to perform
- Acquire and validate data during the course of the test and save it for future processing
- Troubleshoot the hardware configuration

The varied requirements associated with the software, combined with the desire for a rapid prototype development methodology, made LabVIEW™ a natural code development platform. The flexibility and modularity of LabVIEW code enabled the developers to continually test and modify the hardware or the analysis algorithms, without significant schedule or cost impacts related to software development.

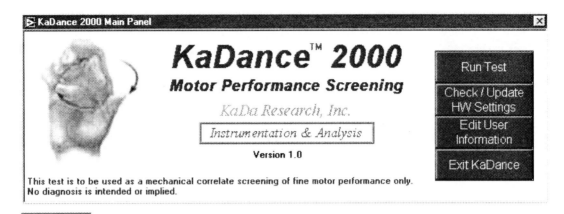

Figure 8-27.
Main Panel from which all KaDance testing, data entry, and troubleshooting is originated.

Certification

Developing the KaDance software application as a component of a commercial medical device required VBI Development Company to adequately establish development and qualification processes. VBI turned to the OverVIEW™ library developed by TimeSlice, Inc. (www.tslice.com). OverVIEW provided an excellent means of tracking development progress and various software metrics (such as VI memory or block diagram nodes). The Unit Test components significantly reduced the time and effort necessary to conduct a rigorous verification and validation program. Finally, the reporting capabilities included within OverVIEW enabled the entire software development to be properly documented in an easily accessible HTML format. All of these features combine to provide the traceability required by the U.S. Food and Drug Administration.

Conclusion

Formulation of commercial software designed for use in a regulated environment, such as within a medical device, can be a daunting task. The development process can be significantly extended by stringent test, verification, and documentation requirements. Additionally, there is the desire for a rapid prototyping capability in order to reduce the time to market. For the KaDance 2000 software, the combination of LabVIEW and OverVIEW proved to be a valuable integrated package, enhancing prototype development and minimizing the impact associated with verification and validation.

For more information, contact Jon B. Olansen, KaDa Research, Inc. 10701 Corporate Dr., Suite 100, Stafford, TX 77477; e-mail jolansen@kadaresearch.com

Part V
Healthcare Information Management Systems

Medical Informatics

9

Managing Disparate Information
ActiveX
ActiveX Data Objects (ADO)
Dynamic Link Libraries
Database Connectivity
Integrated Dashboards

Physicians face the challenge of coping with the increasing demands of managed care, government regulation, better informed healthcare consumers, and the accelerating pace of medical research. Over the past decade, many providers have adopted sophisticated information technology solutions for automating the administrative and financial aspects of group practices, yet most physicians still rely on manual processes and paper charts to manage clinical information and workflow. This practice is clearly inadequate for the demands of today's complex healthcare environment.

Fortunately, technology is one of the fastest changing components within the healthcare information environment. New hardware, advanced communications technologies, and new methodologies are enabling process changes that will dramatically impact the way healthcare is delivered in our hospitals and communities. Today, healthcare organizations can harness the power of information to provide comprehensive, actionable information at the point of decision making. The result of leveraging information technology will be to improve mission-critical clinical, financial, and customer-satisfaction outcomes.

Numerous studies have shown that physicians control over 80% of a healthcare enterprise's controllable costs and resources. Therefore, it is intuitive that providing physicians with the right information, at the right time, in the right way will help them provide the highest quality care possible. It is also worth noting that studies have shown that the highest quality care is often the least expensive care. To this end, in preparing this section of the book, we have drawn upon the insights and experiences of a wide variety of healthcare professionals, including physicians, nurses, computer and information scientists, biomedical engineers, medical librarians, academic researchers, educators, and administrators. Collectively, we will provide examples and identify opportunities of how virtual instrumentation can play an important role in enterprise-wide, clinical management, financial management, strategic decision support, medical simulation, resource planning management, and enterprise application integration solutions for healthcare organizations.

Defining Medical Informatics

Many definitions of the term *medical informatics* have been put forth. It has been defined as "the field concerned with the cognitive, information processing and communication tasks of medical practice, education and research, including the information science and technology to support these tasks" (Greenes & Shortliffe, 1990). Put more simply, medical informatics can be defined as the process of transforming data into information in healthcare. There has been an exponential growth in the use of computers in medicine in recent years, and this has led to rapid expansion of the medical computing industry. It is essential to recognize that the focus is on information management and not computer technology. Informatics ranges from a literature search on a specific medical problem to looking up a laboratory result in a clinical data repository to data warehousing in order to help decide which of two treatments for a specific disease may lead to a better outcome for a patient. It encompasses basic and clinical sciences as well as applied fields in biomedical science (see Table 9-1).

Medical informatics thus implies a much broader range of applications than just those used by healthcare providers. There can be patient-specific information such as a clinical data repository that stores lab results, but there is also knowledge-based information such as reference material found in the research literature. Recent increased consumer demand for access and control of medical information must also be considered. This chapter will focus on computing in clinical situations and will try to illustrate with examples how computers and virtual instrumentation solutions can be used in healthcare.

Table 9-1. *Examples of Applied Medical Informatics or Healthcare Computer Applications.*

- Electronic Medical Record (EMR)/Computerized Patient Record (CPR)
- Decision support
- Information retrieval
- Imaging/Telemedicine
- Medical education
- Consumer health information systems
- Public health information systems
- Telemedicine
- Bioinformatics
- Outcome analysis
- Patient monitoring

Computers in Medicine

The explosion of information available in healthcare has made the practice of medicine integrally tied to the management of information. The practice of tearing apart journals and filing each article according to some classification system in order to manage reference material has become obsolete with the availability of on-line searches linked directly to full text articles. Yet healthcare has lagged behind other industries in the adoption of computer technology as part of routine daily care, in part because of the resistance of clinical staff, especially physicians. This, in turn, is partly because of poorly designed systems and applications that do not easily fit current patterns of workflow without forcing major changes to occur first. Information used in healthcare comes from many sources and one of the most difficult challenges is to deliver this information in a timely fashion, in a format that can easily be used.

However, forces directly and indirectly related to the practice of medicine are forcing change to occur. One example of this is the recent report on medical errors by the Institute of Medicine (IOM; Kohn, Corrigan, & Donaldson, 1999). In November 1999, the Institute of Medicine and the National Academy of Sciences released a 223-page report highlighting the prevalence and consequence of medical errors. The report put forth some disturbing figures. Avoidable medical mistakes kill anywhere from 44,000 to 98,000 people a year, the eighth leading cause of death in the United States, above that of highway accidents, breast cancer, or AIDS.

"These stunningly high rates of medical error, resulting in deaths, permanent disability and unnecessary suffering are simply unacceptable in a medical system that promises first to 'do no harm,' " says William Richardson, chair of the committee that wrote the report and president and CEO of the W. K. Kellogg Foundation, Battle Creek, Michigan. "Our recommendations are intended to encourage the healthcare system to take the actions necessary to improve safety. We must have a system that makes it easy to do things right and hard to do them wrong."

The report called for major reform of the U.S. healthcare delivery system in order to cut the incidence of medical mistakes by at least half within the next five years. The report offers a wide range of reforms and, most notably, recommends that healthcare organizations implement "knowledge-based information systems" to reduce adverse drug events (ADEs) and prevent confusion and delays due to illegible handwritten orders and other situations that negatively impact both the quality and cost of care.

Although some have challenged the numbers in the IOM report and have questioned its methodology, few contest the basic premise that medical errors in general—and medication errors in particular—can and should be reduced.

Another example is The Leapfrog Initiative, which was formed in 2000 by several Fortune 500 companies and other large healthcare purchasers in order to recognize and reward healthcare providers that make improvements in patient safety and quality of care (www.leapfroggroup.org). These businesses provide health insurance to approximately 25 million Americans. The group's goal is to use employer purchasing power to influence hospitals. The group will aggregate performance information on major providers of healthcare into comparative value ratings for their employees, as well as educate employees about the importance of comparing performance of providers.

The group plans to motivate healthcare providers through the use of incentives—patient volume, price, and public dissemination of information about performance. They have already identified initial safety standards, two of which are related to informatics: The first is computerized physician order entry as a means to impact adverse drug events. They would require physicians to enter medication orders via computer with decision support software in place to prevent prescribing errors. The hospitals would have to demonstrate that the system can intercept at least 50% of common serious prescribing errors and that physicians acknowledge reading the directions to any override.

The second standard is evidenced-based hospital referral where elective treatment would be guided to hospitals with superior outcomes. Although hospital volume for certain procedures is one criterion for selection, the group prefers risk adjusted hospital-specific outcomes as direct measures of performance.

Some of the rationale for these standards is that (1) there is scientific evidence that these standards would significantly reduce errors, (2) their implementation is feasible, and (3) purchasers and consumers can easily ascertain their presence or absence.

Another impetus for further implementation of informatics in healthcare is the Joint Commission on Accreditation of Healthcare Organizations (JCAHO; www.jcaho.org/standards_frm.html). The JCAHO evaluates and accredits nearly 19,000 healthcare organizations and programs in the United States. It is considered one of the main accrediting bodies for hospitals and other healthcare delivery organizations, and hospitals must go through a survey every three years in order to remain accredited. New patient safety standards from the JCAHO went into effect in July 2000 and several of these specifically

Table 9-2. *JCAHO Information Management Patient Safety Standards.*

- The hospital plans and designs information-management processes to meet internal and external information needs.
- The information management function provides for information confidentiality, security, and integrity.
- Uniform data definitions and data capture methods are used whenever possible.
- Decision makers and other appropriate staff members are educated and trained in the principles of information management.
- The transmission of data and information is timely and accurate.
- Adequate integration and interpretation capabilities are provided.
- The information management function provides for the definition, capture, analysis, transformation, transmission, and reporting of individual patient-specific data and information related to the process and outcome of the patient's care.
- The information management function provides for the definition, capture, analysis, transmission, and reporting of data and information that can be aggregated to support managerial decisions and operations, performance improvement activities, and patient care.
- The hospital provides systems, resources, and services to meet its needs for knowledge-based information in patient care, education, research, and management.
- The information management function provides for the definition, capture, analysis, transmission, and reporting or use of comparative performance data and information, the comparability of which is based on national and state guidelines for data set parity and connectivity.

addressed information management issues. These elements are provided in Table 9-2.

A survey agenda will typically include an evaluation of the hospital's information management processes as well as interviews with the chief information officer (CIO), director of the library, and director of medical records. Although the standards do not detail specific computer technology, the requirement for data and information management implies the need for fairly sophisticated systems with the application of many principles of medical informatics. The intent of these standards is clearly to improve patient care and safety and improve disease management by measuring outcomes.

The healthcare industry has been compared, albeit unfavorably, to the aviation industry because of similarities such as extensive training, complex skills, dependency on clear communication of information, and a high-stakes setting (Leape, 1994). If the rate of adverse events in the aviation industry approached the 1–5% rate reported in healthcare, not many people would use airplanes as a mode of transportation. There are strong incentives for making flying safe and airplane manufacturers and airlines have focused on safety in aircraft and systems designs. Systems are designed with the idea that errors and failures are inevitable and are built to absorb them with multiple buffers and redundancies. Procedures are standardized and specific protocols must be followed on each trip. Pilots are required to undergo training and evaluation of their response to problems including evaluation of their performance in simulators at regular intervals. There is evidence that human errors often occur as a result of a failure in communication rather than inadequate knowledge or skill level. Pilots and air traffic controllers use standard terminology when communicating with each other.

In comparison, error prevention has not, until recently, been a primary focus of medicine. When errors are discovered, the focus is often on identifying the individual on whom to place blame and not on the search for the root cause of the underlying system failure. The healthcare system approach is to rely on clinicians not to make errors rather than assume they will and design the system to absorb them. The presence of standardization varies widely from location to location—regions of the country, hospital to hospital, or even different areas within a hospital. Although education and training in medicine exceeds that in aviation, the idea of testing performance has never been well received. The fear of malpractice is enough to deter routine in-depth investigation of errors, and there is no FAA equivalent to examine "near misses." Another challenge is in the area of standard clinical terminology or a controlled medical vocabulary. The lack of such a standard leads to poor communication and makes development and implementation of other computer applications more difficult. The most widely accepted standard terminologies are International Classification of Diseases (ICD 9) and Current Procedural Terminology (CPT) coding, which exist for the purposes of billing and not patient care. The codes were developed by the Health Care Financing Administration (HCFA) to assist in the assignment of reimbursement amounts to providers by Medicare carriers and for other purposes such as epidemiological reporting. Coding is usually done on discharge by the medical records department, and the coding is often optimized for billing purposes for the hospital and not care of the patient. Many other classification systems exist, but none have been widely implemented.

Electronic Medical Record

The paper record began as a way for physicians to record observations they made on patients that they could use to remind them of important details the next time they saw the patient. It has since evolved into a multidisciplinary chart used by all caregivers that are involved with a patient during his or her hospitalization. The information is typically recorded in a poorly organized fashion. The chart does not allow for adequate communication among care providers. In order to find a specific note on the paper chart, one needs to have the right volume of the patient's record and hope it was filed in the appropriate section. The focus is on documentation to meet regulatory requirements or justify billing charges and not to meet healthcare needs of the patient. A common scenario is for the patient to arrive in clinic without any medical record available to provide the necessary information. The medical record is woefully inadequate for outcome analyses or clinical trials of treatment because of the poor documentation.

The medical record is where diagnostic, therapeutic, and administrative information dealing with patient care is documented. Its uses include (1) serving as a recollection of current and past medical care including the rationale on which the care is based, (2) providing a means by which healthcare providers communicate about patient care and as a basis on which medical decisions are made, (3) providing the basis for quality assurance programs, and (4) serving as a source of data for research studies.

The patient record should ideally include all the information created and acquired during a patient's lifetime of encounters with the healthcare system. This data includes narrative text such as the history and physical exam, numerical data such as vital signs (heart rate and blood pressure), and laboratory studies images such as X rays and endoscopy pictures, and signal data such as an electrocardiogram. Numerical measurements are easy to computerize while a narrative description of the patient is much more difficult. The importance of this data being accurate can be appreciated when one considers that 80–90% of the time a correct diagnosis can be made from the history and physical exam alone. A computerized patient record has the potential to enhance the quality and availability of these characteristics of the medical record. In a truly integrated delivery network, the system may even provide information about what authorization is needed from the patient's insurance carrier for the level of care being provided.

Although there has been a resurgence of primary care and family practice, much medical care today is still provided by specialists. Instead of a single general internist managing the heart, lung, and kidneys, a patient may have care provided by a cardiologist, pulmonologist, and nephrologist, as well as a number of nonphysician healthcare providers such as a visiting nurse, physical therapist, and social worker. With the increased number of providers involved in a patient's care, the effectiveness and efficiency of communication among them has decreased. There is great potential for an electronic medical record to improve communication. Consider a patient who is seen in the emergency room in the middle of the night who needs a follow-up visit a week later. If the primary care physician could be notified by e-mail at the time the patient is seen and given details of the visit, not only would a follow-up visit be more likely to occur but also the effectiveness of that encounter would be greatly enhanced. Patients have also increasingly demanded e-mail access to their physician in order to communicate about nonurgent issues. Some physicians even allow patients to schedule appointments over the Internet. Furthermore, connectivity to the patient's home would allow monitoring of important data such as blood pressure or glucose levels without the need for office visits.

Computerized Physician Order Entry

Many computerized patient record systems include a module for computerized physician order entry (CPOE). Studies have shown the ability of CPOE to reduce medication errors and related ADEs (Bates et al., 1999). CPOE is undoubtedly the process in medicine with the greatest influence over the quality and cost of care. Inadequate access to clinical information is one of the principal barriers that physicians encounter when trying to increase efficiency and maintain quality of care. There is a need for financial and clinical outcome data to decide on cost-effective treatment. Physicians are often told things such as "You're spending too much money in the ICU." However, with current information systems, the question "What am I spending it on?" can never be answered exactly. It is often not the expensive procedure performed infrequently but the routine tests done hundreds of times a week that are responsible for driving up cost.

Much of the interest in patient safety is focused on medication errors, which are reported to be the largest cause of adverse hospital events (Bates et al., 1998). The current workflow is that a physician writes an order in the paper chart, which must be transcribed. This step is responsible for many of the errors because of issues such as poor legibility resulting in the wrong drug or dose being administered. By reducing ADEs, computerized order entry can reduce costs and length of stay. CPOE may also cut cost by reducing unnecessary variation in care by providing best practices and up-to-date information (Evans et al., 1998). The physician might also be presented with a formulary specific to the patient's insurance carrier when prescribing medications.

Another advantage of the electronic medical record is that it allows different clinicians to view data in different ways, transforming it into information. Different clinicians can more easily view data to fit their needs. One example of this would be a patient summary screen that many systems provide. This may include things such as a problem list, medication list, most recent laboratory results, caregiver list, and allergies.

Monitoring trends in an electronic medical record is much simpler than in the paper record. Electronic documentation of problems such as high blood pressure or high blood sugar assists the physician in anticipating future health problems that may arise from these conditions. Tracking standard preventive measures such as required school immunization or tetanus immunization is also easier.

Another trend is for the medical record to shift from a record of care for an acute episode of illness to more of a repository for an individual's lifetime healthcare. This can more easily be observed in health maintenance organizations (HMOs) where there is one record that many providers access instead of each provider having its own medical record for the patient. This is only truly successful when the medical record is in an electronic form and not a paper format, which often doesn't arrive until days after the patient's visit. This also allows the advantage of easier access to information and analysis of data.

Decision Support

Decision support can be defined as any computer program designed to help healthcare providers make clinical decisions. The application may assist in decisions related to diagnosis or therapy. In either situation the computer must have the correct medical knowledge and specific patient data in order to give case-specific advice. The level of decision support may vary from simply flagging abnormal laboratory values to making recommendations on the most effective therapy for a specific disease. The physician is usually allowed to overrule the advice by acknowledging that she or he has read it. Some simple examples of how decision support can help provide cost-effective care

are alerts for duplicate orders (two of the same test ordered) and suggestions of less expensive medications that are equally effective (an alternative antibiotic that has equal coverage is available).

Another excellent example is knowledge-based orders, which is the combination of a CPOE application with decision support functionally. This type of system provides synchronous decision support as the clinician is entering orders. It can suggest the best drug dosing based on age, height, and weight and can use available laboratory data to suggest correct dosing based on liver or renal function. The system may even go so far as to check an antibiotic ordered against the microbiology lab results to ensure the correct antibiotic has been ordered. This type of system obviously has great potential for helping to provide high-quality cost-effective care.

Information Retrieval

A large amount of research has been conducted on information retrieval. In its simplest form, this process involves researching pertinent literature to help determine appropriate diagnostic or therapeutic options. Index Medicus was an index of journal articles indexed by author and subject and printed in bound volumes that included title and reference information. The National Library of Medicine developed an electronic version but availability was usually limited to a medical library because of limited computing power and storage. More recently, with widespread access via the Internet, searching for information can be done from anywhere, at anytime, and can now immediately provide the full text of the article. Applications have been developed that allow for natural language queries instead of previously used specialized terms. This makes it possible to obtain reference material to help take care of a patient from a workstation on the nursing unit within minutes of evaluating a patient. In addition, many textbooks now have electronic versions, which can also be examined with an electronic search engine. By searching trusted medical sites, up-to-the-minute treatment protocols or research protocols along with their entry criteria can also be discovered in a timely fashion. With the advent of the World Wide Web, vast quantities of medical information are available from a variety of sources. One of the biggest challenges is that although the medical information is more accessible, there is no peer review process to determine which is the high-quality trusted information that should be used.

Medical Imaging

Imaging in medicine now goes far beyond the images obtained in the radiology department, although X rays are still the primary image we think of when talking about medical imaging. Plain two-dimensional radiographs are created when X rays are generated by an X-ray tube that is directed at an area of interest on the patient. A fluorescent screen behind the patient produces light, which is then captured by film to produce an image. The primary uses of these types of images have been for diagnostic purposes or documenting progression of disease or response to treatment. These films are typically stored in a file room and limited to use by a single viewer at a time. However, there is often demand for the films to be available for review by multiple clinicians simultaneously as well as availability in the operating room and at teaching conferences.

Newer systems allow radiological images to be captured and stored as digital images, while still allowing the images to be printed on film if necessary. A major advantage of digital X-ray images is that they can be easily manipulated and the use of computers has led to the developments of complex applications that would otherwise be unavailable. Computerized Axial Tomography (CAT) scans, Magnetic Resonance Imaging (MRI), and Digital Subtraction Angiography (DSA) are but a few examples of these types of applications.

Digital images can be manipulated to bring out details that might otherwise have been missed or would have required further imaging to be seen. Global processing can take a series of two-dimensional images from a CAT scan or MRI and create a three-dimensional reconstruction as shown in Figure 9-1. This is extremely useful for planning surgery and radiation treatment. A surgeon can select the best surgical approach to avoid injury to critical structures. Radiation therapy can be planned so as to avoid irradiation of sensitive surrounding tissues. The goal would be to deliver the maximal radiation dose to a tumor while minimizing exposure to the surrounding tissues.

Digital images can also be transmitted more rapidly and easily than conventional images. In telemedicine expert physicians in tertiary care centers can view a digital image and advise local physicians on the best plan of care without having to move the patient many miles away. For example, cardiac catheterization images could be transmitted to a regional referral center for evaluation of complex cases.

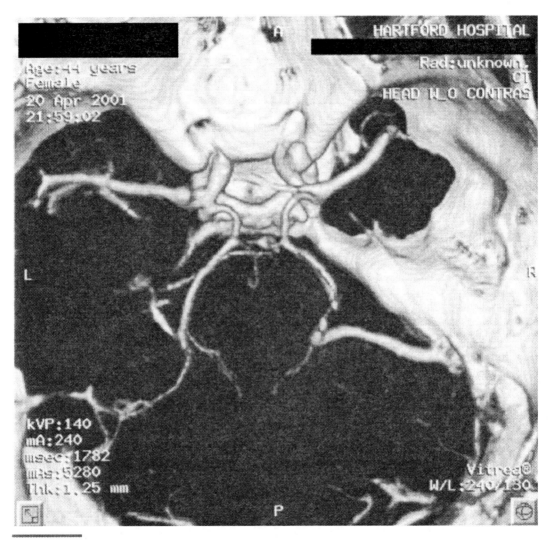

Figure 9-1.
Cerebral arteries at the base of the brain. Three-dimensional reconstruction from a series of two-dimensional CAT scans. (Courtesy of Hartford Hospital.)

One of the major obstacles is still the high cost of replacing film systems with digital systems. In order to get the full benefit of a digital system, viewing equipment must be easily accessible throughout the hospital and not just in the radiology department. This need for widespread availability has made the cost of acquiring digital systems prohibitive for many facilities.

Another challenge is the large amount of storage required for archiving digital image data. A plain chest X ray may require 10–15 megabytes of storage. A large tertiary care referral center with complex imaging such as the three-dimensional images mentioned previously could easily need hundreds to thousands of terabytes of storage. Although a major advantage of digital imaging is rapid viewing throughout the hospital, it is also a drawback since it requires the availability of a high-speed network. Remote viewing is not practical over dialup connections of even 56 kbps because of the large size of the files. However, increasing availability of cable modems and DSL is overcoming this obstacle.

Other types of medical imaging are also being captured digitally and, therefore, are also available for inclusion in the medical record. For example, gastroenterologists performing endoscopic evaluation of the upper or lower gastrointestinal tract can now obtain digital images and store them such that they can be part of a clinical data repository. This can be helpful in conveying information to a surgeon who may need to resect a tumor or to document progression of a disease. Widespread availability of digital cameras and video has allowed for images to be captured in just about any situation where such information could be useful.

Patient Monitoring

The use of computers with increased power and memory has revolutionized the acquisition, display, and processing of physiologic data. Care of critically ill patients requires minute-to-minute monitoring of physiologic functions such as neurologic, cardiac, pulmonary, and renal function. Areas such as the intensive care unit and operating room benefit from the continuous measurement of patient data to help guide management. Bedside point-of-care testing has decreased turnaround time for receiving results of critical laboratory testing. Figure 9-2 demonstrates a typical intensive care unit setup.

The monitor typically displays ECG and pressure waveforms but is quickly becoming the central point for data acquisition and processing. The monitor can now integrate data from other bedside devices such as a ventilator, intelligent intravenous pump, electronic urimeter, and other monitoring devices. Nurses spend a significant amount of time charting physiologic data even in the ICU. Because of its ability to integrate various sources of data, the monitor can now perform tasks such as computer-based charting, which will

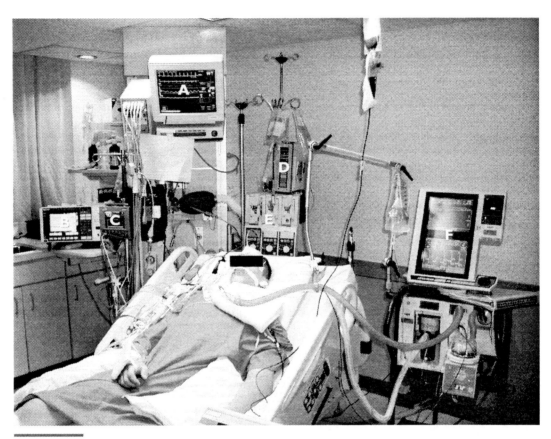

Figure 9-2.
Typical intensive care unit setup. (A) Patient monitor that displays ECG, invasive or noninvasive arterial blood pressure, pulmonary artery pressure, and intracranial pressure or other waveforms. (B) Cardiac output computer that displays cardiac output and arterial and mixed venous oxygen saturation. (C) Sedation monitor. (D) Intracranial pressure monitor. (E) Infusion pumps. (F) Ventilator.

free up the bedside nurse to devote more time to actual patient care. The monitor can provide automatic data entry into the ICU flowsheet along with the calculation of sophisticated derived variables that provide another level of information for guiding therapeutic decision making. A major challenge will be to determine the frequency and quantity of data to record. For example, Figure 9-3 illustrates how ECG arrhythmia monitoring systems use computer algorithms to analyze the ECG and alert caregivers of potentially life-threatening arrhythmias.

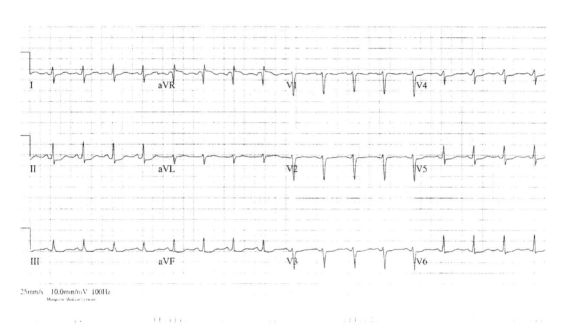

55 years	Vent. rate	102	BPM	SINUS TACHYCARDIA
	PR interval	142	ms	NONSPECIFIC ST AND T WAVE ABNORMALITY
	QRS duration	96	ms	ABNORMAL ECG
Loc:22	QT/QTc	363/473	ms	
	P–R–T axes	63 89 24		
	BP	140/44		

Unconfirmed

25mm/s   10.0mm/mV 100Hz

Figure 9-3.
ECG with computer-interpreted analysis (note "unconfirmed" comment indicating computer interpreted reading not yet confirmed by cardiologist).

Medical Education

Computer-assisted learning is widely available and applied at all levels of medical education from medical school through residency training and on to continuing medical education for practicing physicians. With advances in technology, the multimedia capabilities of current systems make possible the use of sound, video, and interactive instruction. In addition, most information is available outside the classroom or library, even at the patient's bedside.

Practicing physicians have limited time available to attend conferences to keep up to date in their specialty areas. Physicians can continue their medical education via the Internet by downloading on-line lectures using streaming video or slideshows.

The Visible Human Project is one example of how computers are changing medical education. This project, begun in 1989 by the National Library of Medicine, is the creation of a complete anatomically detailed three-dimensional representation of normal male and female human bodies. The aim of the project is to create a digital image dataset of human cadavers in CT, MRI, and cryosection images at 1-mm intervals. These images serve as a reference for studying human anatomy that is especially important given the common use of CAT and MRI scans for patient care.

Medical Simulation

Another area where computers are having a significant impact is in the area of medical simulation. For the past two decades, high-risk industries such as aviation, nuclear power, and the military have recognized that optimizing team resources and performance in the face of challenging environmental conditions requires an initiative to understand the individual and team cognitive psychology. The objective of war games is to modify crew training, promote leadership skills, improve one's assistant and organization skills, and be confronted by a rare but potentially lethal scenario in a safe environment.

Military leaders use war game scenarios to prepare troops and to experiment with eventualities. This recreation of real-life scenarios places trainees in problem situations that allow instructors and mentors to gauge their responses, provide valuable feedback on their performance, and suggest ways to improve their communication skills, reaction time, processing, decision making, and leadership skills. Simulated war settings allow trainees to gain valuable experience in developing their ability to fulfill their battlefield responsibilities.

In 1979, a thorough analysis of 60 airline incidents was reviewed to delineate the basis for the events. Data from cockpit voice and flight data recorders were analyzed and disclosed lethal decision-making errors by individual crewmembers or inadequate teamwork and cooperation among the crew. Further, using detailed simulator studies of the flight crew reconfirmed these findings. Human performance was recognized as the critical issue in aviation safety: Statistics suggest that flight crew actions were causal in more than 70%

of incidents. Technological improvements at the interface between the operator and review equipment did not reduce flight crew errors. Another lesson gleaned from these analyses was that able and well-trained individuals were by no means a guarantee against disaster. Instead, it became readily apparent the highly trained crewmembers had to work together as a team to provide optimal handling of a crisis. As a result of these findings, the aviation industry adapted a training philosophy of Crew Resource Management (CRM). In CRM training, crews are instructed to manage crises as well as manage their individual and collective resources to work together optimally as a team.

Simulation has been used to teach medicine for hundreds of years. As early as the sixteenth century, mannequins were developed to teach obstetrical skills in order to reduce high maternal and infant mortality rates (www.laerdal.com). Simulation has been also incorporated in the education of those involved in emergency and critical care management. For many participants, this training may be the only opportunity to refresh their skills. Many would argue that the recreation of a realistic clinical model with these courses is limited.

Early experience in anesthesia training led to increasing attention being focused on CRM training that was pioneered by Gaba and colleagues (1994) at Stanford University. During a CRM workshop, four to six healthcare providers participate in a variety of realistic event simulations followed by a debriefing session to videotaping of the workshop events and review of individual and team performance. It has become evident that simulator team training plays an important role outside the operating room and can be used to simulate trauma resuscitation, basic and advanced airway management, the administration of conscious sedation, cardiopulmonary resuscitation, malignant hyperthermia, and ICU crises. As a result of a collaborative effort between anesthesia, surgery, emergency medicine, and nursing and clinical support staff, crisis management courses have been developed.

As a training tool, a simulator has the ability to present rare medical, anesthesia, or surgical case scenarios repeatedly so the student can rehearse and practice diagnostic and therapeutic skills without risk to a real patient. Constructive feedback during the debriefing sessions is given via tutorials from the instructor and review of audio/video tapes. Simulation adds additional strength as a tool for practicing task prioritization, team coordination, and communication. Society expects medical personnel to undergo extensive training before they touch patients. Simulation allows novices and those with limited initial training to participate and learn without risk to patients. Acquiring knowledge and critical thinking and decision-making skills and dealing with stress are essential components of one's education and training. A simulation can free the participant from the fear of medico-legal consequences. The freedom to develop one's critical thinking of how to handle a crisis can be accomplished without the fear of compromised patient safety, decreased productivity, or loss of one's license and disciplinary action. Errors during simulation merely lead to peer disapproval, possible embarrassment, and a lower test score.

In simulation-based training experience, the participant has the opportunity to personally compare and contrast the clinical use of various medical devices and medications and treatment modalities and practice psychomotor skills in a dynamic clinical setting that allows them to troubleshoot clinical problems in real time. When compared to listening to a lecture, reviewing a videotape, reading a book chapter, or attending a practice session with a static mannequin, the opportunity to achieve effective training is potentially far superior using simulation technology.

Managing Disparate Information

Information used by physicians comes from many disparate systems—laboratory, transcription, radiology, pathology, and others. Figure 9-4 illustrates a window containing disparate clinical applications from a typical clinical workstation that can be found at many hospitals.

In order to put together a truly integrated electronic medical record, there must be interfaces written for each system. In addition to the clinical systems, administrative and financial systems integration is also needed. The nurse manager of a unit might need census data, severity of illness data, and the operating room schedule for the next day in order to justify changes in staffing levels. In order to access the various applications, a clinician would have to leave one in order to access another. A physician might be looking up results in the clinical information system (CIS) and come across a complete blood count suggestive of anemia. If he or she wanted to look up reference material to help diagnose a specific type of anemia, the physician would have to leave this application and start the medical library application to access reference material.

The ability to integrate disparate ancillary information systems and databases is not a trivial endeavor. Software developers and system integrators face many challenges, including strict performance requirements, multiple

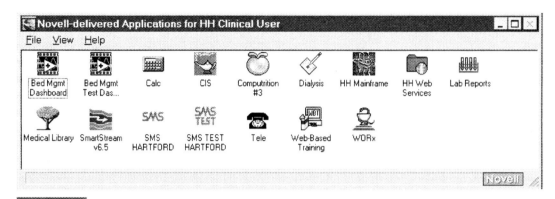

Figure 9-4.
Typical clinical workstation desktop with multiple applications. Bed Management Dashboard (application for assigning patients to specific beds); CIS (clinical data repository, which contains lab results, radiology results, and transcriptions of dictated procedures); Computrition (application used by dietary to monitor nutritional status); Dialysis (database of dialysis patients with pictures of how their dialysis access is constructed); HH Mainframe (used by nursing unit secretaries for ordering labs); HH Web Services (access to hospital intranet, which houses hospital policies and procedures, departmental home pages, drug information); Lab Reports (application that prints a list of a physician's patients and a summary of the last 24 hours of lab results); Medical Library (access to medical library intranet, which has access to Medline searches, full text on-line journals, and textbooks); SmartStream (application for resource management including human resources and materials management); SMS Hartford (Admission, Discharge, Transfer [ADT], and patient financials); Telephone listings; Web-Based Training (Intranet application used for training clinicians and other users on applications); WORx (pharmacy application). (Courtesy of Hartford Hospital.)

measurement and analysis features, shorter development cycles, and demands to integrate information throughout the enterprise. This means that applications must be able to integrate with e-mail, the Internet, ancillary and legacy information systems, and other products like Microsoft Word or Excel. Additionally, the pressure to quickly create applications that are more Windows-like and add features common to many business applications continues to grow. In general, the average time to complete a project has decreased from 12 to 9 months. Fortunately, over the past few years, many technologies have been developed that facilitate the ability to integrate data sets from legacy systems. These technologies include ActiveX®, Component Object Model (COM), ActiveX Data Objects (ADO), Dynamic Link Libraries (DLLs), and database tools such as Open Database Connectivity (ODBC). Each of these technologies will be briefly described.

ActiveX

ActiveX is becoming the de facto way of integrating software components in the Windows environment. Within the LabVIEW environment, ActiveX controls are ActiveX components providing a service to the application in which they are used. They can be embedded in ActiveX containers like Visual Basic or LabVIEW.

The ActiveX for Healthcare Committee (AHC; www.mshug.org/activex/) has enabled a variety of healthcare organizations to achieve interoperability between disparate healthcare information systems. AHC's work has resulted in the development of commercial software applications and the first commercial middleware application based on ActiveX messaging components that enable plug-and-play interoperability. ActiveX for Healthcare messaging components are based on Microsoft's component object model (COM) technology and provide a low-cost, fast, and easy way to integrate multiple applications.

ActiveX for Healthcare messaging components are a set of COM-based components that provide a standard interface to Health Level 7 (HL-7) messages. These objects can be sent to other applications using the Microsoft DCOM protocol or other standard transport mechanisms. A flexible architecture gives users a migration path that makes it easier to connect existing systems and applications from different vendors. For example, vendors can use a custom (non-DCOM) protocol such as TCP/IP sockets or serial connections to easily integrate legacy systems with new technology. This technology benefits healthcare technology users by providing a low-cost, easy-to-implement solution to connect applications from a variety of vendors—a major challenge in today's healthcare industry. More information about HL-7 messages is available at www.hl7.org/.

ActiveX Data Objects (ADO)

The ActiveX Data Objects (ADO) is Microsoft's strategic, high-level interface to all kinds of data. ADO provides consistent, high-performance access to data, whether you're creating a front-end database client or middle-tier business object using an application, tool, language, or even an Internet browser. ADO is a single data interface for one- to n-tier client/server and Web-based data-driven solution development.

Dynamic Link Libraries

One of the most common ways to share code today is through the use of a dynamic link library (DLL) or shared library. A DLL is a code library that is linked to programs when it is loaded or run rather than as the final phase of compilation. This means that the same block of code can be shared between several different tasks or applications rather than each one containing copies of the routines it uses.

LabVIEW capitalizes on this fact by allowing you to call external code in the form of DLLs on Windows platforms and shared libraries on Macintosh and UNIX platforms. Standard application development environments such as Microsoft Visual C++, Visual Basic, and National Instruments Measurement Studio can create DLLs and shared libraries for use in LabVIEW. Using the LabVIEW Call Library function, you integrate the external code directly into your Block Diagram. To simplify the process of integrating external code into LabVIEW, the function names and prototypes exported by the DLL are automatically displayed when you configure the Call Library function.

Database Connectivity

If you need to integrate databases into your measurement and automation applications, add-on tools are available. These database tools deliver high-level, easy-to-use functions for integrating local and remote databases quickly and easily into your LabVIEW programs. With these tools, you can automatically insert data into and select data from databases, as well as perform a host of other common database operations. For programmers desiring advanced database functionality and flexibility in their applications, the Database Tools also allow you to execute SQL statements from within LabVIEW applications.

Open Database Connectivity (ODBC) is a widely accepted application programming interface (API) for database access. It is based on the Call-Level Interface (CLI) specifications from X/Open and ISO/IEC for database APIs and uses Structured Query Language (SQL) as its database access language.

Integrated Dashboards

One of the most difficult challenges in healthcare is to deliver accurate information in a timely fashion and in a format that can be easily used. This has led to the concept of the physician dashboard. The ideal dashboard would be a browser-based portal providing secure access via the Internet. This would allow access from anywhere—the hospital, the physician's office, or home. All applications would be accessible from the same front end that would integrate data and present the information customized to fit the individual physician's needs. Patient information, scheduling, and e-mail are just a few examples of applications accessible directly on the dashboard. A physician could communicate test results with patients via e-mail and simultaneously direct them to trusted consumer health sites on the Internet for more detailed reference information. Patient eligibility and precertification requirements could be determined via direct connection to their insurance carrier. These types of integrated dashboards are only now being developed but hopefully will provide the knowledge and resources to further improve healthcare.

REFERENCES

ActiveX [On-line]. Available: http://www.mshug.org/activex/

Bates, D. W., Leape, L. L., Cullen, D. J., Laird, N., Petersen, L. A., Teich, J. M., Burdick, E., Hickey, M., Kleefield, S., Shea, B., Vander Vliet, M., & Seger, D. L. (1998). Effect of computerized physician order entry and a team intervention on prevention of serious medication errors. *JAMA, 280*(15), 1311–16.

Bates, D. W., Teich, J. M., Lee, J., Seger, D., Kuperman, G. J., Ma'Luf, N., Boyle, D., & Leape, L. (1999). The impact of computerized physician order entry on medication error prevention. *JAMIA, 6*(4), 313–21.

Evans, R. S., Pestotnik, S. L., Classen, D. C., Clemmer, T. P., Weaver, L. K., Orme, J. F. Jr., Lloyd, J. F., & Burke, J. P. (1998). A computer-assisted management program for antibiotics and other anti-infective agents. *NEJM, 338*(4), 232–38.

Gaba, D. M., Fish, K. J., & Howard, S. K. (1994). Crisis management in anesthesiology. New York: Churchill Livingston.

Greenes, R. A., & Shortliffe, E. H. (1990). Medical informatics. An emerging academic discipline and institutional priority. *JAMA, 263*(8), 1114–20.

Health Level 7 [On-line]. Available: http://www.hl7.org/

JCAHO Standards [On-line]. Available: http://www.jcaho.org/standards _frm.html

Kohn, L. T., Corrigan, J., & Donaldson, M. S. (1999). Institute of Medicine (U.S.) Committee on Quality of Health Care in America. *To err is human:*

Building a safer healthcare system. Institute of Medicine. Washington, DC: National Academy Press.

Laerdal Medical Corporation. Stavanger, Norway. Tel. 47–51 51 17 00. Available: www.laerdal.com.

Leape, L. L. (1994). Error in medicine. *JAMA, 272*(23), 1851–57.

Leapfrog Initiative [On-line]. Available: http://www.leapfroggroup.org/

Independent Solution Articles

EWICUM: An Early Warning Intensive Care Unit Monitoring System

by Jan Olav Høgetveit, Ph.D. student, Institute for Surgical Research, Department of Physics, University of Oslo, Norway

The Challenge: Physicians in the intensive care unit (ICU) acquire an enormous amount of data from each patient every day. Because time is limited, it is important to get a fast and exact overview of the patient condition in order to initiate the correct treatment as early as possible.

The Solution: A LabVIEW-based system that acquires data from several external units has been established. Each time a new result arrives, the system calculates several clinical physiological scores, visualizes them, and presents them graphically together with other selected early warning markers in a specially designed Early Warning window.

Introduction

The project consisted of developing a user-friendly computer application for collection, management, and presentation of data in an ICU. ICU patients are surrounded by advanced technical equipment producing thousands of measurements every day. A standard PC with a Windows 95 operating system with two serial ports and a Novell network connection were used together with an application fully written in LabVIEW 5.0. All results from surrounding equipment and relevant laboratories are available on-line and are visualized and presented graphically on a bedside computer.

Design Challenges

Evaluation of critically ill patients is an extremely difficult task that is definitively best performed by a skilled physician. An enormous amount of data is available for each patient, and the physician or nurse has very limited time to spend on each patient. We strongly believe there is an advantage to allowing a computer system to collect data, store it, visualize it, and show it in suitable presentations.

Computer knowledge among IC personnel varies from the computer expert to the unskilled first-time user. A major challenge was to design a system simple enough for the inexperienced and challenging enough for the expert. Still, the most important thing is to allow the first-time user to do all necessary work without too much training. LabVIEW is especially well suited for constructing simple and elegant graphical user interfaces, and it was an easy choice when evaluating different programming tools.

Figure 9-5 illustrates the fundamental setup for the complete EWICUM (Early Warning Intensive Care Unit Monitoring) system. A computer beside the patient is connected to a patient monitor, a ventilator, and the hospital network. The patient monitor (Siemens Sirecust 9600XI) and the ventilator (Siemens Servo Ventilator 300) are connected through a standard RS-232 serial interface.

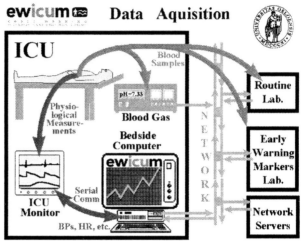

Figure 9-5.
Connections between EWICUM and external medical equipment.

Measurements of specially selected proteases from blood samples are important early warning markers in the EWICUM system. Normally, an ACL Futura spectrophotometer measures the samples, and this equipment is able to communicate directly with the main hospital network. Measurements can also be carried out in a Cobas Bio spectrophotometer. This is an older machine and needs an acquisition system between the machine and the hospital network, which we hooked up with the aid of a Macintosh Sill computer equipped with earlier LabVIEW version 2.0 (Sveen, Borthne, & Aasen, 1994).

The EWICUM System

The EWICUM system is an easy-to-use, plug-and-play ICU monitoring system. When started up, all functions and windows in the system are available through a row of buttons on top of the screen. These buttons can be accessed either by using an interactive touch screen or by clicking with a computer mouse. Windows principles make it easy for users to design a screen according to their wishes by selecting, moving, and resizing windows. Each patient in the unit is supported with a bedside computer, which is connected to a standard Novell network. As soon as a new test parameter is finished somewhere in the hospital, all results are collected in the system, distributed to the correct patient, and stored on a suitable network server. Every time a new result has arrived, new clinical scores are calculated and presented. A clinical scoring system is a way of giving a quantitative description of a patient's degree of illness by defining a normal interval for a set of parameters and giving an increasing score when a parameter falls outside of this interval. The scores are summarized for all parameters and adjusted for age and chronic health points. This quantitative sum gives a simple overview of the degree of illness for the patient. The higher the score, the sicker the patient. All scores are presented graphically and are visualized in order to allow critical care personnel to detect negative developments in the clinical course at the earliest possible stage.

The Early Warning Window

One of the available windows in the EWICUM system is specially designed for early warning of negative evolutions (Figure 9-6). A picture of a human is shown in the upper left corner, covered with easily recognizable organs. Each organ represents one of the six major organ systems: neurological, cardiovascular, pulmonary, renal, hepatic, and hematological. All organs are given a color code to indicate their specific organ function score. Organ dysfunction scores are scoring systems especially designed for measuring and evaluating organ failure and dysfunction. A gray-colored organ represents a normally functioning organ system, while blue through yellow to a blinking red organ indicates increasing dysfunction. The icons are placed on buttons. When pushed, all test measurements and parameters related to the selected organ system are shown. Parameters can easily be transferred to the graph by clicking on the wanted parameter with the mouse. This is performed by an invisible array placed on top of the table. The system knows the order of the parameters in the table, and it is a simple task to compensate for scrolling the table and finding which parameters to show in the graph. The graph contains icons for representing important events; icons indicating surgery and cardiac arrest are shown in Figure 9-6.

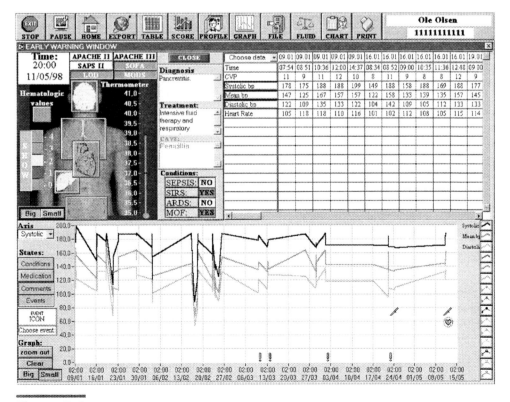

Figure 9-6.
The Early Warning window.

Events and comments are presented with specific icons where additional information can be accessed when an icon is pushed. Technically, the icons are pasted on buttons gathered in a cluster. Some position calculations make it possible to present correlating information in a pop-up window according to the user's wishes.

Conclusion

The EWICUM system combines advanced medical research with a presentation system based on LabVIEW. It has some very important advantages:

- It provides a fast and intuitive overview of large amounts of data.
- On-line data acquisition gives immediate response.
- Built-in early warning markers provide extra care.
- With a very low training threshold, the system is easy to use.

These advantages make the EWICUM system a very suitable and powerful application, well fit for the ICU. Unlike many other programming tools, all design was done from top to bottom. We started designing a user front panel, modified and tested it, and made all programming and lower structure design in the end. This is an invincible design principle for a demanding environment such as an ICU where work and time pressures are the rule, not the exception.

Reference

Sveen, O., Borthne, K., & Aasen, A. O. (1994). Computer acquisition, analysis, and presentation of results from a Cobas Bio analyzer. *Comput. Biol. Med., 24*, 493–504.

Executive Dashboards

10

Overview

By providing fast and easy access to financial, clinical, and quality metrics, virtual bio-instrumentation applications described herein provide caregivers, administrators, and support staff the ability to perform sophisticated analyses to develop accurate models and forecasts and make timely, data-driven decisions. This capability is perhaps best summarized by Bill Gates, chairman and CEO of Microsoft, who states in his latest book:

> I have a simple but strong belief. The most meaningful way to differentiate your company from your competition, the best way to put distance between you and the crowd, is to do an outstanding job with information. How you gather, manage, and use information will determine whether you win or lose.
>
> *Bill Gates*, Business @ the Speed of Thought

The Need for Real-time Performance Measurement

This chapter brings virtual bio-instrumentation further into the realm of information management by demonstrating how VBI can effectively leverage other technologies including the Internet, machine vision, ActiveX components (such as interactive agents), and databases. Specific examples will highlight connectivity to patient information systems, computerized maintenance and management systems (CMMS), and business intelligence and decision support applications.

Each application will consist of detailed descriptions of how virtual instrument solutions have been conceived and developed to meet specific end-user requirements within the biomedical and healthcare arena. Case studies will demonstrate how executive dashboards have been developed to help a wide variety of institutions manage and deploy information. These examples will range from dashboards that support general operations to customized instruments that help hospitals manage fluctuating patient census and bed availability to dashboards designed to enable insurance companies to identify best practice patterns, accurately price products, and increase market share. Collectively, these applications support better, faster, and data-driven decisions.

Data Management

Today's enterprises create vast amounts of raw data, and recent advances in storage technology, coupled with the desire to use this data competitively, have caused a data glut in many organizations. The healthcare industry in particular is one that generates a tremendous amount of data. Tools such as databases and spreadsheets certainly help manage and analyze this data, however, databases, while ideal for extracting data, are generally not suited for graphing and analysis. Spreadsheets, on the other hand, are ideal for analyzing and graphing data, but this can often be a cumbersome process when working with multiple data files. VIs empower the user to leverage the best of both worlds by creating a suite of user-defined applications that allow the end-user to convert vast amounts of data into information, which is ultimately transformed into knowledge to enable better decision making.

Given the vast amounts of data available from increasingly sophisticated enterprise-level data sources, potentially useful information is often left hidden due to a lack of useful tools. Virtual instrumentation allows organizations to effectively harness the power of the PC. VIs can employ a wide array of technologies such as multidimensional analyses and Statistical Process Control (SPC) tools to detect patterns, trends, causalities, and discontinuities to derive knowledge and make informed decisions. Figure 10-1 illustrates these four stages in which raw data is processed to make actionable decisions.

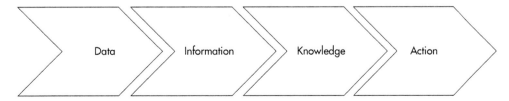

Figure 10-1.
Transforming data into action.

This chapter will discuss several VI applications and tools that have been developed to meet the specific needs of healthcare organizations. We'll pay particular attention to the use of performance indicators and decision support applications that have the ability to trend and forecast various metrics. The use of SPC within virtual instruments will also be demonstrated.

Metrics That Matter—Performance Indicators in Healthcare

Variations in the utilization of health services have been recorded and studied for several decades (Spath, 2000). Comparative data have been available both through state hospital organizations and through their governments. The federal government has had Medicare data available since the early 1970s. These data were available in raw form and, occasionally, in tabular reports. The use of healthcare report cards dates back to the 1980s when several groups, led in part through the efforts of the Health Care Financing Administration (HCFA), began comparing outcomes of different hospitals and using the data to improve healthcare performance. The Medicare MEDPAR files became available shortly after, and, near the end of the decade, the Health Plan Employer Data and Information Set (HEDIS) was created to assist employers in evaluating the value of healthcare provided through health maintenance organizations (HMOs). In the early and mid 1990s, the Joint Commission on Accreditation of Healthcare Organizations (JCAHO) developed and implemented its ORYX program of comparative performance measures for hospitals.

Why Measure?

The rationale for performance measurement is that any organization requires objective feedback about its own performance. This information can be used internally to support quality improvement activities and externally to demonstrate its accountability to the public, regulators, and advocacy groups. Measurement can support claims of quality and provide an early warning of developing problems that could lead to serious errors. It can verify the effectiveness of performance improvement activities and corrective actions and help to identify areas of excellence. It is also very useful in comparing performance with peer organizations using the same measures within the same measurement system.

Consumer and purchaser concerns about healthcare quality have escalated dramatically. Both public and private payers are demanding data to support claims of quality. Performance measurement has gained widespread use in the healthcare industry to reduce subjectivity and reliance on feel or emotion. The absence of quantitative data consigns people to the expression of opinions based on their sense that the organization's rate for performing a procedure or achieving a certain outcome is too high, just right, not high enough, much better since so-and-so left, or much worse than years ago. Often the phrases "It's my sense that this is the reason for . . ." or "My experience is that . . ." are the basis upon which decisions are made. The impression and potential inaccuracy of such judgments may lead to flawed decision making and reduced ability to identify real opportunities to improve performance and outcomes.

The healthcare industry has also begun to use data as a comparison tool. Many organizations compare their current performance against a known entity, usually themselves. They may compare current performance against past or historical performance, or against an internal target that has been determined from such. All acute healthcare organizations accredited by JCAHO now require participation in a program titled ORYX, which is designed to use comparisons with other organizations and promote national benchmarks. Integrating performance measurement into the accreditation process enhances the value of their certification by confirming the link between accreditation and outcomes of care. Many organizations have used comparative patient satisfaction data in this manner for years.

The most widespread use of performance measurement is just that—to measure performance. Data is used to set priorities, targets for improvement, and the maintenance of excellence. It guides the appropriate use of resources and is used for monitoring key points of reference. Without valid and reliable measurement, it's very difficult to understand if improvement is warranted.

Driving the Need to Measure Quality

There are several factors that are driving the growing need to measure individual organization performance and to compare health systems. Financial management is most obvious. Measurement is required to understand both the payment model changes brought about through managed care and the impacts of cost reduction. Purchasers are seeking value for their healthcare dollar, and there is still a need to eliminate unnecessary and costly variations in disease management strategies. Measurement is commonly used in the management of the business of healthcare to ensure that regulatory mandates and accreditation requirements are met, that marketing efforts are aligned to the correct populations, that compliance programs are successful, and to differentiate an organization from its competition. The patients are the principal drivers of increased performance measurement. The need to measure satisfaction and improve patient outcomes is paramount to all organizations' missions as are the establishment of effective disease management strategies.

Measures, therefore, should encompass data from customers, employees, stockholders, suppliers, and communities and should originate from two directions: from the vision, mission, beliefs, and strategic plan of the organization, and from the direct measures needed daily in the operational units to monitor and adjust performance (Godfrey, 1997).

Data

Data can be either existing or new in nature and fall into several types of indicators. Most common are the *aggregate data indicators* that measure things or events that are expected to happen with some frequency. They are based on collection and aggregation of data about many events or phenomena. If the aggregate values fall along a continuous scale, the data is referred to as a continuous variable indicator. A *rate-based indicator* is aggregate data expressed as a proportion or ratio (JCAHO, 1993). Aggregate data indicators often measure a process or an outcome and can serve as early warning systems for either desirable or undesirable events. Aggregate data is useful for making comparisons, trending, and analysis.

Unique data indicators, as their name implies, measure events that occur very infrequently or only once. A sentinel event, defined as "an event that always triggers further analysis and investigation and is undesirable" (JCAHO, 1993), would be an example of this indicator type.

Balanced Scorecards

Too often, financial measurements have dominated the performance measurement systems and reports in organizations. However, in the past few years, many organizations have discovered the value of clear and well-balanced measurement systems deployed throughout their companies (Godfrey, 1997).

These balanced scorecards, or executive dashboards, display discrete data from a variety of sources in one report or on one screen.

Health system executives are increasingly seizing upon this out-of-industry practice—monthly performance dashboards—to keep tabs on their growing enterprises. Based on the concept of automobile or aircraft dashboards, performance dashboards employ a limited number of key indicators and graphics to provide the user with user-friendly snapshots of an operation's overall performance.

Monitoring the ever-increasing expansion of health systems through mergers, acquisitions, and growth has placed a drain on executive time. The traditional approach to monitoring performance by reviewing discrete reports from individual operating units or services has produced an executive information overload and has been shown to be a contributing factor to poor financial performance among hospitals and health systems. The reams of randomly collected performance information—on which executives all too often rely to monitor overall performance—obscure negative trends and interrelationships that would, if seen together, provide an early warning signal of a downturn.

The monthly snapshots of overall performance afforded by dashboards provide a promising antidote to the problems caused by executive information overload, and, although dashboards do not guarantee remedial action, a growing number of executives rely on such condensed overviews of their institution's performance to identify problems. The most successful dashboards combine the following features:

- They contain a balanced cross-section of operational, clinical, and satisfaction metrics in addition to traditional financial indicators. This cross-section of metrics increases the tempo of problem recognition and prediction, which impacts the bottom line, as financial downturns are often produced by problems in other areas.

- Dashboards have a limited number of total indicators. Although more indicators may be necessary to diagnose a problem (drilling down or peeling away the layers of the onion), close monitoring of only 15 to 30 representative measures are needed to identify potential problems effectively.

- Dashboards must express performance targets and variances. Documenting predetermined targets on the dashboard, and variance from the targets, spotlights mounting problems, describes incremental rate of change, and highlights conflicting indicators often characteristic of poor information.

- A final key feature of successful dashboards is the graphical display of all indicators. Converting raw numbers into charts facilitates the recognition of trends and interrelationships among changes within the indicators.

Barriers

Developing effective and useable dashboards can be an extremely time- and resource-consuming task for an organization. There are many barriers to success. Some of the common pitfalls seen in dashboard development are size— attempts to include more than 20 to 30 top-line indicators undermine the summary purpose of a dashboard and confuse the development. The report becomes too detailed. As there is so much that one could track, the principle issue across the short term is to decide where to focus indicator measurement efforts. The ideal is to focus clinical tracking and improvement efforts on high-visibility, high-cost negative events, such as major complications, in 15 to 20 high-profile, high-volume diagnostic related groups, or DRGs.

Another frequently encountered pitfall is attempting to make the development process democratic by satisfying the needs of multiple managers and executives, all with differing agendas. The primary audience for most dashboards should be the CEO, COO, and the board of directors. The dashboard should be designed with this group in mind.

Unwillingness to change or modify indicators can be another problem. The dashboard should not be seen as a fixed end-product. Healthcare is an evolving, complex enterprise, and dashboard indicators need to reflect this. Finally, dashboards should not be tied to the establishment of a strategic planning process. Strategic planning is often a long, cumbersome process. Dashboards need to be viable and responsive.

Information must be current and timely. It is very difficult to institute meaningful improvement processes on information that may be as much as one year old. The frequency of sampling should be determined beforehand and be based on data availability. Quarterly samples are common in healthcare. How the sampling will be done and by whom are barriers that must be overcome in many organizations, as this process is often viewed as more of a chore than one of great excitement. One great barrier to creating comparative data has been the lack of standardized, well-defined measurements. Another challenge to overcome is describing how data are to be reported and used to improve the quality of care. That is, who will receive the measures, and who is accountable for improvement?

PIVIT™—Performance Indicator Virtual Instrument Toolkit

Many of the examples presented in this chapter are derived from an application suite called PIVIT™. The Performance Indicator Virtual Instrument Toolkit is an easy-to-use data acquisition, analysis, and presentation product. PIVIT was created in LabVIEW and developed specifically in response to the wide array of information and analysis needs throughout the healthcare setting.

PIVIT applies virtual instrument technology to access, analyze, and forecast clinical, operational, and financial performance indicators. Some examples include applications that profile institutional indicators (i.e., patient days, discharges, % occupancy, average length of stay [ALOS], revenues, expenses, etc.), and departmental indicators (i.e., salary, nonsalary, total expenses, expense per equivalent discharge, DRGs, etc.). Other applications of PIVIT include 360-degree Peer Review, Customer Satisfaction Profiling, and Medical Equipment Risk Assessment.

PIVIT can access data from multiple data sources. Virtually any parameter can be easily displayed from standard spreadsheet and database applications (i.e., Microsoft Access, Excel, Sybase, Oracle, etc.) using Microsoft's Open Database Connectivity (ODBC) technology. Furthermore, multiple parameters can be profiled and compared in real-time with any other parameter via interactive polar plots and three-dimensional displays. In addition to real-time profiling, other analyses such as Statistical Process Control can be employed to view large data sets in a graphical format. SPC has been applied successfully for decades to help companies reduce variability in manufacturing processes. These SPC tools range from Pareto graphs to Run and Control charts. Figure 10-2 illustrates the Main Menu of PIVIT.

The PIVIT application takes advantage of the latest technology to communicate to users. Microsoft Agent technology is behind a pop-up animated help tool that communicates a message, indicates an alarm condition, or helps the user solve a problem or point out a discrepancy that may have otherwise gone unnoticed. Intelligent agents employ a text-to-speech algorithm to actually

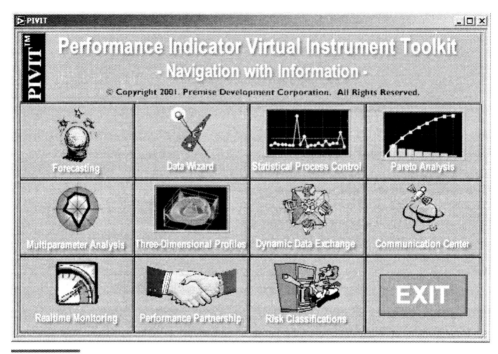

Figure 10-2.
The Performance Indicator Virtual Instrument Toolkit (PIVIT™) Main Menu.

speak an analysis or alarm directly to the user or recipient of the message. In this way, on-line help, user-support, and important messages or alerts can be presented to the user as the need arises. These messages can also be provided in multiple languages. PIVIT's multimedia module can also alert designated personnel by sending an automated message via e-mail, fax, pager, or mobile phone.*

Trending, Relationships, and Interactive Alarms

Figure 10-3 illustrates a virtual instrument that interactively accesses institutional and department-specific indicators and profiles them for comparison. Data sets can be acquired directly from standard spreadsheet and database applications (i.e., Microsoft Access, Excel, Sybase, Oracle, etc.). This capabil-

---

*A functional demo of the PIVIT application has been provided on the accompanying CD-ROM (VBI CD-ROM/sections/Informatics/PIVIT).

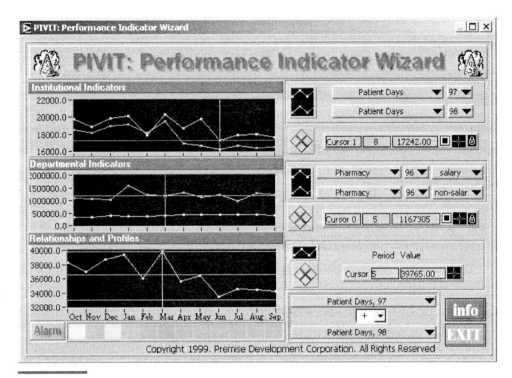

Figure 10-3.
PIVIT—Performance Indicator Wizard. (Courtesy of CRC Press.)

ity has proven to be quite valuable with respect to quickly accessing and viewing large data sets. In the past, multiple data sets contained within a spreadsheet or database had to be selected and then a new chart of this data could be created. Using PIVIT, the user simply selects the desired parameter from any one of the pull-down menus, and this data set is instantly graphed and compared to any other data set.

Interactive threshold cursors use color to dynamically highlight when a parameter is over or under a specific target. Displayed parameters can also be ratios of any measured value, for example, Expense per Equivalent Discharge or Revenue to Expense Ratio. The indicator color will change based on how far the data value exceeds the threshold value (i.e., from green to yellow to red). If multiple thresholds are exceeded, then the entire background of the screen (normally gray) will change to red to alert the user of an extreme condition.

Trending and Forecasting

Figure 10-4 is an example of PIVIT's ability to profile historical trends and project future values. Forecasts are based on user-defined history (i.e., Months for Regression); the type of regression (i.e., linear, exponential, or polynomial); the number of days, months, or years to forecast; and if any off-set should be applied to the forecast. These features allow the user to create an unlimited number of what-if scenarios and allow only the desired range of data to be applied to a forecast. In this example, rather than project the next 12 months based on the previous 56 months of data, only the past 12 months of data was selected. In addition to the graphical display of data values (in this example Patient Days were used), historical and projected tables are also pro-

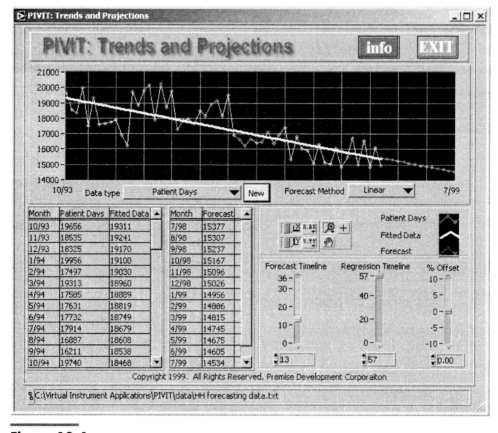

Figure 10-4.
PIVIT—Trends and Projections screen.

vided. These embedded tables look and function very much like a standard spreadsheet.

Forecasted data elements can also be exported from the PIVIT application into standard spreadsheets and databases for distribution and further analysis. Graphs can also be exported into standard analysis or presentation applications such as Microsoft Excel or Microsoft PowerPoint.

Multiparameter Displays

Figure 10-5 illustrates how multiple parameters can be profiled and displayed. In this example, actual versus budgeted numbers are plotted along eight dedicated axes. This polar plot allows the user to easily see which values fall short of or exceed their respective targets. Specific quadrants of the polar chart can be zoomed in on for detailed analysis of individual values. In

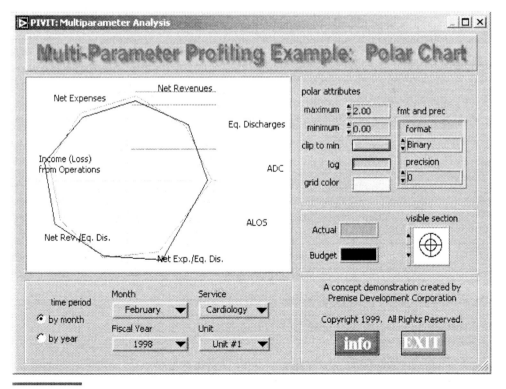

Figure 10-5.
PIVIT—Multiparameter Profiling example.

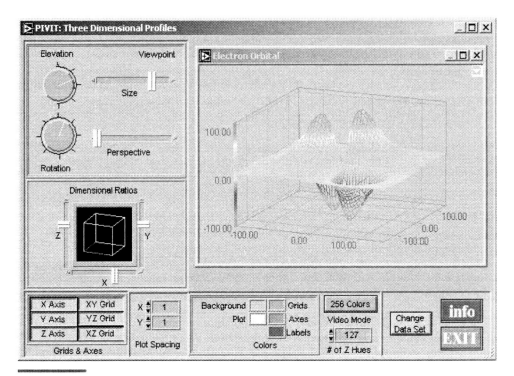

Figure 10-6.
Example of 3D profiling of data.

addition, each polar plot snapshot or slice can be plotted along a vertical axis to generate a three-dimensional profile of the selected parameters over time.

The ability to sift through hundreds of gigabytes of data to discover relationships and patterns that are not obvious, either because the relationship is obscure or because the volume of data to be sifted through to find these patterns is simply overwhelming, is invaluable. In some cases, when large amounts of data are displayed in three dimensions, patterns may be revealed that were not initially apparent. Figure 10-6 illustrates how data can be profiled in three dimensions.

Data Modeling

Figure 10-7 illustrates another example of how virtual instrumentation can be applied to financial modeling and forecasting. This example graphically

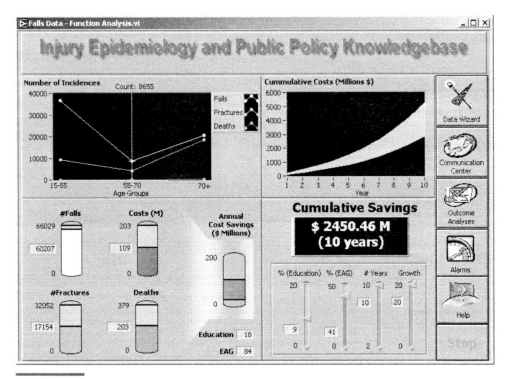

Figure 10-7.
Injury epidemiology and public policy knowledgebase. (Courtesy of CRC Press.)

profiles the annual morbidity, mortality, and cost associated with falls within Connecticut. Such an instrument has proved to be an extremely effective modeling tool due to its ability to interactively highlight relationships and assumptions, and to project the cost or savings of employing educational and other intervention programs.

Virtual instruments such as these are not only useful with respect to modeling and forecasting, but, perhaps more importantly, they become a knowledgebase in which interventions and the efficacy of these interventions can be statistically proven.

By clicking on the Outcome Analyses button, the user can correlate an intervention or an expected outcome to the actual outcome. Note that each data point is actually a value that has been extracted from a database or spreadsheet. This capability exists not only for linear relationships but also for nonlinear relationships, as shown in Figure 10-8.

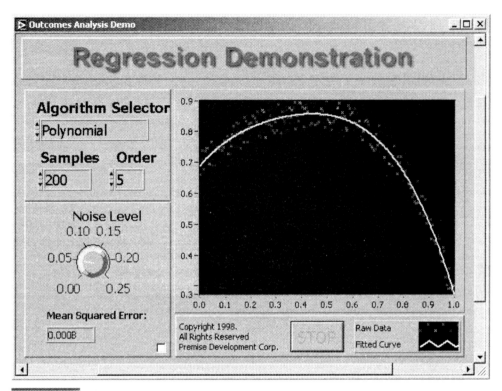

Figure 10-8.
Outcome analyses and regression demonstration.

The regression demonstration illustrates how virtual instruments can identify trends and relationships among otherwise ambiguous data. Linear, exponential, and polynomial curve fitting algorithms, along with a host of other advanced analysis tools, allow the user to find relationships that would not necessarily be evident.

The example program in Figure 10-9 shows how virtual instrumentation can employ standard Microsoft Windows technology (in this case Dynamic Data Exchange [DDE]) to transfer data to commonly used software applications such as Microsoft Access or Microsoft Excel. It is interesting to note that in this example, the virtual instrument can measure and graph multiple signals while at the same time send this data to another application that could reside on the network or across the Internet. It is important to note that in addition to utilizing DDE, virtual instrumentation can use other protocols for interapplication communication. These range from simple serial communication to TCP/IP and ActiveX objects.

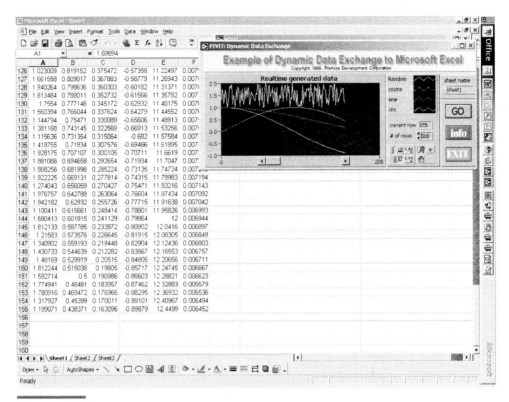

Figure 10-9.
Example of Dynamic Data Exchange (DDE) for interapplication communication.

Figure 10-10 illustrates PIVIT's Communication Center. This application shows various methods by which the user can communicate information throughout an organization. The Communication Center can be used to simply create and print a report or it can be used to send e-mail, faxes, messages to a pager, or even leave voice-mail messages. This is a powerful feature in that information can be easily and efficiently distributed to both individuals and groups in real-time. Additionally, an animated help tool, in this example Merlin the Wizard, can communicate directly with a user at the keyboard.

In addition to real-time profiling of various parameters, more advanced analyses such as Statistical Process Control (SPC) can be employed to view large data sets in a graphical format. SPC has been applied successfully for decades to help companies reduce variability in manufacturing processes. Statistical Process Control has enormous applications throughout healthcare. For example, Figure 10-11 is an example of how Pareto analysis can be

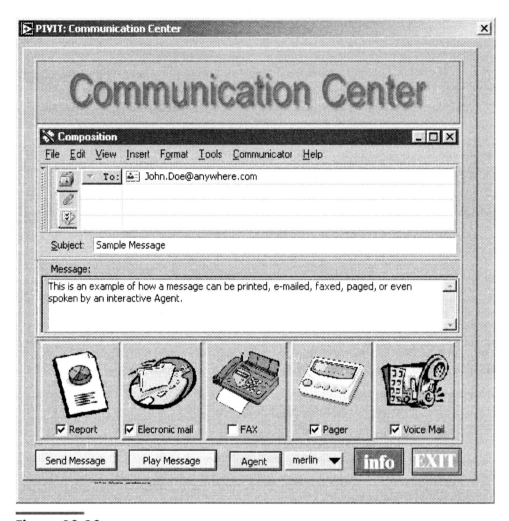

Figure 10-10.
PIVIT's Communication Center.

applied to a sample trauma database of over 12,000 records. The Pareto chart may measure frequency or percentage depending on front panel selection, and the user can select from a variety of different parameters by clicking on the pull-down menu. This menu can be configured to automatically display each database field directly from the database. In this example, 12 different fields (i.e., E-code, DRG, Principal Diagnosis, Town, Zip Code, Payer, etc.) can be selected for Pareto analysis. Other SPC tools include run charts, control charts, and process capability distributions. Figures 10-11 through 10-14 il-

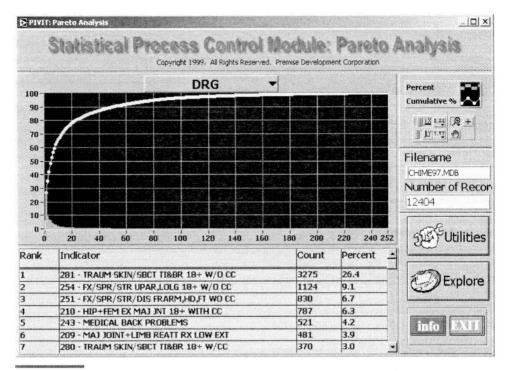

Figure 10-11.
Statistical Process Control—Pareto analysis of a sample trauma registry. (Courtesy of CRC Press.)

lustrate some of these custom virtual instruments that apply Statistical Process Control to a variety of data sets.

Figures 10-12 through 10-14 show how SPC processing can occur in real-time. Control chart points are plotted as sample data points are acquired. Control chart limits are automatically calculated after three samples are acquired, and after # Samples for Limits samples are acquired. Out-of-control points are automatically excluded from the control limit calculation. If the Calc Mode is Manual, the user can press the Calculate Control Limits button at any time to calculate control limits. If the Calc Mode is Auto, control limits are automatically calculated whenever a process shift is detected.

Pressing the View Control Limit Calculation button will display the most recent control chart limit calculation and the points that were found to be out of control. This demo automatically updates a histogram as each sample is acquired. Pressing the Compute Process Capability button computes and displays the process capability based on the last # Samples as shown in Figure 10-13.

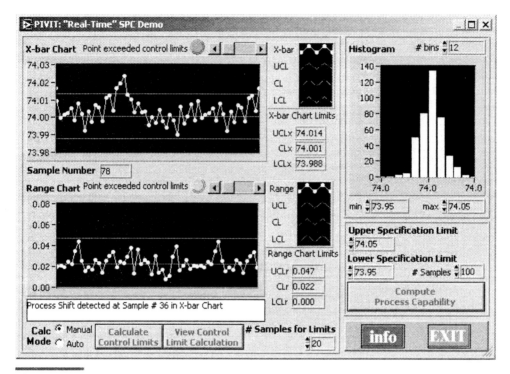

Figure 10-12.
Statistical Process Control—Real-time SPC application.

Figure 10-14 is an example of an instrument that plots the results from the most recent control chart calculation. This example also highlights out-of-control points that were excluded from the control limit calculation on the control chart.

Medical Equipment Risk Criteria

Figure 10-15 illustrates a virtual instrument application that demonstrates how four static risk categories (and their corresponding values) are used to determine the inclusion of clinical equipment in the Medical Equipment Management Program at Hartford Hospital.

Each risk category includes specific subcategories that are assigned points, which when added together, yield a total score that ranges from 4 to 25. Considering these scores, the equipment is categorized into five priority levels

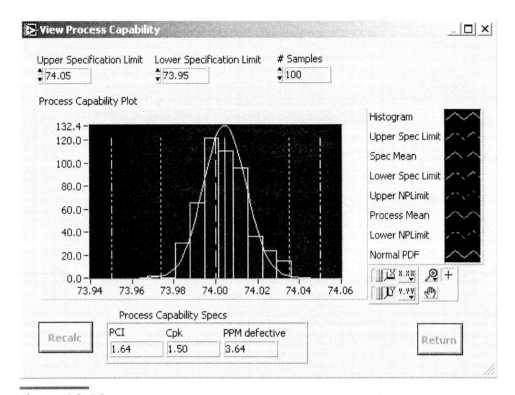

Figure 10-13.
Statistical Process Control—View Process Capability window.

[high, medium, low, hazard surveillance (gray) list, and non-inclusion]. The four static risk categories are

1. **Equipment Function (EF):** Stratifies the various functional categories (i.e., therapeutic, diagnostic, analytical, and miscellaneous) of equipment. The specific rankings for this category are listed in Table 10-1.

2. **Physical Risk (PR):** Lists the worst case scenario of physical risk potential to either the patient or the operator of the equipment (see Table 10-2).

3. **Environmental Use Classification (EC):** Lists the primary equipment area in which the equipment is used (see Table 10-3).

4. **Preventive Maintenance Requirements (MR):** Describes the level and frequency of required maintenance (see Table 10-4).

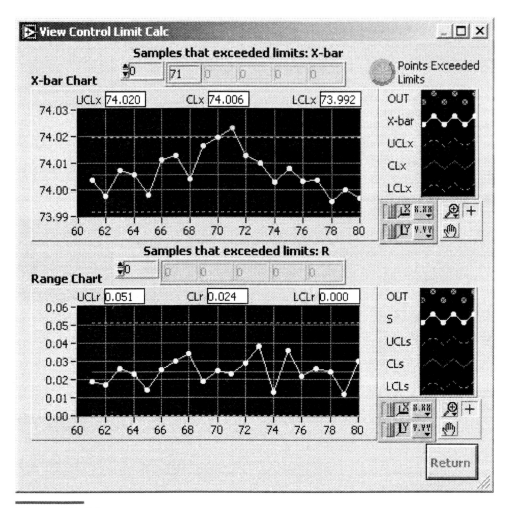

Figure 10-14.
Statistical Process Control—View Control Limit example.

The aggregate static risk score is calculated as follows:

$$\text{Aggregate Risk Score} = EF + PR + EC + MR$$

Using the system described in Tables 10-1 through 10-4, clinical equipment is categorized according to the following priority of testing and degree of risk:

- **High Risk:** Equipment that scores between and including 18 to 25 points on the evaluation system. This equipment is assigned the highest risk for testing, calibration, and repair.

Table 10-1. *Equipment Function Ranking.*

	Risk Category I: Equipment Function (EF)
Point Score	Function Description
10	Therapeutic—Life support
9	Therapeutic—Surgical or intensive care
8	Therapeutic—Physical therapy or treatment
7	Diagnostic—Surgical or intensive care monitoring
6	Diagnostic—Other physiological monitoring
5	Analytical—Laboratory analytical
4	Analytical—Laboratory accessories
3	Analytical—Computer and related
2	Miscellaneous—Patient-related
1	Miscellaneous—Nonpatient related

Table 10-2. *Physical Risk Ranking.*

	Risk Category II: Physical Risk (PR)
Point Score	Description of Use Risk
5	Potential patient death
4	Potential patient injury
3	Inappropriate therapy or misdiagnosis
2	Equipment damage
1	No significant identified risk

Table 10-3. *Environmental Use Classification Ranking.*

	Risk Category III: Environmental Use Classification (EC)
Point Score	Primary Area of Equipment Use
5	Anesthetizing locations
4	Critical care areas
3	Wet locations/Labs/Exam areas
2	General patient care areas
1	Nonpatient care areas

Table 10-4. *Preventive maintenance ranking.*

	Risk Category IV: Preventive Maintenance (MR)
Point Score	PM Frequency
5	Monthly
4	Quarterly
3	Semi-annually
2	Annually
1	Not required

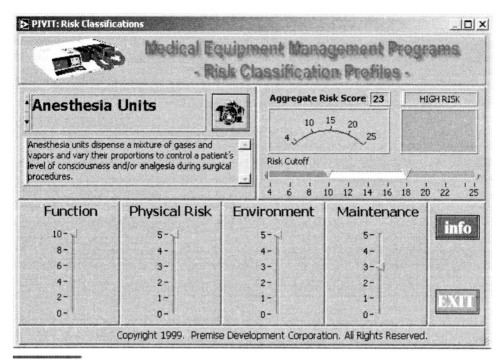

Figure 10-15.
Medical equipment risk classification profiler. (Courtesy of CRC Press.)

- **Medium Risk:** Equipment that scores between and including 15 to 17 points on the criteria's evaluation system.

- **Low Risk:** Equipment that scores between and including 12 to 14 points on the criteria's evaluation system.

- **Hazard Surveillance (Gray):** Equipment that scores between and including 6 and 11 points on the criteria's evaluation system are visually inspected on an annual basis during the hospital hazard surveillance rounds.

- **Non-inclusion:** Medical equipment and devices that pose little risk and score less than 6 points may be deleted from the management program as well as the clinical equipment inventory.

Future versions of this application will also incorporate various dynamic risk factors such as user errors, mean-time-between failure (MTBF), device failure within 30 days of a preventive maintenance or repair, and the number of years beyond the American Hospital Association's recommended useful life. These metrics are accessed from the computerized maintenance management system (CMMS).

Peer Performance Reviews

The virtual instrument shown in Figure 10-16 has been designed to easily acquire and compile performance information with respect to institution-wide competencies. It has been created to allow every member of a team or depart-

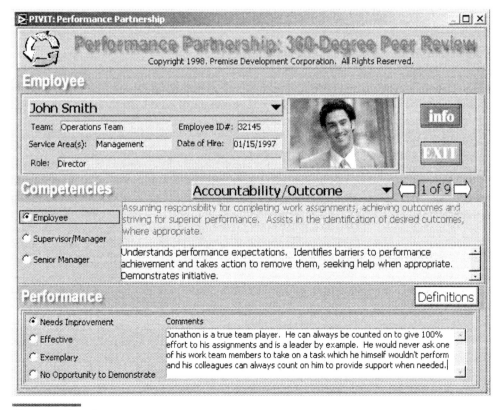

Figure 10-16.
Performance reviews using virtual instrumentation. (Courtesy of CRC Press.)

ment to participate in the evaluation of a coworker (360-degree peer review). Upon running the application, the user is presented with a sign-in screen where he or she enters a unique username and password. The application is divided into three components. The first (top section) profiles the employee and relevant service information. The second (middle section) indicates each competency as defined for employees, managers, and senior managers. The last (bottom) section allows the reviewer to evaluate performance by selecting one of four radio buttons and also provide specific comments related to each competency. This information is then compiled (with other reviewers) as real-time feedback. Figure 10-17 illustrates this process.

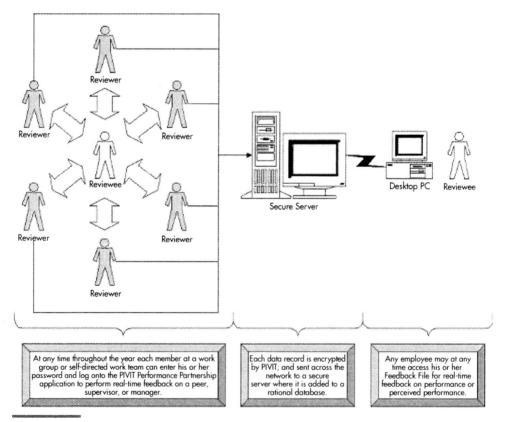

Figure 10-17.
PIVIT's peer performance review process.

Executive Information Dashboard

The Executive Information Dashboard is an easy-to-use business intelligence (BI) application that has been created using virtual bio-instrumentation technology. It has been designed to allow administrators, managers, and knowledge workers to easily access, analyze, and display real-time information from databases, spreadsheets, and ancillary information systems. This capability provides a cross-functional view of enterprise health to decision makers across the company. Decision makers can easily move from big-picture analyses to transaction-level details while at the same time share this information throughout the enterprise to derive knowledge and make timely, data-driven decisions. The VBI technology enables users to see trends and relationships in their data directly from their desktop PCs. Users can easily delve into vast amounts of data to discover not only where relevant information can be found but also why specific data shifts, discrepancies, or changes occur.

By utilizing instruments based on visual computing, users can easily understand data significance. Given the graphically intuitive data query and navigation tools, users do not need advanced statistical or mathematical backgrounds to see for themselves what their data represents. Everyday users can access, analyze, and share their data to quickly find information in places where they might not have thought (or been able) to look. Findings are displayed in clear, animated graphical and multidimensional reports and presentations to solve everyday business problems. Figure 10-18 shows the Main Menu of a prototype Executive Information Dashboard at Hartford Hospital.

The Executive Information Dashboard applies virtual instrument technology to the assessment and analysis of clinical, operational, and financial performance indicators. We've already provided examples of some of the institutional and departmental indicators. Virtually any parameter (from any fiscal year) can be easily accessed and displayed. Furthermore, any parameter can be profiled and compared (i.e., added, subtracted, multiplied, or divided) with any other parameter. In the following examples, we illustrate some of the analysis and presentation capabilities of the Executive Information Dashboard, and, in particular, a sub-VI of the dashboard called the Bed Management Digital Dashboard. We begin with the Building Profile Wizard as shown in Figure 10-19.

This instrument interactively profiles the infrastructure specifications for each building within a healthcare system simply by clicking on a specific building. It can model annual and cumulative costs of maintenance and utilities.

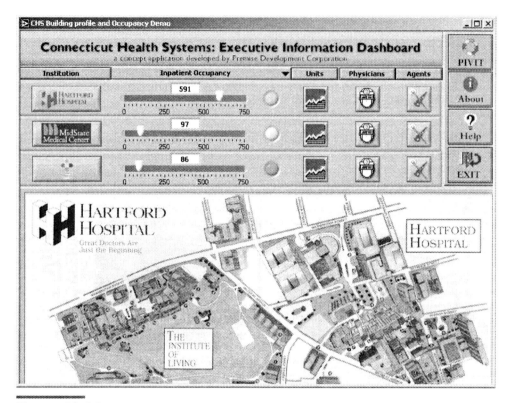

Figure 10-18.
The Executive Information Dashboard Main Menu.

The Building Profile Wizard can also display and model the revenue and net revenue of rental income. Finally, this module can perform cost/benefit analyses of maintaining existing infrastructure versus building a new facility.

The Bed Management/Census Control Dashboard

Overview

The Bed Management Dashboard (BMD) is an easy-to-use BI product. It has been designed to allow administrators, clinicians, managers, and knowledge workers to easily access, analyze, and display real-time patient and bed avail-

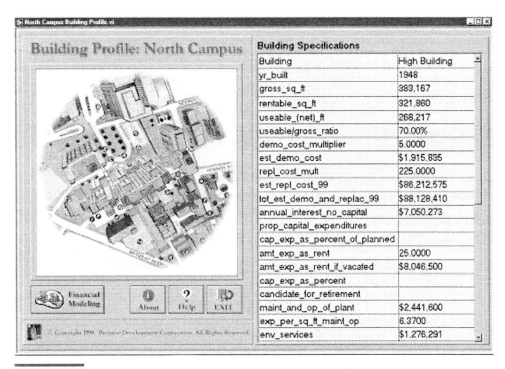

Figure 10-19.
The Building Profile Wizard.

ability information from ancillary information systems (i.e., Admission/Discharge/Transfer systems), databases, and spreadsheets.

An Air Traffic Control Tower for Beds

In many ways, the Bed Management Dashboard is much like an air traffic control tower. For instance, this application is real-time and mission critical. It must handle both scheduled and emergency events. The BMD is an on-line system that is typically deployed hospital-wide. It assists with the clinical and business decision process that occurs when a patient needs to be assigned to a specific bed location. The system comprises intuitive modules to display and provide alerts on quantitative results in real-time. Additionally, it creates a robust data mining application or knowledgebase to analyze historical trends and predict future events. Collectively, this system provides organizations with an array of enabling technologies to

- Schedule/reserve/request patient bed assignments;
- Assign/transfer patients from the emergency department or other clinical areas such as Intensive Care Units, Medical/Surgical Units, Operating Rooms, Post-Anesthesia Care Unit, and so on;
- Reduce/eliminate dependency on phone calls to communicate patient and bed requirements;
- Reduce/eliminate paper processes to manage varying census levels;
- Apply SPC and Six Sigma* methodologies to manage occupancy and patient diversion; and
- Provide administrators, managers, and caregivers with accurate and on-demand reports and automatic alerts via pagers, e-mail, phone, and intelligent software agents.

Table 10-5 summarizes the major problems that the BMD addresses.

The following diagrams illustrate some of the capabilities of this dashboard. Figure 10-20 profiles in real-time the current occupancy of any care unit. This data is accessed via standard Health-Level 7 (HL-7)* interfaces to the Admission/Discharge/Transfer (ADT) system. The occupancy profile can be displayed to show total occupancy as well as subset populations such as outpatient (day surgery) and emergency room activity. Further detail can be obtained by selecting a particular unit and clicking on the Occupancy Details button.

Figure 10-21 shows the Patient Distribution Module, which allows the user to select a particular campus or facility and then drill down within that facility for more specific information. For example, by clicking on the virtual elevator, the user can easily see the real-time patient population distribution, referring physicians, or DRG patterns throughout a healthcare facility. Interactive alarms and on-line agents can also be configured to alert the user of unusual developments according to the criteria they set.

---

*Six Sigma is a process of collecting data and investigating the data to see how we can reduce the number of defects. Sigma is standard deviation from the ideal target. Six Sigma is a vision of quality that equates with only 3.4 defects per million opportunities for each product or service transaction.

*HL-7 is an industry standard data format for health-related information.

Table 10-5. *The Bed Management Dashboard provides solutions to many problems.*

PROBLEM	SOLUTION
ADT systems are not designed to provide sufficient clinical information for appropriate patient placement.	BMD integrates all required information in real-time (i.e., monitor required, negative pressure room, etc.).
Lack of accurate bed availability information can result in lost admissions and excessive wait times.	BMD provides the information necessary to significantly improve the efficiency of patient placement and discharge.
Inefficient communication occurs while searching for the appropriate bed for a patient.	BMD automates the notification process via dashboard, pager, e-mail, etc.
23+ hour observation of outpatients who occupy inpatient beds without payer authorization is costly.	BMD has built-in automated alerts via intelligent agents, pagers, etc.
It is difficult to access meaningful historical, current, and predictive data.	BMD provides data mining and decision support tools to create useful information from data.

Hospital Summary

Administrators and others who need to see a global view of the hospital status primarily use the screen shown in Figure 10-22. A table and pareto chart profile the various units and the current status summaries of each unit. These patients can also be rolled up into services, or grouped by physician. In addition, patients can be aggregated in many other ways such as time of admission, length of stay, or admitting diagnosis. Hospital-wide summary values are always displayed for easy access to the current status of the hospital.

The Pareto Analysis Module is an interactive and real-time application that dynamically sorts and ranks each facility, unit, or care area by bed size, occupied bed, percent occupancy, available bed, or percent availability. This information is displayed both graphically and in a tabular format. The user can sort any of the variables simply by clicking on the column heading. Although data values are constantly changing, historical profiles and trends can be easily recalled through the Indicator Trending Module.

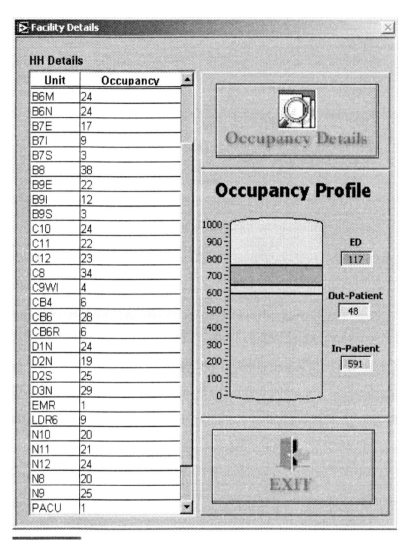

Figure 10-20.
Real-time occupancy profile by care unit.

How the Bed Management Dashboard Works

The Bed Management Dashboard is accessible via a web browser or via a client application. The supporting architecture of the BMD system is a standard N-tier server-based system, which depending on the end users' needs,

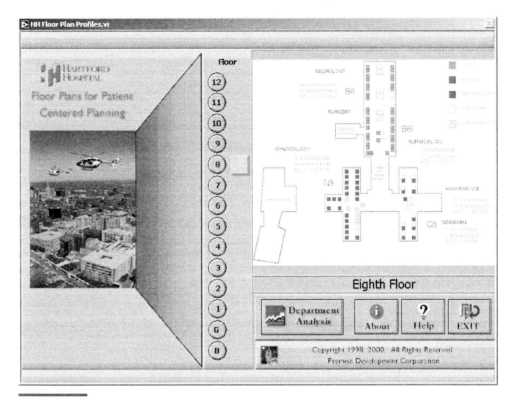

Figure 10-21.
Real-time Patient Distribution Module.

consists of one or more of the following: a web server, an application server, a database server, and an ADT interface server.

Key information from the hospital's ADT system is automatically fed to the BMD via a Health Level Seven (HL-7) data stream. The system has also been designed to accept inputs from patient monitoring and nurse call networks. Figure 10-23 illustrates the flow of information from the ADT system to the user's desktop. The system typically receives up to 12,000 transaction messages a day that are parsed into appropriate data elements by an HL-7 parser and are stored on a database server.

The login authentication module accesses a hospital's central user login repository to validate the user's login information. This module enables single-user sign-on (SSO) capability by providing a user with the same username and password used by the other applications that authenticate to the central user login repository.

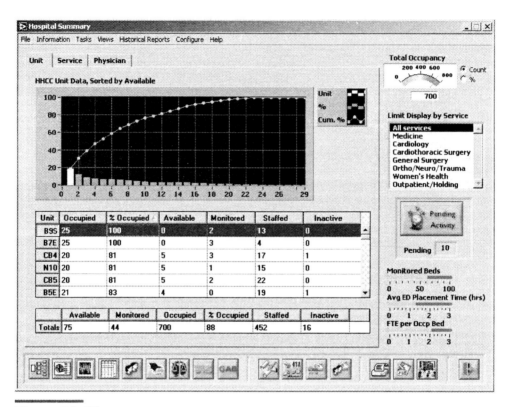

Figure 10-22.
Real-time Pareto Analysis Module of hospital census.

The integrated data mining and report module consists of a number of standard reports that take advantage of the data warehouse nature of the BMD. These reports include, but are not limited to, historical census report, real-time census report, bed manager report, physician discharge report, and discharge compliance report. In addition, access to the data by industry-standard report writers such as Crystal Reports and Microsoft Access give technical users the ability to create complicated or special reports as desired. For example, the BMD offers extensive ad hoc reporting capabilities ranging from length of stay (LOS) and Care Day metrics to asset management analyses (on beds and patient monitor utilization) to biosurveillance reports that have been requested by state and federal organizations such as the Department of Public Health and the Centers for Disease Control and Prevention.

The utility and configuration module gives system operators the ability to administer BMD users (e.g., add new users, modify new or existing user se-

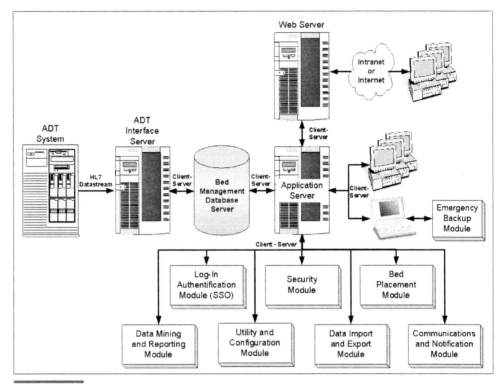

Figure 10-23.
System diagram. The BMD consists of four major components: the ADT Interface, the Database Server, an Application Server, and a Web Server. The system has been designed to interface with any ADT system that provides an HL-7 interface and utilizes TCP/IP as a communication protocol.

curity settings, etc.); administer service, unit, rooms and beds (e.g., add or modify clinical services, units, rooms and beds, and their interrelationships); define automated alert thresholds; and configure unit floor plan diagrams.

The embedded backup utility module consists of a local version of the bed management database that is constantly updated. Access to this database gives users a self-contained and mobile version of the system that can be used in the event of catastrophic failure of the system hardware or network hardware or in the event of a crisis that removes the users from direct access to the hospital network.

One of the most important features of the BMD is that it reformats information from the ADT system and presents it to the clinical user in a more user-friendly and process-oriented manner. Dynamic and interactive graphical

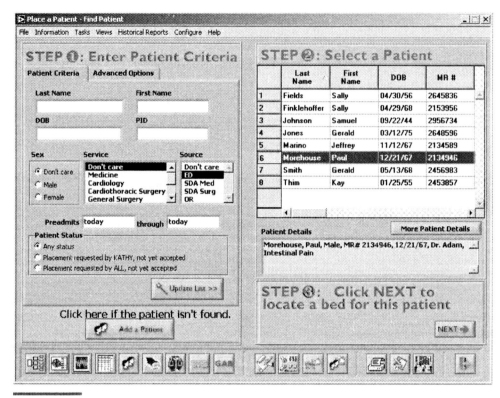

Figure 10-24.
Place a Patient—Find Patient.

presentations of data are used extensively. Figure 10-24 illustrates how all the patients from a given admitting source, such as the emergency department, can be displayed and selected from a dynamically sortable Smart Table. The screen in Figure 10-24 is primarily used to select a patient for placement to a hospital bed. The user first enters various criteria to identify the patient and the system then displays all of the patients meeting the specified criteria. Finally, the user selects the patient in question and presses Next to move to another screen where an available bed is located and requested.

Decisions for patient placement can be centralized or decentralized. The tool allows proper communication between the affected parties. Status of decisions is automatically tracked, and a monitoring process can detect and notify key stakeholders of any process delays. Admitting or emergency departments can automatically be notified of decisions, if appropriate. Information is reported on-line in screen views of data tailored to the needs of a particular class of system user. Unit personnel can view both detailed information as well as summary roll-ups of their patients. Administrators and program directors can view data over a wider scope that encompasses multiple units, services, or physicians. Longer term, retrospective reporting is also supported, both by user-configured screen-based summaries, as well as by third-party tools, such as Crystal Reports, via an ODBC link to the database. A key feature of this system is its use of intelligent agents. These on-line agents are constantly monitoring and analyzing patient and census information, and they have the ability to detect key system situations, such as a high census in a unit (i.e., no available beds), excessive ED placement time for a particular patient, or delays in responses to placement requests. These agents can be configured to notify individuals via on-screen messages, emails, pagers, or synthesized voice phone messaging.

Once a patient is selected, the Bed Finder screen shown in Figure 10-25 is used to locate a specific bed or unit for the patient. The user first enters vari-

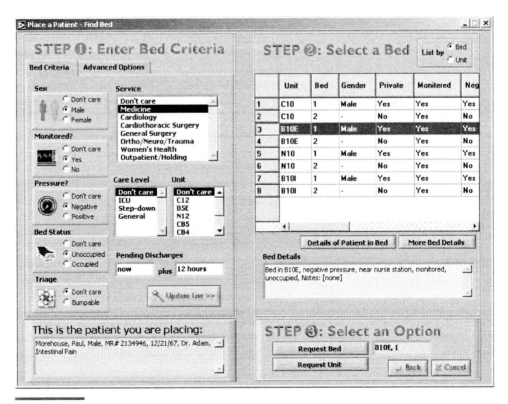

Figure 10-25.
Find a bed.

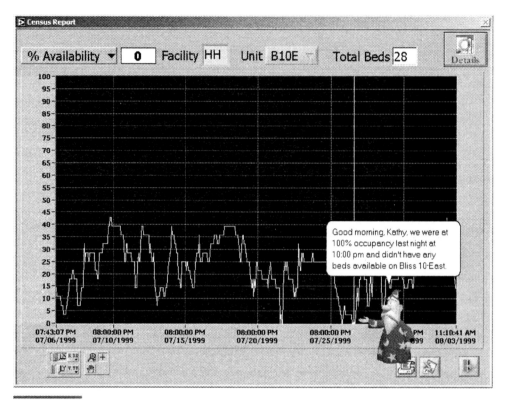

Figure 10-26.
Census report with interactive agents.

ous criteria about the type of bed that is needed (i.e., patient gender, monitor required, etc.). The system then displays all of the available beds that meet the specified criteria. Finally, the user selects a particular bed for the previously specified patient.

The interactive report in Figure 10-26 allows the occupancy trends on a specific unit, group of units, or the entire hospital to be examined over a selectable time interval. The pull-down selector provides options of viewing the data as Occupancy, % Occupancy, Availability, and % Availability. A data pointer line can be dragged across the report to show the actual numeric value in a field at the top of the report. Comprehensive statistical analyses are also available that empower managers and knowledge workers to analyze concrete data and take action on it. Additional drill-down details, print, and export features are also available. The on-line intelligent agent, in this case

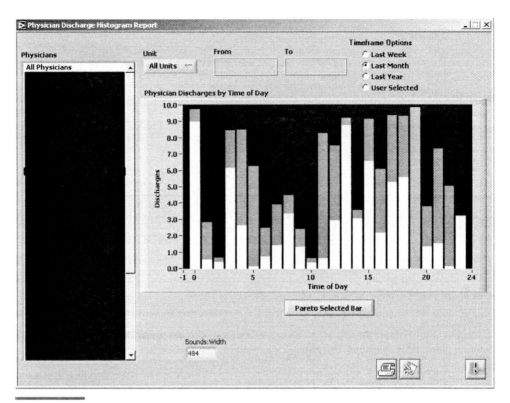

Figure 10-27.
Physician discharge histogram report.

Merlin the Wizard, provides assistance and alerts the user of important alarm conditions as needed.

The interactive report shown in Figure 10-27 rolls up the selected patient discharge activity into a single 24-hour format. Hospital-wide or specific unit data can be selected. Timeframe selection options are provided. Each bar represents the number of discharges for a one-hour period of the day. The overall bar displays the total for all physicians, while the highlighted sub-bar is the value for the specific physician selected on the left.

In addition to pareto analysis, the Bed Management Dashboard offers a complete array of SPC tools including run charts, control charts, and process capability distributions. This module is particularly useful for monitoring and analyzing parameters in real-time such as patient occupancy and throughput, referral and payment patterns, and network activity (see Figure 10-28).

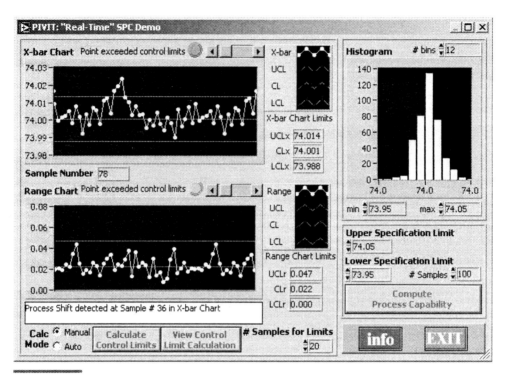

Figure 10-28.
Control charts.

Figure 10-29 shows the pending admissions and transfers that have been officially registered via the hospital's ADT system. The screen highlights any requested bed placements made by the admissions coordinator or other units that wish to transfer patients. The top table shows the amount of pending activity coming in, and the bottom table shows the pending activity out.

In all cases, the screen can show information for a specific unit, a service, or the entire hospital. The user can also limit the display so that he or she will only see the activity that will happen in the next 3 hours, 12 hours, 24 hours, and so on. Also, users can see the details of who made a request and when the request was made.

Figure 10-30 provides a graphical view of an intensive care unit. Each bed is presented as a square using a simplified floor plan view. Spatial relationships between beds and key functions (i.e., Central Nursing Station, etc.) are displayed. Colors on screen are used to indicate the selected attribute of the pa-

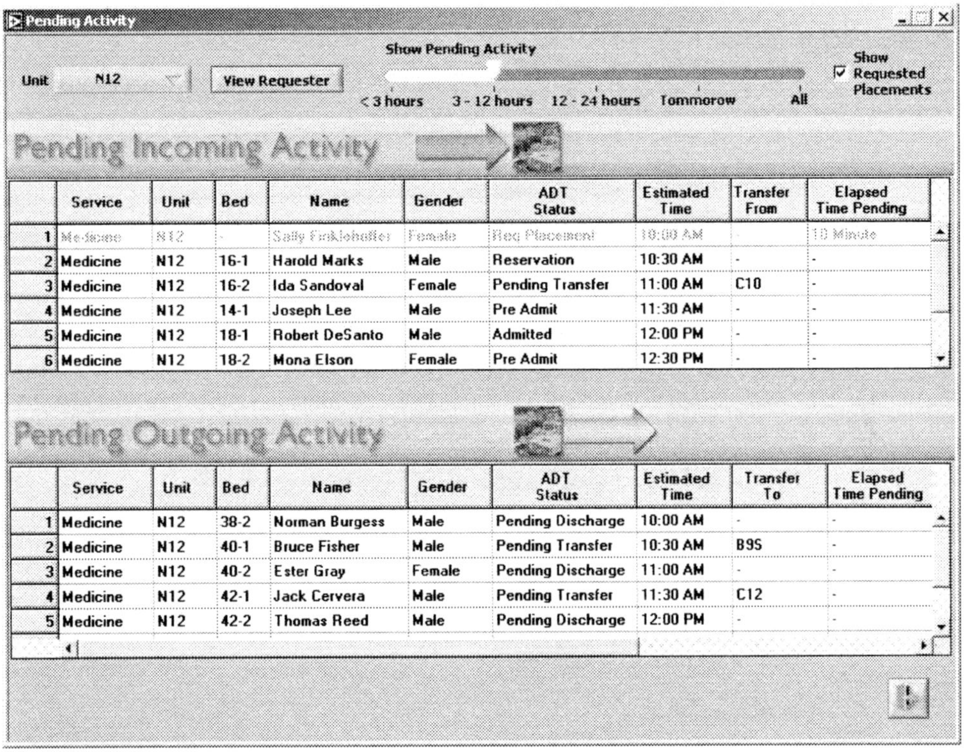

Figure 10-29.
Pending activity.

tient or bed. For example, Figure 10-30 indicates Available Beds, as noted in the current selection shown in the pull-down at the top of the screen. Available beds are green, occupied beds are red, gray beds (which are flashing on screen) are beds with pending discharges and the black bed is closed/inactive.

Many other color-coded options are available via the pull-down selector. These include patient gender (i.e., pink or blue), patient's use of monitors, patient's use of negative pressure, security risk patients, sitters, and so forth. Bed information is also available, including negative pressure, monitoring capabilities, and proximity to the nursing station. In addition, embedded timers indicate how long a patient has been in a bed or how many hours are left until the patient is to be transferred or discharged.

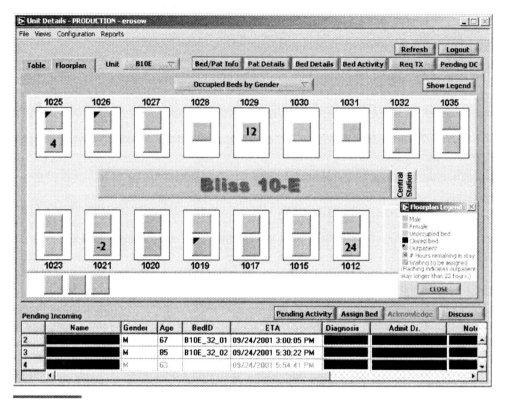

Figure 10-30.
Unit details (floor plan view).

Patient Confidentiality

A full security system is embedded within the dashboard to audit user access and to assign users to definable system roles. These roles restrict both processes and the ability to view or change key data. The system is designed to be fully compliant with the evolving Health Insurance Portability and Accountability Act (HIPAA) regulations.

Summary

Successful organizations have the ability to measure and act on key indicators and events in real-time. By leveraging the power of virtual instrumentation and open architecture standards, the Bed Management Dashboard improves patient placement efficiency and saves time and money by assisting with the clinical and business decision processes associated with patient admissions, transfers, and discharges.

REFERENCES

Godfrey, A. B. (November 1997). Spiders, scorecards, and instrument panels. *Quality Digest, 17.*

Joint Commission on Accreditation of Healthcare Organizations. (1993). *The measurement mandate.* Oakbrook Terrace, IL: Author.

Spath, P. L. (November/December 2000). Are report cards measuring up? *Journal of AHIMA,* 29–35.

The Physician/Network Assessment Tool:
A Business Intelligence and Data Mining Tool
for Healthcare Insurance Claims

by Eric Rosow and Joseph Adam, Premise Development Corporation; Marcia Satlow, M.D, vice president and medical director, The Hartford Insurance Group

Overview

By leveraging standard technologies such as Open Database Connectivity (ODBC) and ActiveX Data Objects (ADO), executive dashboards and virtual instruments can serve as powerful front-end decision support and data mining tools that integrate disparate data sources and allow end-users to see trends and relationships that might otherwise go undetected. This capability allows individuals and organizations alike to make fast, data-driven decisions.

The Challenge: The Hartford Insurance Group (HIG) desired an easy-to-use executive dashboard that could effectively integrate disparate data sets to assess and manage medical outcomes. The data sets consisted of claims, invoices, ICD-9 and CPT codes, network performance indicators, costs, and clinical outcomes. The goal of this integrated dashboard was to allow the HIG to make faster decisions, to identify best-practice patterns, to accurately price products, and to increase market share.

The Solution: A collaborative team of senior executives and information technologists designed and developed a powerful data mining and modeling tool to meet multiple business and technical goals. Specifically, the dashboard has two major functions: (1) easy, immediate data retrieval and display, and (2) data analysis.

Data Retrieval and Display

Data can be retrieved and displayed at differing levels:

- Micro (an individual claim, medical provider, or ICD-9 diagnosis) or
- Macro (entire region, network, injury type).

This ability allows for the examination of specific problems relating to claims or providers and is an essential tool for quality improvement processes. Data display is user-friendly and allows users who are less facile with numbers to visualize numerical data and better understand the significance of simple statistical norms.

Forthcoming in the proceedings of the 2002 Annual HIMSS Conference and Exhibition (Atlanta, GA, January 27–31, 2002). HIMSS is headquartered in Chicago, IL.

Data Analysis

The Physician/Network Assessment Tool is used to analyze treatment patterns (statistics) of cohorts of claims. Basic statistical parameters (recommended by URAC*) have been built into the tool and are automatically displayed. The parameters defining these cohorts can be customized and changed as business needs dictate, as the customer or user wishes, or as the data indicate. The toolkit is user-defined, so queries do not have to be hard coded.

For the more sophisticated analyst, all of the numerical data can be exported into other applications (e.g., Excel, Crystal Reports, etc.) allowing for more complex analysis, interpretation, and display. This type of function can be used to

- Immediately recognize outliers.
- Redefine the cohort into sub-cohorts, allowing the user to further analyze problems as they are uncovered.
- Clarify contributory issues.
- Continue to monitor the effectiveness of changes and case management procedures.

These abilities allow the user to institute corrective actions quickly by appropriately contracting physicians and network groups with the best outcomes.

Variables (Filter Settings)

Some variables and the rationale used for HIG selection for the demonstration are listed next. The variables built into the toolkit were based on a survey from URAC. These variables are used to define cohorts for comparison and analysis.

1. **Medical Costs:** These costs are reported as total, average, median, and percentages.
2. **Individual Claim, ICD-9 Diagnosis, or ICD-9/CPT Combination Costs:** *Individual values* are displayed as a Pareto chart and in a dynamic/sortable table. Values are shown as a percentage of the *total value of the cohort*. In many screens, the *cohort value* for a variable is displayed in relation to the value for *the entire state being analyzed*. This is a useful, immediate indication of relative magnitude (and therefore potential significance) of the variable.

---

*URAC is an independent body that accredits health organizations. It has credibility since it is not bound to the insurance industry.

3. **Service Categories:** These contain groups of related CPT (and other treatment) codes and are used to report utilization and treatment patterns. Examples include
   - Physical Medicine
   - Physical Therapy
   - Chiropractic
   - Surgery/Anesthesia
   - Radiology
   - Hospital Inpatient
   - Other
4. **Utilization/Treatment Patterns:** These describe the overall use of service categories at varying periods, or for the duration of the claim.
5. **Indemnity Costs:** Due to data capture issues we have used these as an equivalent measure of
   - Disability Duration
   - Impairment Ratings

Methods

The Physician/Network Assessment Toolkit maps diagnosis codes with treatment provided (Treatment Statistics screen). HIG's decision to map in this way was based on observations that the most complicated claims have *multiple diagnoses* falling into *many categories*, causing difficulty in gaining consensus on appropriate classification of the claim as a whole. By mapping diagnosis/treatment combinations, treatment patterns for diagnoses can be compared regardless of claim or injury type. Data modeling will reflect changing statistical patterns as data input occurs, the data repository increases, and data integrity improves. The system can also be configured to provide real-time alerts when aberrant patterns occur, to allow immediate corrective action.

HIG's Business Applications

Internal to the Organization

1. **Individual Claims:** All information on a claim can be displayed. Invoice details (to the line item level) are retrievable and can be grouped chronologically and isolated by provider specialty. This gives invaluable treatment information that can be used to track costs effectively and respond to customer queries.
2. **Medical Provider Profiling:** All medical treatments provided on all claims seen by a provider in a chosen cohort can be accessed and displayed. Claims can be sub-grouped (by account if necessary). Treatment provided for specific

time periods (as specific as one-day intervals) on a claim or group of claims can be retrieved (at the invoice level). This information can be used to inform the provider of patterns in comparison to peers (medical provider profiling). This feature has been shown to reduce aberrant behavior.

3. **Provider/Network Classification:** Treatment patterns can be used as a basis for comparison of treatments offered by providers in different networks, at different facilities, at different times in the evolution of claims, or for different cohorts of claims.

4. **Optimal Treatment Patterns for Claim Types:** Treatment patterns associated with the best outcomes for types of claims within specific diagnosis/injury groups can be determined.* This will be achieved through the claim medical/claim actuarial data analysis partnership, supported by our academic consultants.

5. **Comparison of Treatment Patterns to Accepted Standards:** Provider treatment patterns can be compared with National Medical Specialty Association Standards (such as the ACOEM Guidelines), federally generated standards (AHCPR Low Back Guidelines), or Jurisdictional Workers' Compensation Medical Treatment Protocols.

6. **Exclusive Provider Organizations:** Based on outcome analyses, preferred providers can be identified for specific claim types (EPO formation).

7. **Predictive Capabilities:** With the toolkit medical costs for specific injuries can be accurately predicted, enabling the HIG to increase market share for businesses producing those injuries where the HIG has appropriate providers. The HIG can also price its products accordingly.

8. **Suboptimal Outcomes:** Treatment patterns associated with poor outcomes for specific cohorts can be defined.

9. **Claim Classification:** Claims that fit specific patterns can be distinguished early on, and individualized file strategies (i.e., medical, claim, and case management) can be implemented to allow for more efficient allocation of resources. These include
   - Referral to a more appropriate provider or facility where feasible;
   - Assignment to more skilled claim handlers, supporting workers' compensation redesign;
   - More intensive case management oversight; and
   - Defining optimal claims for settlement/disposal.

---

*Data integrity issues (common to the entire industry) affect the accuracy of possible conclusions and limit our ability to institute change but provide an objective basis for relevant QA processes.

External to the Organization

1. **Demand for Performance Measures:** These will become the industry norm and will be demanded by customers. The HIG believes that it will be able to meet that demand more effectively than its competitors.

2. **Innovative Partnerships outside the Insurance Industry:** During the development of this application, several innovative partnerships have been established with academic institutions (medicine and economics). These relationships have provided ongoing support and unbiased guidance in the tool's design and development. This is important in that it distinguishes the toolkit from other analytical products that might be perceived as being tainted by development by the insurance industry or associated vendor products.

3. **Market Potential:** The Physician/Network Assessment Toolkit may have market potential in other areas, including
   - Large self-insured customers,
   - URAC and other oversight organizations,
   - Medical provider organizations, and
   - Medical specialty associations.

4. **Fraud Detection:** The immediate access to detailed information made available by the toolkit provides invaluable support for detection of medical fraud. Information about a provider's activities down to the level of a single day (as defined by invoice) is readily available. A provider's treatment patterns can also be displayed in any desired combination needed (for example, electrodiagnostic testing used with backs).

5. **Legislative/Policy Implications:** Because the toolkit displays data so graphically and persuasively, it empowers observers to understand the failings of the current system. It can therefore be used to demonstrate egregious medical practices to legislative bodies and policy makers, which could drive changes that would benefit workers, employers, competent providers, and insurers.

6. **Quality Assurance/Continuous Quality Improvement:** The development of the Physician/Network Assessment Toolkit has allowed the HIG to quantify problems with data integrity—a challenge that is widespread in many industries. The HIG has used this information to develop QA mechanisms that improve its data reporting, analytical reliability, and business decisions.

Figures 10-31 through 10-39 illustrate some of the features and capabilities of the Physician Assessment Toolkit.

Conclusion

Executive dashboards and virtual instrumentation allow organizations to effectively harness the power of the PC to access, analyze, and share information throughout the enterprise. Various institutions have conceived and developed user-defined solutions to meet specific requirements within the healthcare and insurance industries. These dashboards support gen-

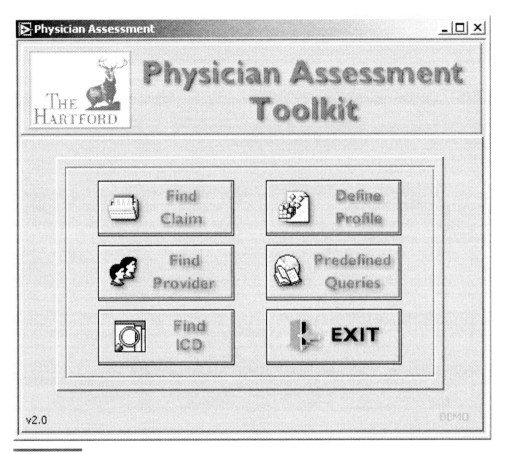

Figure 10-31.
The Physician and Network Assessment Main Menu allows the user to quickly find a claim, provider, or ICD code. In addition, a library of predefined queries have also been established by the Hartford Insurance Group. Also, new searches can be performed by selecting the Define Profile button.

eral operations, help hospitals manage fluctuating patient census and bed availability, and enable insurance companies to identify best-practice patterns, accurately price products, and increase market share. Decision makers can easily move from big-picture analyses to transaction-level details while safely sharing this information throughout the enterprise to derive knowledge and make timely, data-driven decisions. Collectively, these integrated applications directly benefit healthcare providers, payers, and, most importantly, patients.

Note: For more information about the Physician/Network Assessment Tool, please contact info@premisedev.com.

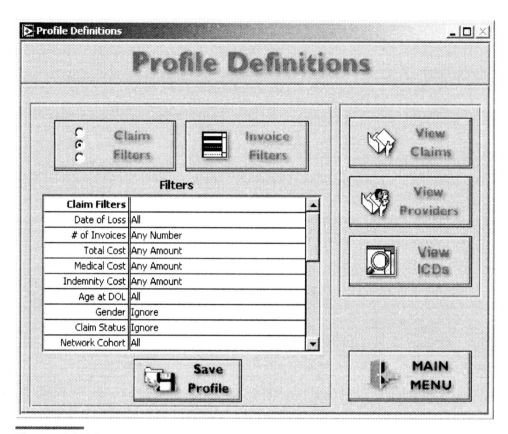

Figure 10-32.
When the Define Profile button is selected, the user can refine the selection criteria by invoking claim and invoice filters. All filter criteria are listed in a scrollable filter table as shown, and at any time the user can view the claims, providers, or ICD codes that match the selected profile definition.

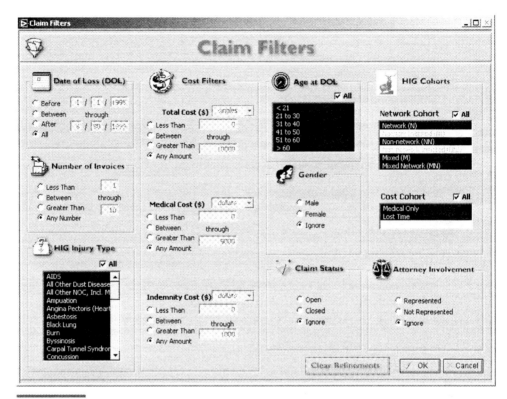

Figure 10-33.
The Claim Filters Module allows users to easily filter multiple databases to identify a desired claim or set of claims by various attributes. These attributes include data of loss, number of invoices, injury type, total cost, medical cost, indemnity cost, policy holder's age and gender, and claim status.

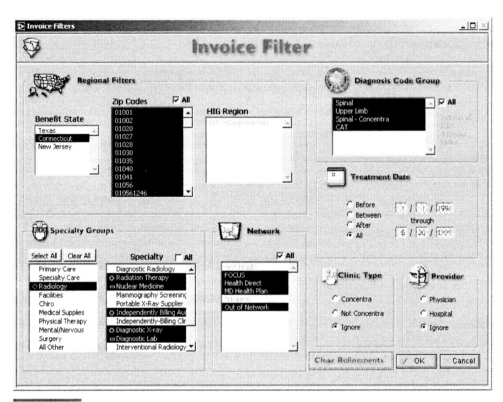

Figure 10-34.

The Invoice Filter Module allows users to easily filter multiple databases to identify a desired invoice or set of invoices. Invoices can be filtered by state and then by zip code, specialty group, diagnosis code, treatment date(s), or provider.

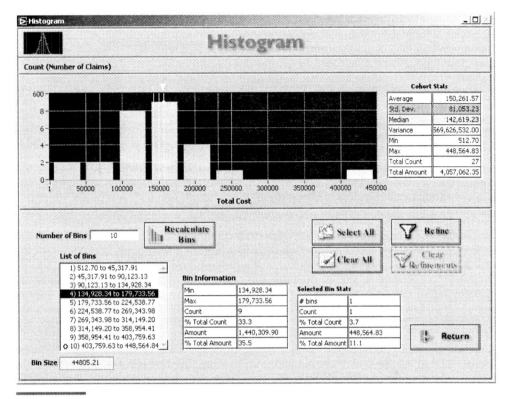

Figure 10-35.
The Histogram Module profiles claims by provider, ICD codes, or cost to easily identify averages, outliers, and distribution patterns.

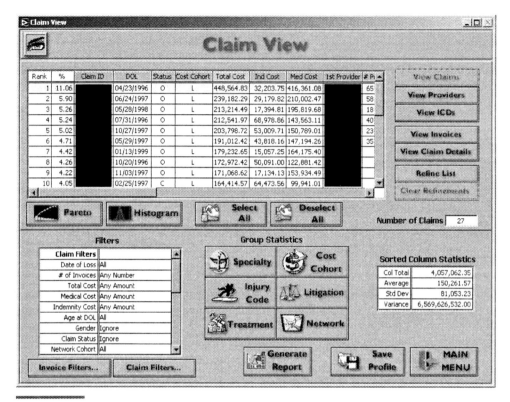

Figure 10-36.
All claims that match the user's criteria are displayed in dynamically sortable tables. Any column heading can be clicked on to sort the entire table into ascending or descending order. In this way, it is easy for the user to quickly identify claims by cost, date, ICD code, status, or provider. The most expensive claim is highlighted in yellow on screen.

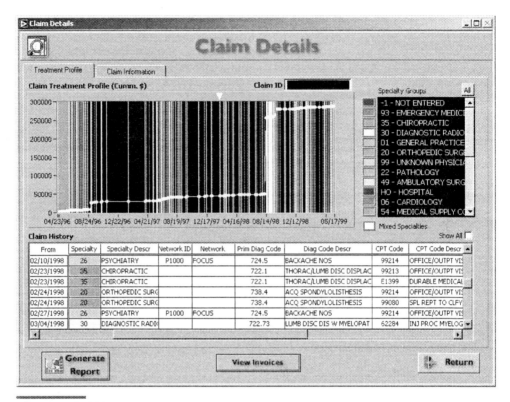

Figure 10-37.
The Claim Details Module chronologically illustrates specific treatments, providers, and costs in both a graphical and table format. This feature allows the user to see at a glance significant patterns, trends, and outliers.

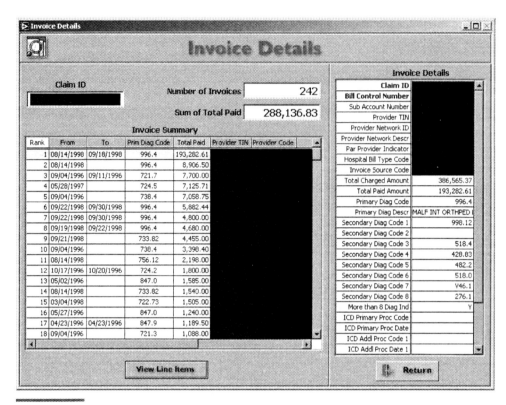

Figure 10-38.
The Invoice Details Module allows the user to drill down multiple levels for any invoice. This module also provides the user with dynamic tables that can be sorted simply by clicking on the column headings.

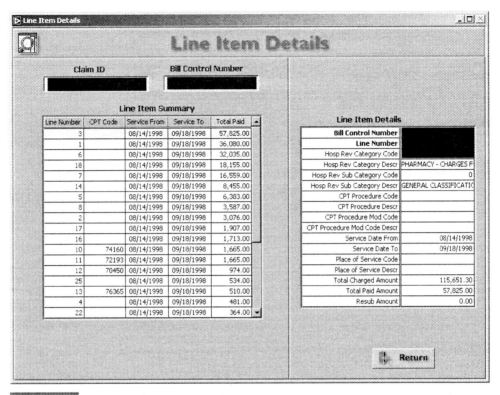

Figure 10-39.
The Line Item Details Module allows the user to drill down to individual line items for any invoice. This module also provides the user with dynamic tables that can be sorted simply by clicking on the column headings.

The Examinator: A LabVIEW-based Exam Maker, Taker, and Grader

by Eric Rosow and Joseph Adam, Premise Development Corporation

Overview

Virtual instrumentation allows hospitals and teaching institutions to conceive, develop, and implement biomedical and educational applications that achieve enhanced clinical and technical interpretations and result in significant cost reductions.

In a collaborative approach, physicians, researchers, instructors, and biomedical and software engineers have developed a virtual instrument training and evaluation system that can be easily customized to address the educational needs of practically any discipline.

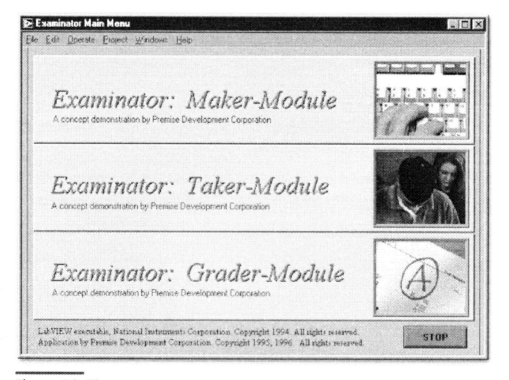

Figure 10-40.
The Examinator Main Menu.

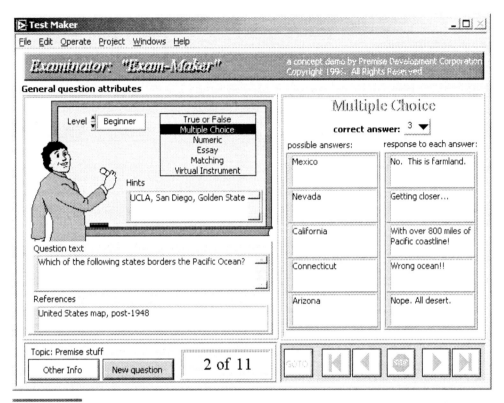

Figure 10-41.
Sample interface for the Exam-Maker Module. This panel illustrates how the Examinator allows the instructor to create questions. In this example, a multiple-choice question has been created with interactive responses.

Educational Virtual Instrumentation

Teaching any form of instrumentation in an allied health program, such as clinical laboratory science, is an expensive endeavor. The average cost of a tabletop spectrophotometer ranges from $3,000 to $5,000 per instrument, and the cost of an automated diagnostic chemical analyzer would likely exceed the scope of an educational department's budget.

The virtual instrument *Examinator*™ allows each trainee to obtain hands-on experience of sorts on a wide range of instruments. The system allows the instructor to design, create, modify, and distribute an electronic training program. The trainer may define a series of questions: true or false, multiple-choice, numeric, matching, or essay (i.e., case studies). Most importantly, the trainer may include virtual instrument questions.

Virtual instrument questions display an image of the actual instrument. The knobs, buttons, and indicators on the instrument have the same functionality as on the real-world instrument

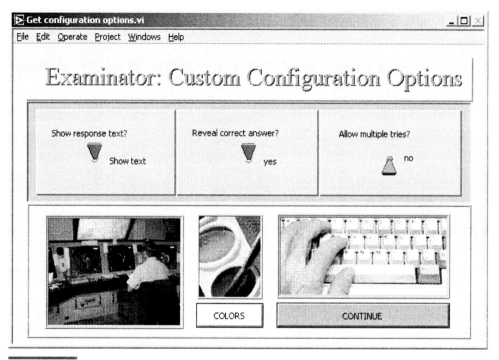

Figure 10-42.
The Examinator can be configured to provide hints and feedback, reveal the correct answer to a question, or allow the student multiple tries at achieving the correct answer. In this way, the Examinator serves as both an educational application as well as an evaluation tool.

and the trainee may manipulate them in order achieve the desired results. The user is presented a task to perform (e.g., calibrating the instrument) and provided with hints using interactive, multimedia message boxes. The computer monitors all actions of the trainee and can judge the correctness of an action based on which parameter was changed, the degree to which the parameter was changed, and the time elapsed between actions.

The software has three major components:

1. An educator-defined program to define sequence of the training program,
2. An interactive user interface to facilitate the execution of the training program, and
3. A module to provide feedback on performance and suggest areas for improvement and additional study.

Note: LabVIEW's graph object was used to allow the user to draw lines between selections of the Matching question.vi.

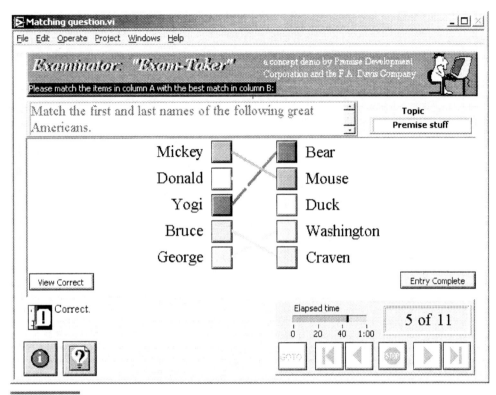

Figure 10-43.
Sample interface for the Exam-Taker Module. This panel illustrates how a student would interact with various questions during an exam or practice exercise. In this example, the student must match the first and last names of five characters. Once the matches have been made, the application provides immediate feedback to the user. In addition, if the Exam Taker Module has been configured to allow the user to view the correct answers, he or she could simply click on the View Correct button to see the correct sequence.

Conclusion

Educational software continues to play a significant role in training and self-directed learning. Virtual instrumentation applications developed for educational purposes can be an extremely powerful and cost-effective learning and evaluation tool. By leveraging the open architecture of industry-standard computers, educational virtual instruments offer a learning experience for students that might not otherwise be feasible due to access or cost constraints.

Note: For more information about the The Examinator, please contact info@premisedev.com.

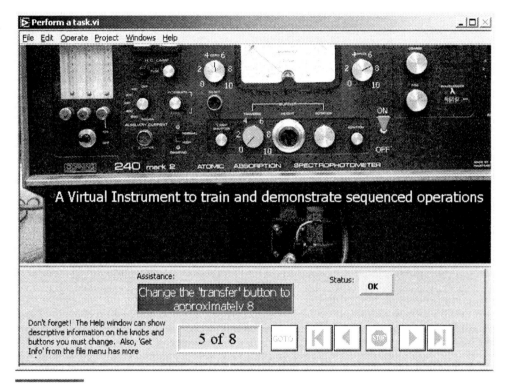

Figure 10-44.

Example of a virtual instrument. This panel illustrates how the Examinator teaches or tests a student on the use of an instrument or medical device. This task-based approach is a unique feature of the Examinator.

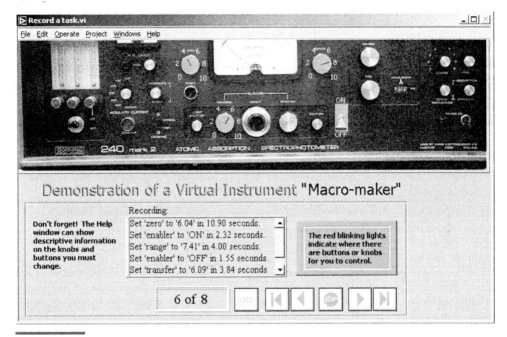

Figure 10-45.

Example of a Virtual Instrument Macro-Maker. This panel illustrates how the Examinator's Macro-Maker Module allows an instructor to create a customized test sequence for a student. In this way, an unlimited number of test conditions and simulations can be created for a student.

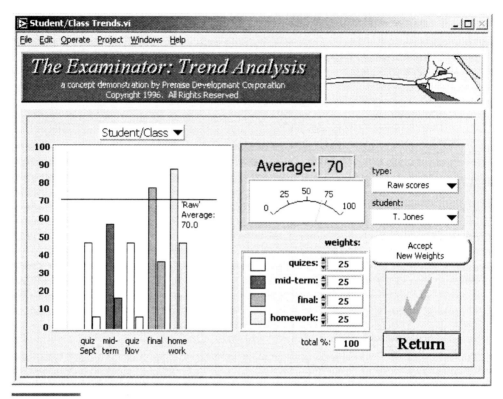

Figure 10-46.

Example of profiling and trending. This panel illustrates how the Examinator can profile students, teachers, exams, and classes. The instructor can interactively assign various weights to exams, quizzes, homework, and so on for a final grade.

Part VI
Advanced Topics

Mathematical Modeling / Simulation of Physiologic Systems

11

Cardiovascular System Modeling and Simulation
>   Theoretical Basis
>   Model Development
>   Heart Model
>   Circulatory Model
>   Experimental Measurements
>   Computational Aspects
>   Parameter Estimation and Sensitivity Analysis
>   CardioPV Program Development

Pulmonary Mechanics Modeling and Simulation
>   Theoretical Basis
>   Model Development

Integrated Cardiopulmonary Dynamics Modeling and Simulation

Some of the most significant benefits the LabVIEW™ environment offers to the biomedical student or professional is the means to readily integrate complex analysis capabilities with on-line data collection schemes and comprehensive, interactive data display capabilities. Indeed, these specific benefits greatly enhance the tasks normally associated with mathematical modeling and system identification processes applied to advanced research endeavors. While this development is an evolving effort, this chapter presents several original contributions to the study of cardiopulmonary dynamics under the premise of virtual bio-instrumentation. (Refer to Chapters 3 and 4 for relevant physiologic and data collection descriptions.)

In this chapter, we present a variety of mathematical models and associated software development tools that provide a rich platform for pursuit of integrated cardiopulmonary research:

(1) A closed-loop model has been developed that integrates a dynamic heart model, including ventricular and pericardial coupling, with a lumped-parameter model of both the systemic and pulmonic circulations (Olansen, Clark, Khoury, Ghorbel, & Bidani, 2000; Olansen, 1999). The identified model produces realistic, continuous hemodynamic waveforms around the circulatory loop, as well as reasonable estimates of cardiac output and blood pressure gradients and volume distributions. The control case then serves as a basis for study of pathophysiological states and the impact of direct and series ventricular interaction. Our simulations have demonstrated various effects of septal contraction and pericardial mechanics in modulating ventricular pump function, as well as the effects of changing afterloads on ventricular wall motions and pump function.

(2) A mathematical model that simulates airway mechanics has been developed based on work done previously (Liu et al., 1998; Athanasiades et al., 2000), including a new, specialized component model structure that specifically describes the dynamic behavior of lung tissue in spontaneously breathing human subjects over full capacity maneuvers as well as brief maneuvers of variable frequency (Athanasiades et al., under review). Both model versions have been incorporated into the **PulmPV** virtual instruments (VIs), providing a user-friendly graphical interface for operation of the model. This software also enables the use of the parameter estimation routines described in Olansen et al. (2000) as a useful adjunct to the study at hand. Results of the lung tissue study show that the variation of the Kelvin body presented here can successfully predict normal lung tissue dynamics in humans over the full volume capacity and for different breathing frequencies. As such, the model may be used as a component part in models of lung/airway mechanics. Although abnormalities in lung tissue behavior have not been addressed here, the model can be adjusted to account for different pathologies.

(3) The models delineated in points 1 and 2 have been integrated to further pursue the effects of cardiopulmonary interaction. Specifically, an assessment of hemodynamic fluctuations as a result of pulmonary system variations was accomplished. While the results presented herein do not include the study of pulmonary disease processes or the effects on gas exchange, the integrated model does provide a foundation upon which many of these potential research projects can be built.

The **CardioPV** analysis program, which incorporates the complete cardiovascular model, and the **PulmPV** program, which contains the complete airway mechanics model, each apply system identification techniques to estimate the key model parameters and serve as virtual testbeds for assessing the global effects of localized mechanical or hemodynamic alterations. Additionally, the **CardioPulm** software package builds on the capabilities of **CardioPV** and **PulmPV** while incorporating the integrated cardiopulmonary model previously described. This enables the assessment of various clinical interventions of the pulmonary system and their concomitant effects on cardiovascular hemodynamics. Given the demonstrated capabilities of the **CardioPV**, **PulmPV**, and **CardioPulm** software packages, with the incorporated model developments, *virtual modeling* can serve as a useful adjunct in cardiopulmonary research.

These research endeavors of Dr. Olansen have been previously published (Olansen et al., 2000; Olansen, 1999) and are adapted here with the appropriate permissions. The CD-ROM contains the full functioning **CardioPV**, **PulmPV**, and **CardioPulm** models including VI and CIN source code, as well as the Levenberg-Marquardt parameter estimation routines for you to learn from and improve upon.

Cardiovascular System Modeling and Simulation

This section presents the development of a closed-loop model for the study of ventricular interaction and pericardial mechanics on hemodynamics throughout the circulatory loop. This endeavor integrates three distinct areas of modeling: arterial circulation, closed-loop circulation, and heart mechanics (particularly ventricular mechanics).

Theoretical Basis

A comprehensive survey of mathematical models of the circulatory system (Melchior, Srinivasan, & Charles, 1992) has shown that, while characterizing the same biological system, these models vary significantly in their complexity, modeling assumptions, and objectives. Circulatory models range from simple resistive-compliant Windkessel models for the study of the interaction between the left ventricle and its systemic arterial afterload (Burkhoff, Alexander, & Schipke, 1988; Greene, Clark, & Mohr, 1973), to very complex distributed network representations of the systemic vascular tree (Rideout, 1991) used for more detailed studies of the hemodynamics of the arterial system. Several lumped-parameter models of intermediate complexity have been developed to characterize the complete closed circulatory loop (Ursino, 1999; Chung et al., 1997; Sun, Bashera, Lucariello, & Chiaramida, 1997; Santamore & Burkhoff, 1991; Hardy, Collins, & Calvert, 1982; Beyar & Goldstein, 1987; Halperin, Tsitlik, Beyar, Chandra, & Guerci, 1987) and have been employed in studies of a wide variety of physiological phenomena, ranging from the circulatory response to gravitational acceleration (Hardy et al.) to the effects of intrathoracic pressure variations (Beyar & Goldstein; Halperin et al.). Heart pump models range from simple spring-dashpot models (Braunwald, Ross, & Sonnenblick, 1967; Palladino & Noordergraff, 1998) to elastance models of ventricular function (Suga, 1969; Suga, Sagawa, &

Shoukas, 1973) to finite element models of ventricular mechanics (Guccione & McCulloch, 1991). As a practical matter, elastance models are relatively simple in structure and can characterize the pump properties of the atria and ventricles quite well.

In addressing the question of modeling ventricular interaction *in vivo*, the development of an adequate closed-loop model of the circulation is a prime requisite. That is, in the intact circulation, changes in the output of one ventricle eventually affect the input to the other ventricle. All closed-loop circulatory models developed to date can emulate this *series* interaction between the ventricles and their vascular afterloads. However, in addition to this series interaction, ample experimental evidence points to another significant form of interaction called *direct* ventricular interaction (Chung, 1996; Badke, 1982; Bemis, Serur, Borkenhagen, Sonnenblick, & Urschel, 1974; Bove & Santamore, 1981; Little, Badke, & O'Rourke, 1984; Santamore, Lynch, Meier, Heckman, & Bove, 1976; Slinker, Goto, & LeWinter, 1989; Taylor, Covell, Sonneblick, & Ross, 1967), where the state of one ventricle can affect the function of the other via the compliant interventricular septum (SPT). Through this direct form of interaction, the systolic and diastolic properties of both ventricles are interrelated. In addition, direct ventricular interaction is enhanced by the presence of the relatively stiff pericardium, and these two factors can have important consequences on the overall performance of the heart, as well as the circulatory system as a whole. Few models of the circulation treat either direct ventricular interaction (Santamore & Burkhoff, 1991; Amoore & Santamore, 1989) or the influence of the pericardium (Burkhoff & Sagawa, 1986).

Direct ventricular interaction is addressed in some models of the closed-loop circulation (Santamore & Burkhoff, 1991; Amoore & Santamore, 1989) via the use of interaction gains to model the modulation of ventricular pressures. However, these models do not include an explicit dynamic description of the contracting septum and therefore cannot predict the hemodynamic consequences of the temporal (phasic) interaction between the contracting free walls and the septum. Moreover, these same models do not address the role of the pericardium in modulating ventricular interaction and overall cardiac performance. In the present study, we (1) connect mathematical models of the pulmonary and systemic circulations as physiologic afterloads to the dynamic models of the ventricles of the heart, (2) include mathematical models of the passive and active behavior of the atria, and (3) enclose the atria and coupled ventricles in a pericardium. These extensions are based on work previously developed (Olansen et al., 2000; Olansen, 1999; Chung et al., 1997; Chung, 1996).

Model Development

The integrated cardiovascular model considers two major components, cardiac mechanics and circulatory hemodynamics. The structures chosen to develop this lumped-parameter model were adopted with regard to the following considerations: (1) The ventricular model should be capable of describing the continuous direct interaction of the right and left ventricles via the interventricular septum; (2) The atrial models should adequately describe both the passive and active (contractile) behavior of each atrium (the atria are considered separate uncoupled compartments); (3) The heart model should include a description of the pericardial influence on the cardiac mechanics associated with all four chambers; (4) The resistive, compliant, inertial system used to describe the hemodynamics of both the systemic and pulmonary vasculature should be appropriately apportioned with regard to blood volume, reflect appropriate mean pressure levels, and present an appropriate hydraulic input impedance to the ejecting ventricles; and (5) The complete model should be versatile and easily applied to a variety of simulated hemodynamic conditions (normal and pathophysiological).

Heart Model

Ventricular Description. The general form of the ventricular model is based on the work of Chung et al. (1997) on the isolated ventricle. The basic premise upon which the model is built is that the ventricles can be modeled as a three-walled system: the right ventricular (RV) and left ventricular (LV) free walls and the coupling septal wall. This creates three functional volumes as depicted in Figure 11-1A. A septal volume of zero indicates a flat septum; the chamber volumes are then equal to the free wall volumes. $V_{SPT} > 0$ represents septal movement into the RV, whereas $V_{SPT} < 0$ indicates leftward septal motion. This characterization of functional volumes is intended to signify septal wall motion only and does not affect the total chamber volumes unless accompanied by imbalanced input and output blood flows.

The ventricular walls are characterized by time-varying elastance functions that relate instantaneous pressure and volume. These elastance functions produce a smooth transition from a nonlinear end-diastolic pressure-volume curve (EDPVR) to a linear end-systolic pressure-volume relationship (ESPVR), both of which have been established in Chung et al. (1997). Time-varying activation functions $e_v(t)$, which are representative of normalized

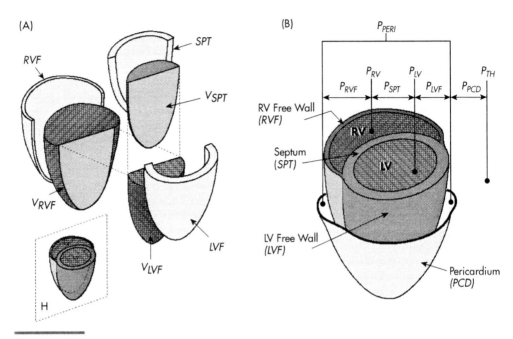

Figure 11-1.
Volume distribution (A) and force balance (B) for model of coupled ventricles within the pericardium. See text for description and nomenclature. (From Chung et al, 1997.)

elastance curves, are used to develop the elastances. A more complete description of the cardiac model equations can be found in Chung et al.

Atrial Description. The ventricular model previously described was extended to include a dynamic description of the atria. For this model, the atrial septum is assumed to be rigid, that is, the atria are uncoupled and have no direct mechanical influence on each other or the ventricles. The atrial free walls are characterized by a time-varying elastance in a fashion similar to the ventricles. In this case, however, the EDPVR and ESPVR have been developed to produce reasonable outputs that correspond to the work of Lau, Sugawa, and Suga (1979) as well as to the data we have collected.

Heart Model Dynamics. Figure 11-1B depicts a schematic diagram showing the components of the heart model. Pressure (force) balances across the appropriate walls, in addition to mass balances between the heart chambers, are used to derive the dynamic interaction of the heart chambers, the septum, and the pericardium. While the atria are modeled as active free walls within the pericardium, the atrial septum is considered rigid for model simplification purposes. The heart chambers are enclosed within an elastic pericardium under open-chest conditions (i.e., intrathoracic pressure variations are neglected; P_{PERI} is referenced to atmosphere).

The mass balances are developed based on the principles of conservation of mass and the continuity equation and result in a series of dynamic flow equations describing the changes in chamber volumes as a function of time:

$$\dot{V}_{LV} \equiv \frac{dV_{LV}}{dt} = Q_{MT} - Q_{AO} \tag{1a}$$

$$\dot{V}_{RV} \equiv \frac{dV_{RV}}{dt} = Q_{TC} - Q_{PM} \tag{1b}$$

$$\dot{V}_{LA} \equiv \frac{dV_{LA}}{dt} = Q_{LA} - Q_{MT} \tag{1c}$$

$$\dot{V}_{RA} \equiv \frac{dV_{RA}}{dt} = Q_{RA} - Q_{TC} \tag{1d}$$

where Q_{MT}, Q_{AO}, Q_{TC}, and Q_{PM} are flows through the mitral, aortic, tricuspid, and pulmonary valves, respectively, and Q_{LA} and Q_{RA} are the flows into the left and right atria. The flows are derived using the generic formula:

$$Q_i = \Delta P_i / R_i \tag{1e}$$

where P_i is the forward pressure gradient across the flow resistance (R_i) encountered at each heart valve and the atrial inlets.

Circulatory Model

The systemic and pulmonary circulations are modeled as physiologic afterloads to the dynamic model of the coupled ventricles located, along with the atria, within the pericardium. Previous studies (Hardy et al., 1982; Tsitlik et al., 1992) have characterized the circulatory loop as a series of compliant hydraulic elements. We also utilize this approach and characterize the circulatory hemodynamics in terms of an apportioned resistive, compliant, and inertial system (Chung, 1996). A hydraulic equivalent circuit model of the circulatory loop is depicted in Figure 11-2.

Each segment of the circulatory loop is modeled via a set of equations describing the relationships between the pressure (P_i, mmHg), volume (V_i, ml),

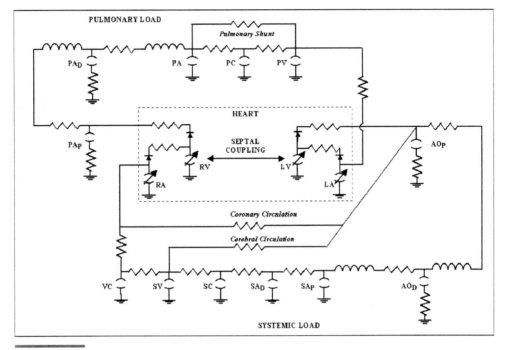

Figure 11-2.
Hydraulic equivalent schematic of the closed-loop circulatory model. Model parameters are listed in Table 11-1 and the Appendix. (Reprinted with permission from Olansen et al., 2000.)

and flow (Q_i, ml/sec) associated with that segment. The equations governing each hydraulic segment are listed next and they contain a number of parameters that describe each segment's resistance (R_i, mmHg – sec/ml), inertance (L_i, mmHg – sec^2/ml), and compliance (C_i, ml/mmHg). The relationships used to characterize each hydraulic element are

$$P_i = P_{TM.i}(V_i) + R_{T.i}\dot{V}_i + P_{EX.i} \tag{2a}$$

$$P_{TM.i} = \frac{1}{C_i}(V_i - V_{0.i}) \tag{2b}$$

$$\dot{Q}_i = \frac{P_{i-1} - P_i - Q_i R_i}{L_i} \tag{2c}$$

$$\dot{V}_i = Q_i - Q_{i+1} \tag{2d}$$

where i indicates the specific element (e.g., AO_P for proximal aorta [aortic arch], VC for vena cava). Some segments of the circulatory loop contain only resistive and compliant elements because, as the diameter of the blood vessels diminishes around the circulatory loop (namely, the arterioles, capillaries, and venules), the volumetric flow likewise declines, reducing the significance of the inertance (L_i) parameter characterizing the blood column. Neglecting this term for the appropriate elements reduces equations 2c and 2d to a single ordinary differential equation.

In the previous derivation, the natural state variables are the volumes of each compliant element (V_i). However, the linear pressure-volume relationship about a given operating volume, given in equation 2b, allows a transformation of variables. Thus, the transmural pressure ($P_{TM,i}$), the pressure across the segment wall, is classified as a state variable. This leads to the following alternative equations:

$$\frac{dV_i}{dP_{TM,i}} = C_i \tag{2e}$$

$$\dot{P}_{TM,i} \equiv \frac{dP_{TM,i}}{dt} = \frac{1}{Ci} \ (Q_i - Q_{i+1}) \tag{2f}$$

In this case, the pressure within a given segment (P_i) can be determined using equation 2a. We assume experimental conditions consistent with the open-chest animal preparation, hence the external pressure ($P_{EX,i}$) is atmospheric.

Experimental Measurements

Pressure measurements from selected sites in the circulatory loop were obtained from open-chest canine preparations in the Center for Experimental Cardiac Electrophysiology, Baylor College of Medicine, Houston, Texas. These anatomic sites include the left ventricle, aortic arch, descending aorta, and femoral arteries. Additional records were obtained from the inferior vena cava, the right atrium, and right ventricle. The pressure recordings from four sites were typically acquired simultaneously, and the ECG was recorded as an independent timing reference.

Pressures were obtained using solid state catheter-tip transducers from Millar Instruments (Houston, TX). Positioning of the transducers within the heart chambers or vessels was verified via X-ray imaging prior to initiating acquisition. The analog recordings were digitally sampled at a rate of 500 Hz, which was sufficient to capture all significant frequency content of the acquired waveform. Data acquisition was accomplished using a mobile PC platform consisting of a National Instruments AI-16E-4 PCMCIA DAQCard and a 266 MHz laptop. Analog signal conditioning was incorporated via an SCXI-1120 module (National Instruments) in order to amplify the incoming signals (ECG 1000; Pressure 10).

The acquisition was controlled via virtual instruments designed and developed within the LabVIEW programming environment, as well as through use of the turnkey program BioBench™. (LabVIEW and BioBench are both products of National Instruments, Austin, TX). The data collection virtual instruments enabled (1) the raw data to be converted to engineering units (mmHg) using previously established calibration curves; (2) continuous real-time display of the acquired data for quality control; and (3) storage of acquired, calibrated data for post-processing. A sample of the data acquired is shown in Figure 11-3A.

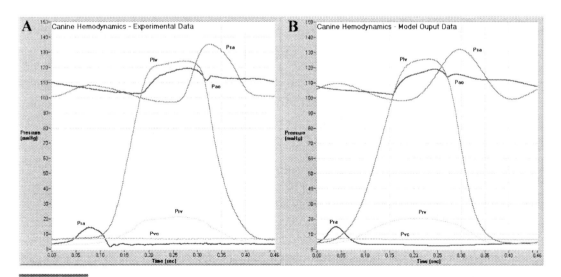

Figure 11-3.

(A) Typical canine cardiovascular pressure data collected during open-chest experiments. (B) Typical model output pressure data produced under similar simulated conditions as the data displayed in panel A (e.g., equivalent heart rate, open chest, intact pericardium). LV = left ventricle, RV = right ventricle, AO = aorta, SA = systemic arteries, VC = inferior vena cava, RA = right atrium. (Reprinted with permission from Olansen et al., 2000.)

Computational Aspects

The model equations were initiated at the time of end-diastole with the atrio-ventricular valves open and the semilunar valves closed. Initial conditions employed for the 32 state variables, some of which were adopted from previous studies (Chung et al., 1997; Chung, 1996), are defined in this mode. The model parameters were divided into static and adjustable sets, with the adjustable parameters determined using the parameter estimation algorithm described next. Nominal parameter values were selected to produce physiologically realistic values, such as cardiac output (CO) and hemodynamic waveforms for a typical 25-kg dog. Pulmonary shunt flow was set at 2% of the mean pulmonary blood flow (Nunn, 1993), whereas mean coronary and cerebral flows were set at 5% and 15% of cardiac output, respectively (Folkow & Neil, 1971; Mountcastle, 1968). The model equations were programmed in C and solved using a variable step-size Runge-Kutta-Merson algorithm with a maximum time step size of 2×10^{-3} sec and an error tolerance of 1×10^{-6}.

Parameter Estimation and Sensitivity Analysis

A nominal set of parameters was first obtained that provided acceptable fits to a variety of indices such as cardiac output (CO), left and right ventricular ejection fractions, mean blood pressures and blood volume distributions around the circulatory loop, and ventricular and arterial pulse pressure waveforms. Model output pressures from a typical control case are depicted in Figure 11-3B, corresponding to the pressures obtained experimentally, and shown in Figure 11-3A. Compare Figures 11-3A and 11-3B, and you will establish that the model output is in general agreement with typical *in vivo* data.

With the baseline hemodynamics established in the control case just described, use was made of parameter sensitivity analysis (Paulsen, Clark, Murphy, & Burdine, 1982; Khalil, 1992) and a parameter estimation scheme (Marquardt, 1963) to achieve better agreement with experimental data. Examinations of the magnitude and time course of the relative sensitivity functions associated with all of the circulatory hemodynamic parameters (i.e., resistances, compliances, and inertances) were evaluated for the purpose of ranking the degree of sensitivity of the set (as described in Olansen, 1999). This analysis revealed that the most sensitive parameters were R_{TAO}, C_{AOP}, and L_{AOP} of the systemic circulation and R_{TPA}, C_{PAP}, and L_{PA} of the pulmonary circulation. The resistance parameters R_{SAD} and R_{SC} are also quite sensitive, as they are largely responsible for establishing the mean aortic and systemic arterial pressures that are directly compared with the data.

Based on the dynamic sensitivity analysis, the model parameters classified as most sensitive are included in the nonlinear least-squares parameter estimation algorithm. This step significantly enhances convergence of the algorithm by constraining parameter variations to those that can bring about the greatest change in the system variables. The nature of the closed-loop circulation allowed initial estimation of the parameters to be performed on each circulation (i.e., systemic and pulmonic) separately followed by an integrated fine-tuning. This primarily enabled the ventricular systolic profiles and hemodynamic means to be established prior to identifying the parameters responsible for the oscillatory perturbations in the pressure waveforms. Figure 11-4 depicts the user interface for operation of the parameter estimation

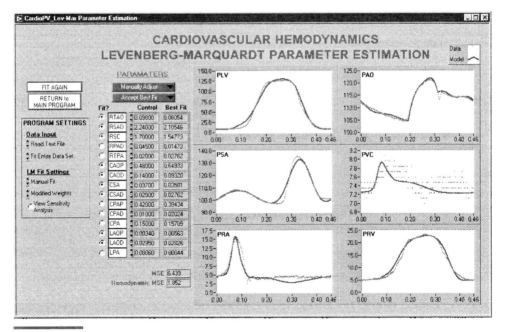

Figure 11-4.
Front panel of **CardioPV** parameter estimation routine depicting typical results of identification analysis. Parameters are defined in Table 11-1 and the Appendix. *MSE* is the mean square error of all compared data sets. *Hemodynamic MSE* disregards errors in pressures within heart chambers. (Reprinted with permission from Olansen et al., 2000.)

algorithm with typical results. These results show good agreement for specified pressures around the cardiovascular loop. Table 11-1 summarizes parameter values determined for each of the experiments conducted. These techniques are described in greater detail in Olansen et al. (2000).

It should also be noted that heart rate is a significant factor in determining the circulatory hemodynamics. As such, the heart rate established from each data set was utilized in determining the adjustable parameters. Since the heart rate varies with respiratory, baroreceptor, and other influences, only small data intervals incorporating three to four beats of roughly equal RR intervals (i.e., the interval of time between the peaks of the R wave in the ECG) were used at one time in the identification scheme. The effects of varying heart rate on the circulatory hemodynamics are discussed later.

By producing physiologically realistic outputs, this model allows the study of ventricular interaction through the septum (direct), as well as via the circulatory loop (series) under simulated *in vivo* conditions. Once model parameters were identified, simulations were conducted to examine a number of additional problems that are often referenced in the literature. The results of these simulations are discussed next.

Table 11-1. *Key parameter values identified via the described parameter estimation routine for three separate canine experiments. Units for resistance terms are mmHg − sec/ml; for compliance terms ml/mmHg; and for inertance terms mmHg − sec^2/ml.*

Parameter	Description	Typical Values	Identified Values Dog 1	Dog 2	Dog 3
R_{SAD}	Resistance: systemic arterioles	2.2400	2.1050	1.7498	2.1162
R_{SC}	Resistance: systemic capillaries	1.7000	1.5477	1.1444	1.1012
R_{PAD}	Resistance: distal pulmonary artery	0.0450	0.0147	0.4263	0.0454
R_{TAO}	Viscoelastic element: prox. aorta	0.0980	0.0805	0.0853	0.0813
R_{TPA}	Viscoelastic element: pulm. artery	0.0200	0.0276	0.0300	0.0197
C_{AOP}	Compliance: proximal aorta	0.4800	0.6493	0.3745	0.6818
C_{AOD}	Compliance: distal aorta	0.1400	0.0932	0.0105	0.0270
C_{SA}	Compliance: large systemic arteries	0.0370	0.0358	0.0250	0.0318
C_{SAD}	Compliance: systemic arterioles	0.0280	0.0276	0.0308	2.5876
C_{PAP}	Compliance: proximal pulm. artery	0.4200	0.3943	0.6025	0.2277
C_{PAD}	Compliance: distal pulm. artery	0.0100	0.0202	0.0593	0.3647
C_{PA}	Compliance: pulmonary arterioles	0.1500	0.1571	0.0007	0.6688
L_{AOP}	Inertance: proximal aorta	0.0034	0.0056	0.0234	0.0450
L_{AOD}	Inertance: distal aorta	0.0295	0.0282	0.0506	0.0277
L_{PA}	Inertance: pulmonary artery	0.0006	0.0004	0.0002	0.0001

CardioPV Program Development

The **CardioPV** program, developed using the LabVIEW programming environment, is a compilation of original VIs designed both as the user and data interface to the model operation. Sophisticated graphical displays highlight the user interface while the **CardioPV** software can accept data from numerous sources, including direct acquisition via a data acquisition (DAQ) card for on-line operation. The model C code described earlier is integrated into the **CardioPV** program as an external call from LabVIEW. **CardioPV** also provides a continuous graphical feedback of the model operation, enhancing user understanding of the model performance as well as aiding any debugging efforts that may be required.

The computational flow diagram shown in Figure 11-5 depicts the various stages that are included within the software package. Data can be acquired directly, allowing the software analysis to be utilized during an ongoing experiment. Data can also be read from a number of previously recorded file types. In these cases, the model is run with the average heart rate (HR) determined from the data and will run for the same elapsed time as included in the data segment for comparison purposes. When data is linked in the program, the option to directly compare model output using the aforementioned parameter estimation and sensitivity analysis routines becomes available. This capability greatly enhances the functionality of the model. The model can also be run without reference to experimental data by supplying a user-specified HR and simulation elapsed time.

The screen panel shown in Figure 11-4 provides the user with multiple options to completely control the parameter estimation process, if so desired. Yet, its default configuration is sufficient to enable novice users to typically obtain satisfactory results. The sensitivity analysis is conducted in the background and can be reviewed during performance of the parameter estimation routine.

Once acceptable parameter values are obtained, simulations can be continuously conducted for a number of the alterations or pathophysiologies. In these cases, all model parameter values are held constant except for those changes necessary to simulate the desired intervention. The model will run continuously until it reaches a steady state output, at which time mean pressure, volume, and flow data around the loop, as well as a variety of cardiac indices such as cardiac output (CO) and ejection fraction (EF), are calculated. These features allow easy manipulation of model parameters, which help to provide a quick, convenient, and thorough analysis of ventricular

CardioPV Software Package Computational Flowchart

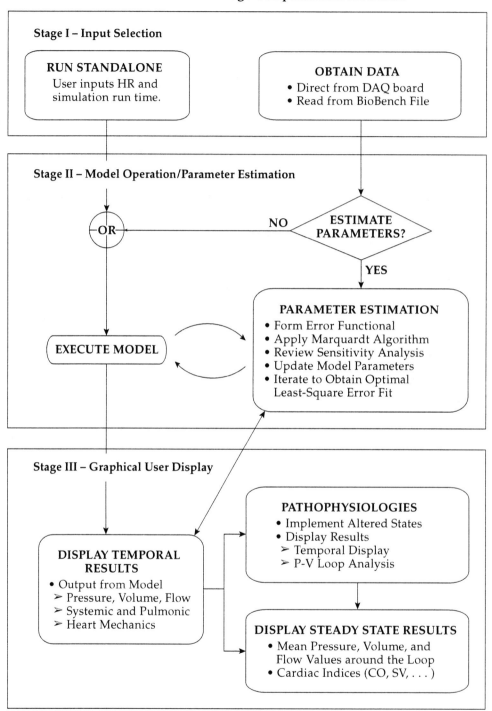

Figure 11-5.
Computational flow diagram of the **CardioPV** software package. (Reprinted with permission from Olansen et al. 2000.)

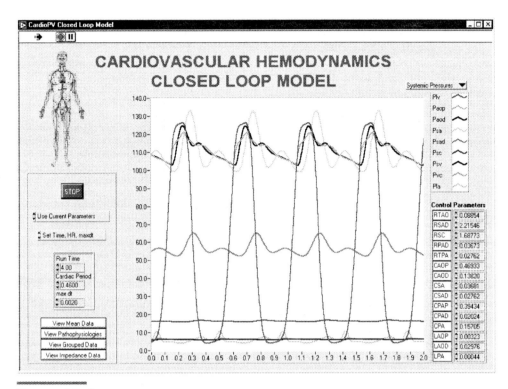

Figure 11-6.
Front Panel of **CardioPV** program. The user sets the model options on the left, while control parameter values are depicted on the right. The user selects the output set to view via the selection box at the top right.

interaction, pericardial influence, and normal hemodynamics as well as the simulation of a wide variety of pathophysiological conditions or clinical interventions.

From the main **CardioPV** front panel shown in Figure 11-6, the user can select the set of model outputs that are of interest. The outputs are grouped together as systemic pressures, volumes and flows, pulmonic pressures, volumes and flows, as well as septal, free wall, and elastance dynamics. Upon starting the top level VI, the model will run with the default parameter values. (If the user has selected Estimate Parameters, the Levenberg-Marquardt algorithm will be started first.) Once the model has run, the output is available for viewing. As a reminder, the parameter values included in the version on the CD-ROM are for a typical 25-kg dog, not a human.

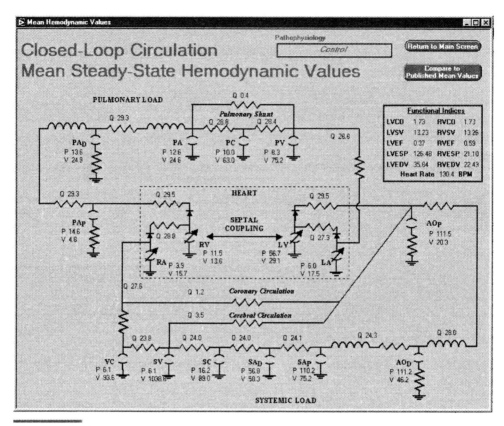

Figure 11-7.
Representation of the mean hemodynamic data and associated cardiac function indices output from the closed-loop **CardioPV** model.

The mean values and cardiac indices also become available as depicted in Figure 11-7. This screen allows the modeler to view the steady state results of the model throughout the entire loop in one snapshot. These values can then be compared with typical values published in the literature, as exemplified by Figure 11-8. There are a number of additional output views and formats that **CardioPV** is capable of producing, many of which are used to depict specific research results in Olansen et al. (2000) and all of which the user can view by running the VIs from the accompanying CD. These include model outputs for a variety of pathophysiologic conditions and the ability to group related conditions as well as estimations of cardiac output impedance.

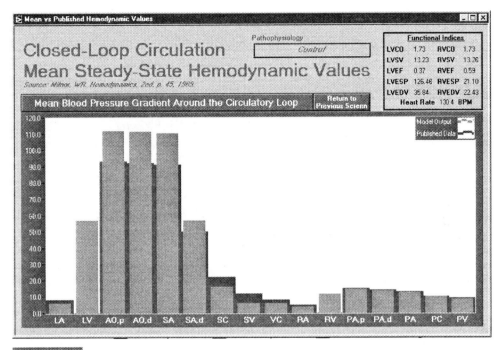

Figure 11-8.
Representation of **CardioPV** model outputs compared with typical data previously published in the literature.

Pulmonary Mechanics Modeling and Simulation

This section focuses on the means of collecting pulmonary function data, and using that data to verify the development of theoretical models. Specifically, the particular model addressed here first deals with a study of the viscoelastic properties of lung tissue and is then incorporated into a more comprehensive nonlinear model of pulmonary mechanics.

Theoretical Basis

Lung Tissue Viscoelastance. The mechanical properties of lung tissue are described by the relationship between transpulmonary pressure (alveolar pressure minus pleural pressure) and the volume of air contained in the alveoli (lung volume). This relationship is nonlinear and reveals both elastic and nonelastic behavior. In particular, under dynamic conditions, the pressure-volume (P-V) curve exhibits *hysteresis*, whereby the inspiratory and expiratory paths do not coincide (inspiration involves higher pressures than expiration). The area enclosed by the hysteretic loop equals the energy dissipated during each breathing cycle. This suggests the existence of a nonelastic component in the tissue, commonly referred to as the lung tissue resistance. Similarly, dynamic compliance quantifies the elasticity of the material and is measured as the slope of the line connecting the extrema of the hysteretic loop. Under static conditions, the loop collapses onto a single, nonlinear P-V curve, the slope of which equals the static compliance of the lung tissue. Lung tissue also exhibits *stress adaptation* (Sharp, Johnson, Goldberg, & van Lith, 1967), that is, the pressure differential across it changes value over time under constant volume conditions. Equivalently, the volume changes even after the pressure has been fixed, a phenomenon termed *creep* (Suki, Barabasi, & Lutchen, 1994). It has been suggested that stress adaptation in the tissue can be attributed to the same mechanism responsible for the hysteretic loop observed under dynamic conditions (Hughes, May, & Widdicombe, 1959). The properties described earlier are characteristic of viscoelastic behavior.

Experimental studies (Sharp et al., 1967; Bachofen, 1968; Bachofen & Hildebrandt, 1971; Hildebrandt, 1970; Lutchen et al., 1994), mostly on anesthetized animals, show that the dynamic behavior of lung tissue exhibits the following properties: (1) Dynamic compliance decreases as the frequency increases, (2) Tissue resistance drops with increasing frequency, and (3) The static P-V curve is nonlinear and local incremental compliance varies with volume.

The linear viscoelastic solid, or Kelvin body, consisting of a compliance in series with a parallel combination of another compliance and resistance (often called the Maxwell body), is capable of simulating many aspects of tissue behavior, including the dependence of tissue resistance and compliance upon the frequency of breathing. Jonson et al. (1993) employed this model in mimicking data collected from healthy anesthetized humans ventilated in the linear range of total respiratory system behavior (lung tissue plus chest wall). Svantesson, John, Taskar, Evander, and Johnson (1996) proposed a variation of the Kelvin body that employs nonlinear viscoelastic components and demonstrated that this model was able to predict respiratory system mechanics in ventilated rabbits.

Other modeling schemes have been introduced in the literature as well. Among them is the work of Hildebrandt (1970), who proposed a composite model of linear viscoelastic and plastoelastic elements. The latter are required to account for the static hysteresis that some experimenters have discovered in lung tissue (i.e., P-V hysteresis under zero flow conditions). The need to address respiratory mechanics properties in nonlinear ranges and higher frequencies has guided research to more advanced modeling efforts. These include distributions of Kelvin bodies (Hantos, Daroczy, Csendes, Suki, & Nagy, 1990), viscoelastic and plastoelastic elements (Stamenovic, Glass, Barnas, & Fredberg, 1990), nonlinear viscoelasticity (implemented with a Volterra series; Suki & Bates, 1991), and coupled dissipative and elastic processes (hysteresivity; Fredberg & Stamenovic, 1989).

Variations of the Kelvin body have been employed in the past to characterize lung tissue properties in ventilated dogs (Bates, Brown, & Kochi, 1989), but most studies have employed this model to simulate the total respiratory system (lung and chest wall) in anesthetized humans (Jonson et al., 1993), ventilated rabbits (Svantesson et al., 1996), and anesthetized dogs (Similowski et al., 1989) and cats (Kochi, Okubo, Zin, & Milic-Emili, 1988).].

Unlike these works, this study confines the application of the model to the simulation of human lung tissue dynamics only. As discussed in Chapter 4, transpulmonary pressure data is collected in naturally breathing human subjects, and, thus, the lung tissue is isolated from its neighboring structures. Then, the data is used to run the model (experimentally generated driving waveform), identify its parameters, and make comparisons, both for dynamic and frequency-dependent variables. This study establishes that this simple model fits experimental data well, both over the nonlinear range of breathing volumes and also over a range of breathing frequencies. This model is subsequently used as part of a larger model of the respiratory system that includes characterizations for airway, thorax, and diaphragm dynamics.

Pulmonary Mechanics. Similar to the numerous modeling endeavors undertaken to describe the cardiovascular system, a wide variety of modeling efforts have been applied to the study of pulmonary mechanics, dynamics, and gas exchange. For the present study, we consider only the modeling of airway mechanics. Previously, detailed distributed models of airway mechanics have been developed to describe pressures and flows in each airway generation (Elad & Kamm, 1989; Pardaens, van de Woestijne, & Clement, 1972; Thiriet, Bonis, Adedjouma, Hatzfeld, & Yvon, 1987; Wada, Seguchi, & Tanaka, 1991). The complexities of these models enable them to mimic both tidal breathing and a variety of maneuvers, such as the forced vital capacity (FVC) maneuver. However, the complexities also make it difficult to estimate the distributed parameters of the model from data routinely collected in a PFT. Conversely, lumped parameter models, which are adequate to describe airway dynamics, yet sufficiently simple to be useful for clinical applications, have also been developed (Golden, Clark, & Stevens, 1973; Jackson & Milhorn, 1973; Olender, Clark, & Stevens, 1976; Tomlinson, Tilley, & Burrows, 1994). Lumped models of this type have been applied to the study of various breathing maneuvers, from tidal breathing to panting to the FVC maneuver.

One objective of this type of model is to gain better insight into normal respiratory function for potential applications in clinical settings. The assessment of work of breathing (WOB) is one common clinical reason for studying pulmonary mechanics. WOB measurement finds application in the assessment of respiratory failure and in ventilatory management, as an indicator of the metabolic energy of breathing. Recently, with the advent of new ventilatory modes for the critically ill patient, assessment of breathing effort has received revived interest among clinicians. The application of this model to the estimation of the WOB and respiration energetics is discussed in Athanasiades et al. (2000).

Model Development

The pulmonary mechanics model includes nonlinear characterizations of airway resistance, chestwall compliance, and lung tissue viscoelasticity. The model features separate resistive coefficients for the upper, middle, and small airways; a compliant characterization of partially supported airways; a modified Kelvin body that describes the dynamic compliance of lung tissue; and a static compliance for the chest wall. The mathematical models employed here are based on our respiratory models developed in Liu et al. (1998), Athanasiades et al. (2000), and Athanasiades et al. (under review). Read these articles for more detailed theoretical descriptions of these models.

A physical representation of the full model, depicting its individual components, appears in Figure 11-9A, along with its pneumatic analog in Figure 11-9B. To allow for a complete accounting of the work associated with breathing and for a better response to maneuvers of variable frequency, the following modifications are introduced to the original model:

(1) A nonlinear compliance C_{CW} is used to represent the combined elastic behavior of the chest wall and diaphragm, such that the energy stored in these structures can be accounted for;

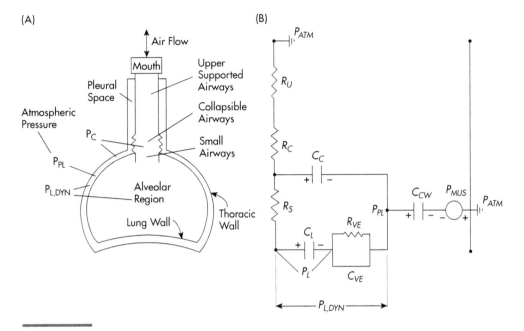

Figure 11-9.
(A) Physical model of respiratory system; (B) Pneumatic analog of pulmonary mechanics model. (Reprinted with permission from Athanasiades et al., 2000.)

(2) As a result of this modification, the signal driving the model is no longer pleural pressure P_{PL}, but rather the pressure P_{MUS}, which describes the net equivalent effect of respiratory muscle activity; P_{MUS} is directly influenced by the respiratory controller in the brain and its waveform is easily reproduced in simulated maneuvers; and

(3) A viscoelastic structure (Kelvin body) replaces the original pressure-volume characterization of lung tissue. The modifications are treated in more detail in the following paragraphs.

Model functions and parameters for all subjects are summarized in Athanasides et al., 2000. Parameter values of the full nonlinear model have been tuned to match experimental records of volume and flow in the FVC maneuver, as it is rich enough to excite most system dynamics. In general, parameter values comply with specifications given in Liu et al. (1998). In addition, they remain unchanged for all simulated maneuvers in a single subject.

Lung Tissue Viscoelastance

A pneumatic representation of the lung tissue model appears as the shaded portion in Figure 11-10. It employs a combination of a nonlinear compliance C_L, of volume V_A (lung volume), in series with a linear compliance C_{VE}, of volume V_{VE}, and linear resistance, R_{VE}. The compliance, C_L, describes the relationship between V_A and the pressure drop across the tissue wall, under static conditions. Experimental evidence shows that this pressure-volume relationship is nonlinear (Sharp, et al., 1967; Bachofen, 1968); it is modeled herein as a logarithmic function of volume. Logarithmic expressions such as this have been used previously to characterize elastic structures of the lung (Liu et al., 1998).

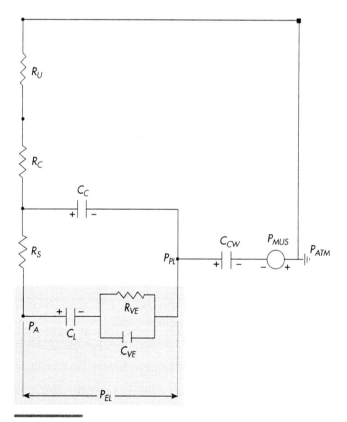

Figure 11-10.
Pneumatic representation of pulmonary mechanics model. Shaded region describes lung tissue viscoelastance.

The total pressure drop across the shaded elements of Figure 11-10 is the dynamic elastic recoil of lung tissue, P_{EL}, accounting for both the elastic forces and the forces introduced by any nonelastic behavior. We use our measurements of P_{EL} as the independent forcing function in our simulations. This allows us to isolate the lung tissue structure from the total respiratory system, effectively separating lung tissue from the adjoining airways and chest wall. Figure 11-10 shows a pneumatic representation of the total respiratory system, which includes characterizations for the airways and chest wall, in addition to lung tissue. The model successfully predicts lung volume and flow waveforms in dynamic maneuvers that include tidal breathing, panting, and forced vital capacity efforts (simulations generated with a similar model appear in Liu et al., 1998). We have used this model to simulate our measurement procedure, as explained later. It is presented here to more clearly identify the modeled lung tissue structure in the context of a full lung/airways model.

The volumes V_A and V_{VE}, associated with the compliant elements of the lung tissue model, are the two independent variables that completely describe the dynamic behavior of the pneumatic network shown in the shaded portion of Figure 11-10. The differential equations describing the motion of the system can be derived from the balance of pressures along the branches of the network:

$$\begin{bmatrix} 1 + \dfrac{bC_{ve}(TLC - RV)}{(V_A - TLC)(V_A - RV)} & 0 \\ -1 & 1 \end{bmatrix} \begin{bmatrix} \dot{V}_A \\ \dot{V}_{VE} \end{bmatrix} + \begin{bmatrix} 1/(R_{ve}C_{ve})V_{VE} \\ 1/(R_{ve}C_{ve})V_{VE} \end{bmatrix} = \begin{bmatrix} \dot{P}_{el} \\ 0 \end{bmatrix} \qquad (3)$$

Chest Wall

In previous studies (Liu et al., 1998), the chest wall was assumed rigid and the lungs were driven by the pleural pressure waveform measured via the esophageal balloon technique. Since elastic forces developed in the chest wall and diaphragm expend part of the effort during breathing, we have reformulated our model to include a lumped characterization of the thoracic wall and diaphragm. They are modeled as a series combination of an independent pressure source, P_{MUS}, serving as the dynamic force driving the model, and a passive compliant element, C_{CW}, as shown in Figure 11-10. The volume of air contained in C_{CW} is the chestwall volume, V_{CW}, and the pressure across it is the chestwall elastic recoil, P_{CW}. This is equal to the difference between the total pressure across the chest wall, pleural pressure P_{PL}, and the pressure developed due to the respiratory muscles.

P_{CW} is commonly approximated by a sigmoidal curve (Taylor et al., 1967) that is linear in the range of quiet breathing and saturating at higher and lower volumes. We mimic this relationship again with a logarithmic expression. Numerical values are chosen so that chestwall compliance, $C_{CW} = dP_{CW}/dV_{CW}$, is approximately 0.2 $1/cmH_2O$ in the linear range (volumes of 2.5 to 3 liters) for the subject tested. During tidal breathing the chest wall has mostly outward recoil.

Airways

Functional relationships and parameter values modeling the airways are adopted directly from Liu et al. (1998), except for the resistive characterization of the peripheral airways, which now bear a nonlinear relationship with volume only. The corresponding characterization of peripheral airways in Liu et al. incorporated an effort-dependent term, aimed to achieve the limitation of flow during the expiratory portion of the forced vital capacity maneuver. We replace that mechanism by restricting the airways' elastic recoil P_C to positive values only: The airways can now sustain distension, but not compression.

As formulated in Liu et al., the value of R_S increases exponentially as lung volume V_A decreases. The formula aims to capture the dependence of small airway resistance on lung volume. As the lung inflates, small airways imbedded into the lung parenchyma are stretched open, allowing free passage of airflow (low resistance). At low lung volumes, these airways are constricted, offering large resistance to flow. The value of R_C is inversely proportional to the square of the volume in the airways V_C. The same principle applies here: As the airways become narrower (lower volume) they offer higher resistance to flow, and vice versa. The characterization of R_C was originally developed in Golden et al. (1973) and Olender et al. (1976).

Full Model of Respiratory Mechanics

The equations governing the motion of the system are developed using a Lagrangian analysis (Greenwood, 1977). The dynamic behavior of the system is described completely by generalized coordinates whose number is equal to the number of energy storage elements minus the number of constraints. In the absence of inertial elements, kinetic energy is zero and potential energy is stored in the four compliant compartments. Due to the kinematic constraint $V_{CW} = V_A + V_C + V_D$, however, only three independent generalized coordinates are realized, namely, $q_1 = V_C$, $q_2 = V_A$, and $q_3 = V_{VE}$, depicted in the mechanical representation of the model in Figure 11-11.

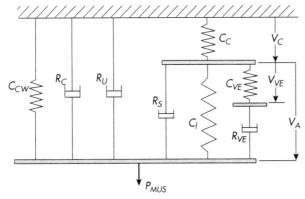

Figure 11-11.
Mechanical equivalent of pulmonary mechanics model. (Reprinted with permission from Athanasiades et al., 2000.)

In matrix form, the resultant equations of motion are

$$\begin{bmatrix} R_V + R_C & R_V + R_C & 0 \\ R_V + R_C & R_V + R_C + R_S + R_{VE} & -R_{VE} \\ 0 & -R_{VE}) & R_{VE} \end{bmatrix} \begin{bmatrix} \dot{V}_C \\ \dot{V}_A \\ \dot{V}_{VE} \end{bmatrix} + \begin{bmatrix} P_C + P_{CW} \\ P_L + P_{CW} \\ P_{VE} \end{bmatrix} = \begin{bmatrix} P_{MUS} \\ P_{MUS} \\ 0 \end{bmatrix} \quad \textbf{(4)}$$

The temporal variation of the generalized coordinates V_C, V_A, and V_{VE} is described by three coupled, nonlinear, ordinary differential equations. P_{MUS} is the independent, time-varying input function resulting from a particular breathing pattern. The first term on the left side of equation (4) represents the dissipative forces introduced by the four resistances in the model while the second term on the left side of the equation represents the elastic forces that characterize the compliant elements. The term on the right side of equation (4) represents the applied force.

Experimental Measurements

Instrumentation and methodologies developed for the measurement of pulmonary mechanics relied on simultaneous pleural pressure and lung volume recordings. The former is approximated by esophageal pressure, measured with a naso-gastric catheter balloon (Milic-Emili, Mead, Turner, & Glauser, 1964). Volume is measured either with spirometry or by integrating volumetric flow from a pneumotachograph. The setups described earlier in Chapter 4 are sufficient to perform this type of experiment. The data collection associated with this modeling endeavor is also described in Chapter 4.

Lung Tissue Model Parameter Identification

Similar to the approach developed for the cardiovascular studies, parameter estimation techniques are employed in the current settings to relate analytical results to experimental data. These efforts to coordinate model prediction with clinical research serve as a potent validation tool for the modeling efforts undertaken.

The equations of motion of the lung viscoelastance model (equation 3) involve five parameters, namely, b, R_{VE}, C_{VE}, TLC, and RV. The latter two are identified for each subject from standard pulmonary function tests (Olansen et al., 1998). Parameter b establishes, in the P-V plane, the slope of the static elastic recoil curve. This parameter is initially identified such that the static recoil fits, in a least-squares sense, measured $P_{EL} - V_A$ data over the full range of lung volumes (collected with the shuttering technique described in Chapter 4). After parameters R_{VE} and C_{VE} are identified (as explained next), the value of b is further tuned (manually) such that model simulations of the dynamic $P_{EL} - V_A$ relationship match experimental data.

Employing the virtual instrumentation capabilities that are the focus of this text, the parameters R_{VE} and C_{VE} can be identified using a nonlinear Levenberg-Marquardt algorithm. This routine is described in detail in Olansen et al. (2000). Typical results from this identification scheme are displayed in Figure 11-12.

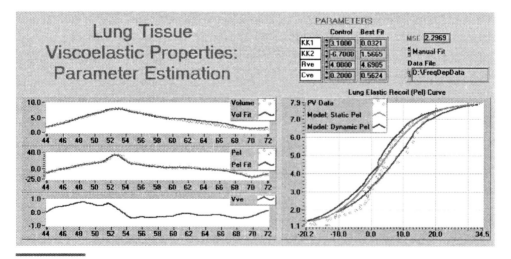

Figure 11-12.
Front panel of the parameter estimation routine as applied to the analysis of viscoelastic properties of lung tissue.

The values of parameters R_{VE} and C_{VE} determine the shape of the $P_{EL} - V_A$ loop in a manner described mathematically as the equation of a rotated ellipse (Athanasiades et al., under review). Figure 11-13 plots the components of the equation while varying parameters R_{VE} and C_{VE}, independently. Effective elastance, $1/C(\omega)$, initially a constant value at $1/C_L$, increases considerably as the value of R_{VE} increases (panel A, solid line). It attains a constant value for large values of R_{VE}. The dependence upon C_{VE} (panel B) is weak, with a modest initial increase that peaks at 0.23 L/cmH$_2$O. The term quantifying the extent of hysteresis (Athanasiades et al., 2000), $\omega R(\omega)$ similarly exhibits a peak at $R_{VE} = 36$ cmH$_2$Osec/L (panel A, dashed line); it subsequently declines to a constant value. The term decays slowly as C_{VE} increases (panel B).

Figure 11-14 plots measurements of the dynamic elastic recoil of lung tissue P_{EL} (filled and empty dots), against volume measurements for three

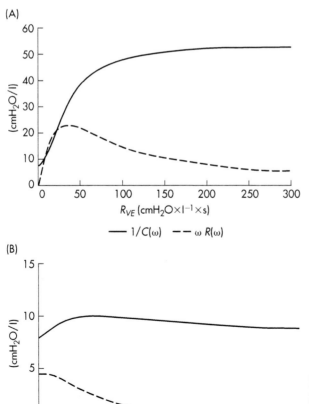

(A)

(cmH$_2$O/l)

R_{VE} (cmH$_2$O×l^{-1}×s)

—— $1/C(\omega)$ – – $\omega R(\omega)$

(B)

(cmH$_2$O/l)

C_{VE} (l/cmH$_2$O)

Figure 11-13.
Sensitivity of the components of the lung tissue viscoelastance model to variations in the parameters R_{VE} and C_{VE}.

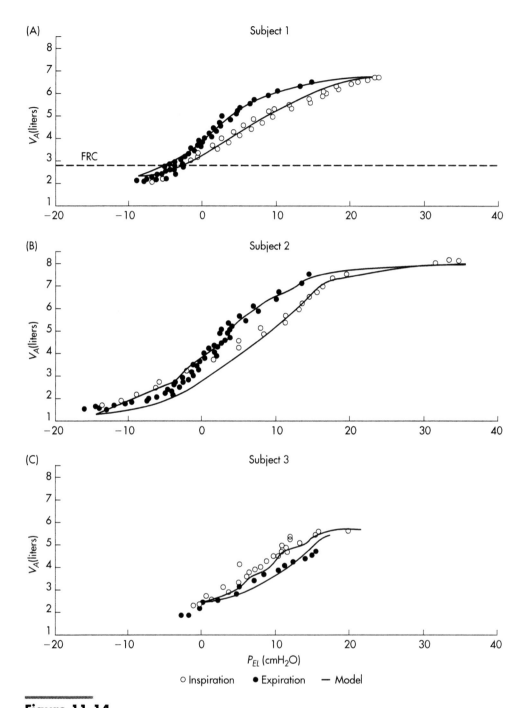

Figure 11-14.
Experimental data and model predictions of lung elastic recoil pressure plotted versus alveolar volume.

human subjects performing a full capacity maneuver. The data exhibit measurable hysteresis in the range of volumes above functional residual capacity (FRC). In subject 1, at mid-tidal volume (4.47 liters), the inspiratory path pressure is approximately 3.5 cmH$_2$O higher than the expiratory path pressure (mid-tidal volume hysteresis, or MTH). A maximum hysteresis of 6.6 cmH$_2$O is observed at approximately 5.8 liters. Close to FRC the inspiratory and expiratory paths cannot be clearly separated. Volumes below FRC correspond to negative values of P_{EL}. There is noticeable variation among the three subjects, both in the extent of the volume excursion (subjects were instructed to perform full capacity maneuvers) and the slope of the hysteretic loops. Model predictions are also plotted in Figure 11-14 and show good agreement with the measurements, in both the general shape of the loop and the extent of hysteresis.

PulmPV VI Development

The **PulmPV** software package provides a sophisticated graphical user interface to the full airway mechanics model described in this chapter. The **PulmPV** program, also developed using LabVIEW, is a compilation of original VIs designed both as the user and data interface to the model operation. Sophisticated graphical displays highlight the user interface while the **PulmPV** software can accept data from numerous sources, including direct acquisition via a DAQ card for on-line operation. The model C code described earlier is integrated into the **PulmPV** program as an external call from LabVIEW. **PulmPV** also provides a continuous graphical feedback of the model operation, enhancing user understanding of the model performance as well as aiding any debugging efforts that may be required. Figure 11-15 schematically depicts the execution flow of the **PulmPV** software package. Several features such as user selection of appropriate species or options for breathing patterns (e.g., tidal, ventilated, PEEP settings, etc.) were added to the program to demonstrate its adaptability.

The complete airway mechanics model was modified from Liu et al. (1998) and Athanasiades et al. (2000) and incorporated into the **PulmPV** software. The main results taken from this study are associated with identification of parameters that account for changes made to the model. A nominal set of parameters was first obtained that provided reasonable flows and volumes for given P_{PL} variations. Model output pressures from a typical control case are depicted in Figure 11-16. If you compare Figure 11-16 with classical textbook figures, you will see that the model output is in general agreement with typical *in vivo* data.

Pulmonary PV Software Package Computational Flowchart

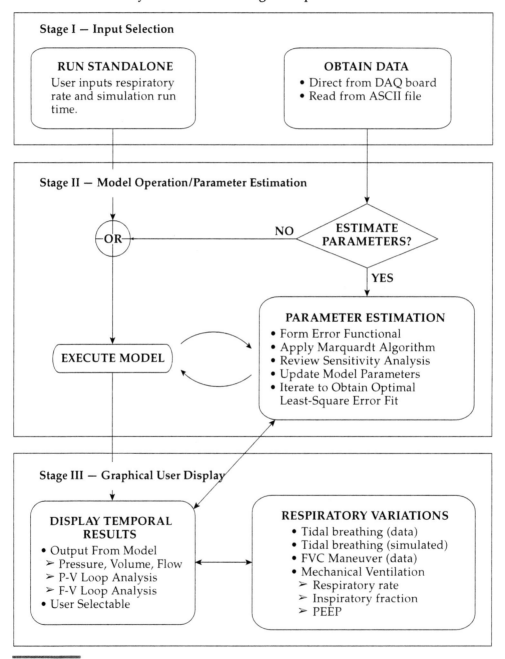

Figure 11-15.
Computational flow diagram for **PulmPV** software.

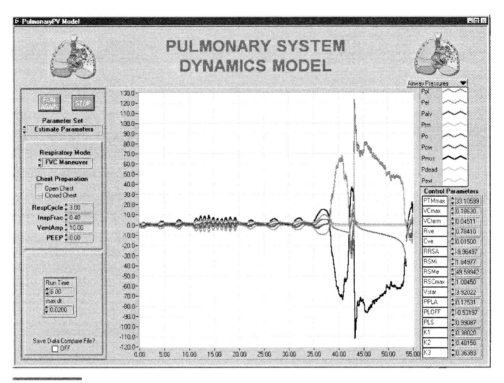

Figure 11-16.
Front panel of **PulmPV** VI depicting typical model output pressures. Parameters are
defined in the Appendix.

PulmPV Parameter Identification

When data is linked in the **PulmPV** program, the option to directly compare
model output using the parameter estimation and sensitivity analysis routines of Olansen et al. (2000) becomes available. This capability greatly enhances the functionality of the model. The forced vital capacity (FVC) maneuver was chosen for the identification analysis due to expected excitation
of most dynamic modes of the respiratory system. By estimating the values of
the adjustable parameters for a specific subject via direct comparison with the
subject's data records, the results of simulated pathophysiologies become directly applicable to that subject.

The screen panel shown in Figure 11-17 provides the user with multiple options to completely control the parameter estimation process, if so desired.
Yet, its default configuration is sufficient to enable novice users to typically

obtain satisfactory results. The sensitivity analysis is conducted in the background and can be reviewed during performance of the parameter estimation routine.

PulmPV also incorporates the ability to perform parameter sensitivity analyses associated with estimation routines. Examinations of the magnitude and time course of the relative sensitivity functions associated with all of the airway mechanics parameters (i.e., resistances and compliances) were evaluated to rank the degree of sensitivity of the set. A typical output of the initial sensitivity analysis for all adjustable parameters is included in Figure 11-18. The data in this figure is quite busy but is intended to give you insight into the determination of sensitive parameters. From Figure 11-18, it is apparent that the parameters PTM_{max} and PLS are the most sensitive in an inversely proportional manner, particularly during large volume excursions. These parameters are included in the descriptions of the transmural pressures across the collapsible airways, P_{TM}, and the lungs, P_{EL}, respectively. On the other hand, alterations in RRSA and V_{star}, which are component parameters of the

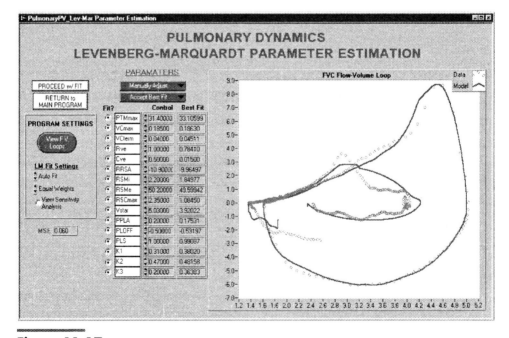

Figure 11-17.
Front panel of **PulmPV** parameter estimation routine depicting typical results of identification analysis. Parameters are defined in the Appendix. *MSE* is the mean square error of all compared data sets.

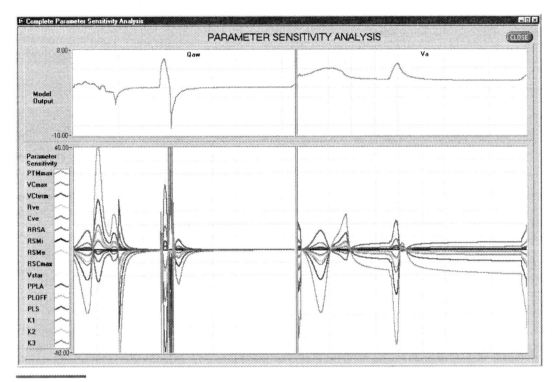

Figure 11-18.
Front panel of **PulmPV Sensitivity Analysis** VI depicting initial sensitivity of all adjustable parameters for subject 1. Parameters are defined in the Appendix.

small airway resistance, R_S, are directly proportional to changes in the lung volume and airway flow.

The current sensitivity analysis is also in agreement with the results of Liu et al. (1998) in that the sensitive parameters have a regional importance relative to the different phases of the breathing cycle. For example, R_C has a significant effect throughout the FVC maneuver, whereas R_S is more evident at low lung volumes, and R_{UAW} manifests its influence on the peak inspiratory and expiratory flow rates. Based on the sensitivity analysis, the most sensitive parameters are included in the nonlinear least-squares parameter estimation algorithm. This enhances convergence of the algorithm by constraining parameter variations to those that can bring about significant change in the state variables.

Figure 11-17 depicts the user interface for operation of the parameter estimation algorithm with typical results for subject 1. Results for another subject are evidenced in Figure 11-19. These results show good agreement for

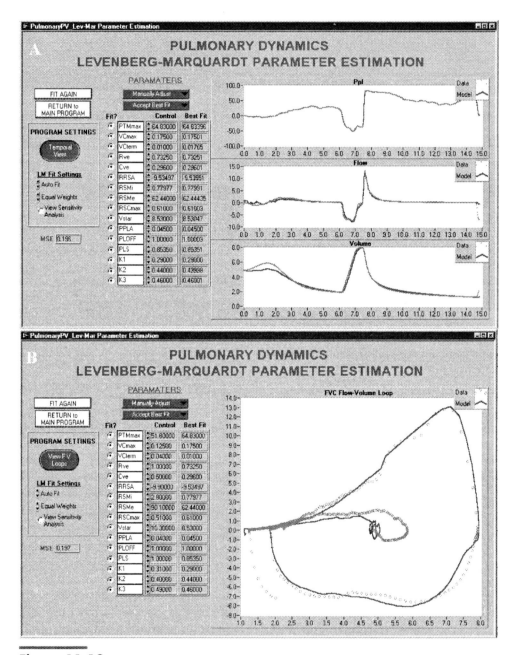

Figure 11-19.
Front panel of **PulmPV Parameter Estimation** VI depicting results for subject 2 in (A) temporal format and (B) flow-volume loop format. Parameters are defined in the Appendix.

specified pulmonary pressures, volumes, and flows. A summary of parameter values determined for each of the experiments conducted is given in the Appendix. These techniques are described in greater detail in Olansen et al. (2000) and Olansen (1999).

Integrated Cardiopulmonary Dynamics Modeling and Simulation

A current thrust in the modeling of biological systems is the development of large-scale integrative models that can help unravel complex interactions between different organ systems. Based on decades of research and analysis of isolated subsystems, computer modeling at the organ level is poised to take its place in quantitative computer-aided medical diagnosis. Previously, subsystem models of the uncontrolled circulation and pulmonary mechanics were discussed. This section pulls these subsystem models into a larger integrated, closed-chest cardiopulmonary model for two specific cases: the dog and the normal human subject. This integrated model gives us the opportunity to study the effects of interaction between the cardiovascular and pulmonary systems in each case. In particular, we examine the alterations in hemodynamics and cardiac mechanics caused by variations in breathing patterns. Tidal breathing, the vital capacity maneuver, and intubated subjects supported on a ventilator are the specific cases considered herein.

Background

Several studies of the interaction of the human cardiovascular and pulmonary systems have previously been reported in the literature. Typically, however, they focus on a particular area of interest, rather than overall cardiopulmonary function. Examples of these studies date back to the 1950s when Coleridge and Linden (1959) and Opdyke and Brecher (1950) studied respiratory effects on ventricular preload (i.e., left and right atrial pressures). Since then, substantial effort has been applied to characterizing systemic afterloads under the influence of respiration. This includes studies of the effects of intrathoracic pressure on left ventricular function (Robotham et al., 1978; Summer, Bromberger-Barnea, Shoukas, Sagawa, & Permutt, 1976; Ruskin,

Bache, Rembert, & Greenfield, 1973) and the effects of positive end-expiratory pressure (PEEP) on cardiovascular function in ventilated patients (Cassidy, Robertson, Pierce, & Johnson, 1978; Fewell, Abendschein, Carlson, Murray, & Rapaport, 1980; Powers & Dutton, 1975; Qvist, Pontoppidan, Wilson, Lowenstein, & Laver, 1975), as well as work by Beyar and Goldstein (1987), who have studied interaction specifically in the performance of cardiopulmonary resuscitation (CPR). The integrated model presented herein is sufficiently comprehensive to enable us to study a variety of interactive phenomena, including the ones delineated previously.

Human Cardiovascular Model Development

Comparative physiology endeavors to associate various physiologic functions across species. Indeed, when we consider conducting animal experiments, the underlying intent is to transform any insight obtained so that it may enhance our understanding of human physiology. We have applied this concept in the current study by appropriately scaling the cardiovascular model developed in the first section. Specifically, it is necessary to increase the volumes and flows around the loop to accommodate the larger fluid content of a normal 70-kg man. Pressure variations about the mean are roughly the same; heart rate is divided by a factor of approximately 2.15. Gross adjustment of the full parameter set is necessary to bring the values of the parameter set into the appropriate range for humans. Parameters used to describe the nominal human subject are listed in Table 11-2 and the Appendix.

Applying the parameter adjustments already mentioned yielded acceptable fits to a variety of indices, such as cardiac output (CO), left and right ventricular ejection fractions, mean blood pressures, and blood volume distributions around the circulatory loop, that are documented as typical in the literature (Mountcastle, 1968; Folkow & Neil, 1971). Table 11-3 documents the mean hemodynamic pressure and volumes around the circulatory loop for both dog and human simulations. Using these new parameters, the model computes hemodynamic waveforms in good agreement with previously published human data (Mountcastle, Folkow & Neil). Figure 11-20 depicts the typical model output that could be compared with standard textbook waveforms from Mountcastle. Comparison of these figures establishes that the model output is in general agreement with typical human data, verifying the models' adaptation to mimic human cardiac function.

Table 11-2. *Typical CardioPulm control parameter values for both canine and human studies.*

Parameter	Description	Typical Values Canine	Typical Values Human
R_{SAD}	Resistance: systemic arterioles	2.1050	0.6560
R_{SC}	Resistance: systemic capillaries	1.5477	0.4977
R_{PAD}	Resistance: distal pulmonary artery	0.0147	0.0097
R_{TAO}	Viscoelastic element: proximal aorta	0.0805	0.0814
R_{TPA}	Viscoelastic element: pulmonary artery	0.0276	0.0176
C_{AOP}	Compliance: proximal aorta	0.6493	0.9730
C_{AOD}	Compliance: distal aorta	0.0932	0.2200
C_{SA}	Compliance: large systemic arteries	0.0358	0.0360
C_{SAD}	Compliance: systemic arterioles	0.0276	0.0470
C_{PAP}	Compliance: proximal pulmonary artery	0.3943	0.3900
C_{PAD}	Compliance: distal pulmonary artery	0.0202	0.0200
C_{PA}	Compliance: pulmonary arterioles	0.1571	0.1571
L_{AOP}	Inertance: proximal aorta	0.0056	0.0052
L_{AOD}	Inertance: distal aorta	0.0282	0.0298
L_{PA}	Inertance: pulmonary artery	0.0004	0.0004

Table 11-3. *CardioPulm model output under control conditions: Mean pressures and volumes at various compartments throughout the closed-loop circulation. The units for pressure and volume are mmHg and ml, respectively.*

Vascular Segment	Description	Canine Pressure	Canine Volume	Human Pressure	Human Volume
AO_P	Aorta (proximal)	111.3	20.3	104.5	49.1
AO_D	Aorta (distal)	111.0	46.2	103.6	135.2
SA_P	Small systemic arteries	110.2	75.2	95.1	224.5
SA_D	Systemic arterioles	56.8	58.3	45.4	175.0
SC	Systemic capillaries	16.2	89.0	14.6	264.4
SV	Systemic veins	6.1	1038.8	5.8	3260.5
VC	Vena cava	6.1	93.6	6.1	310.3
RA	Right atrium	3.9	15.7	4.9	70.6
PA_P	Pulmonary artery (prox.)	14.6	4.8	18.9	18.9
PA_D	Pulmonary artery (distal)	13.6	24.9	17.7	75.0
PA	Pulmonary arterioles	12.6	24.6	12.2	74.7
PC	Pulmonary capillaries	10.0	63.0	11.6	203.3
PV	Pulmonary veins	8.3	75.2	11.3	230.4
LA	Left atrium	6.0	17.5	7.0	88.0

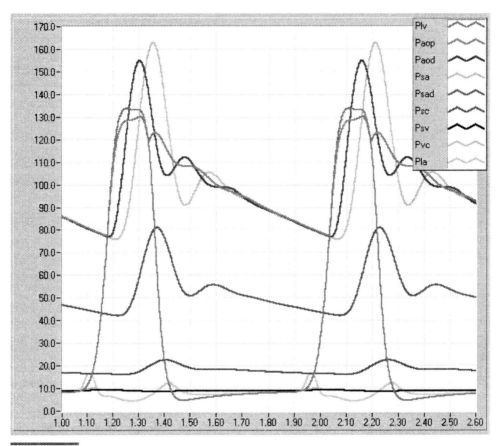

Figure 11-20.
Typical **CardioPulm** model output pressures for human systemic arterial system during respiration.

Deviation from Dog Model

Comparing the pressure data in Figure 11-20 with the canine data depicted earlier, it appears that there is little difference between the species. Indeed, for the most part, pressure profiles are strikingly consistent across the mammalian species (Li, 1996). Mean flow velocities are generally conserved as well among species, with only a slight dependence on body weight (see Li, p. 104). However, the blood volumes and flows associated with those same species vary substantially. For example, a typical 70-kg man will have a heart rate of

60–75 beats/min and a respiratory rate of 12–15 breaths/min, with a cardiac output (CO) of approximately 5 L/min. This means the man must average a stroke volume (SV) of 66–83 ml per heartbeat. A 25-kg dog, on the other hand, could have a resting heartbeat of 120 BPM, a respiratory rate of 24 breaths/min, and a CO of 2 L/min, resulting in an average SV requirement of 16 ml per heartbeat. It is these differences that must be accounted for in the extrapolation of the model from one species to another.

Integration of Cardiovascular and Pulmonary Models

An important change necessary to the process of combining these models is to anatomically close the chest of the animal (i.e., to consider the effects of intrathoracic, or pleural, pressure variations). Specifically, the pressures in the cardiac chambers, the aorta, vena cava, and pulmonary arteries and veins are all referenced to atmosphere in the open-chest configuration, whereas the nonthoracic component pressures are referenced to a relatively constant tissue pressure. Thus, in the first section, those pressures referenced to atmosphere are all considered absolute pressures since the absolute and transmural pressures are equivalent in the open-chest case (i.e., $P_{PL} = P_{Ref} = 0.0$). By closing the chest, however, as in the current assessment, the pleural pressure, P_{PL}, is established and the transmural pressures in the thoracic cavity are measured relative to P_{PL}. The pulmonary arterioles and capillaries are more influenced by alveolar pressure than pleural pressure (Milnor, 1972; Hawkim, Michel, & Chang, 1982), such that P_{ALV} is added to the transmural P_{PA} and P_{PC} to obtain their absolute pressures.

The other prominent feature necessary to integrate the cardiovascular and pulmonary models is the nonlinear effect the variation in lung volume has on the resistance of blood flow through the pulmonary capillaries. As the lungs expand, the alveolar blood vessels are elongated and thinned, resulting in an increase in pulmonary vascular resistance, which we denote by the overall capillary resistance, R_{PC}. The relationship used herein to emulate this physiological phenomenon is

$$R_{PC}(V_A) = R_{PC,0} \left[\frac{V_A}{V_{A,max}} \right]^2 \tag{5}$$

where $R_{PC,0}$ is a constant characterizing the overall magnitude of R_{PC} and $V_{A,max}$ is the value of V_A when R_{PC} is dominated by $R_{PC,0}$.

Computational Aspects

The model equations were initiated at the time of end-expiration with no flow at the mouth and at end-diastole with the atrioventricular valves open and the semilunar valves closed. Measured physiological data of the pleural pressure, P_{PL} (simulated data for use in canine studies), or a generated external pressure waveform data, P_{EXT}, are the input waveform used to run the appropriate model (P_{PL} for the normal model, and P_{EXT} for the ventilator version). Initial conditions employed for the 36 state variables are defined in this mode. Nominal parameter values were selected to produce physiologically realistic values, such as cardiac output (CO), and hemodynamic waveforms for a typical 25-kg dog. Pulmonary shunt flow was set at 2% of the mean pulmonary blood flow (Nunn, 1993), while mean coronary and cerebral flows were set at 5% and 15% of cardiac output, respectively (Folkow & Neil, 1971; Mountcastle, 1968). The model equations were programmed in C and solved using a variable step-size Runge-Kutta-Merson algorithm with a maximum time step size of 2×10^{-3} sec and an error tolerance of 1×10^{-6}.

CardioPulm VI Development

The **CardioPulm** software package provides a sophisticated graphical user interface to the interactive cardiopulmonary model described in this chapter. The **CardioPV** and the **PulmPV** software served as the basis for the development of **CardioPulm**. The **CardioPulm** program is a compilation of original VIs, the top level of which is shown in Figure 11-21, designed both as the user and data interface to the model operation. Sophisticated graphical displays highlight the user interface while the **CardioPulm** software can be adapted to accept data from numerous sources, including direct acquisition via a DAQ card for on-line operation. The model C code described earlier is again integrated into the **CardioPulm** program as an external call from LabVIEW. **CardioPulm** also provides a continuous graphical feedback of the model operation, enhancing user understanding of the model performance as well as aiding any debugging efforts that may be required.

Figure 11-22 schematically depicts the execution flow of the **CardioPulm** software package. In addition to combining the models included in the **CardioPV** and **PulmPV** programs, several software modifications were necessary to create the **CardioPulm** program.

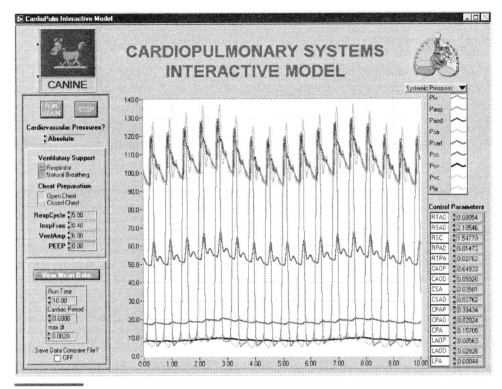

Figure 11-21.
CardioPulm front panel depicting typical results of canine model. Model settings are established via inputs on the left, while parameter values are controlled on the right.

(1) Since combined cardiopulmonary data could not be directly obtained from human subjects in this study, the parameter estimation portions of the software were not needed in the interactive model.

(2) The study of specific cardiovascular pathophysiologies was removed from the combined model as this study required near-term steady state hemodynamics, which are not available with the incorporation of pulmonary dynamics.

(3) Additional features such as user selection of appropriate species to options for breathing patterns (e.g., tidal, ventilated, PEEP settings, etc.) were added.

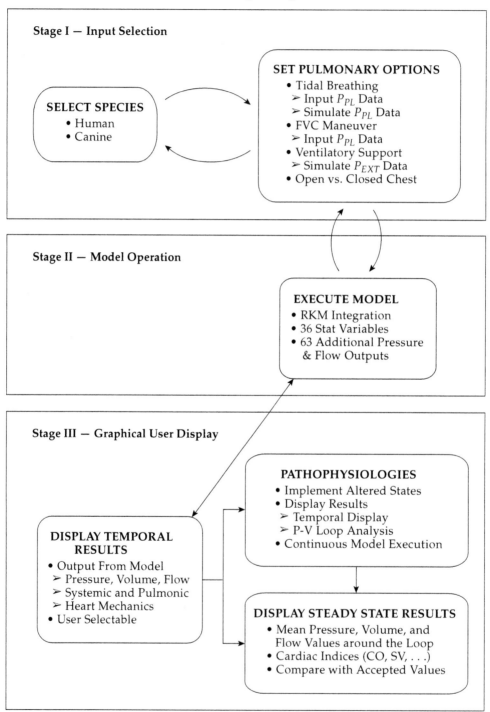

Figure 11-22.
Computational flow diagram for **CardioPulm** software package.

REFERENCES

Amoore, J. N., & Santamore, W. P. (1989). Model studies of the contribution of ventricular interdependence to the transient changes in ventricular function with respiratory efforts. *Cardiovasc. Res., 23,* 683–694.

Athanasiades, A., Ghorbel, F., Clark, J. W., Jr., Niranjan, S. C., Olansen, J., Zwischenberger, J. B., & Bidani, A. (2000). Energy analysis of a nonlinear model of the normal human lung. *J. Biol. Systems, 8*(2), 115–139.

Athanasiades, A., Ghorbel, F., Clark, J. W., Jr., Olansen, J., Niranjan, S.C., Zwischenberger, J. B., & Bidani, A. (under review). Modeling lung tissue properties in normal adult human subjects. *ASME Trans.*

Bachofen, H. (1968). Lung tissue resistance and pulmonary hysteresis. *J. Appl. Physiol., 24*(3), 296–301.

Bachofen, H., & Hildebrandt, J. (1971). Area analysis of pressure-volume hysteresis in mammalian lungs. *J. Appl. Physiol., 30*(4), 493–497.

Badke, F. R. (1982). Left ventricular dimensions and function during right ventricular pressure overload. *Am. J. Physiol., 242.* H611–H618.

Bates, J., Brown, K., & Kochi, T. (1989). Respiratory mechanics in the normal dog determined by expiratory flow interruption. *J. Appl. Physiol., 67*(6), 2276–2285.

Bemis, C. E., Serur, J. R., Borkenhagen, D., Sonnenblick, E. H., & Urschel, S. W. (1974). Influence of right ventricular filling pressure on left ventricular pressure and dimension. *Circ. Res., 34,* 498–504.

Beyar, R., & Goldstein, Y. (1987). Model studies of the effects of the thoracic pressure on the circulation. *Ann. Biomed. Eng., 15,* 373–383.

Bove, A. A., & Santamore, W. P. (1981). Ventricular interdependence. *Prog. Cardiovasc. Dis., 23,* 365–388.

Braunwald, E., Ross, J., & Sonnenblick, E. H., (1967). *Mechanisms of contraction of the normal and failing heart.* Boston: Little, Brown.

Burkhoff, D., Alexander, J., & Schipke, J. (1988). Assessment of windkessel as a model of aortic impedance. *Am. J. Physiol., 255 (Heart Circ. Physiol., 24).* H742–H753.

Burkhoff, D., & Sagawa, K. (1986). Ventricular efficiency predicted by an analytical model. *Am. J. Physiol., 250 (Regulatory Integrative Comp. Physiol., 19).* R1021–1027.

Cassidy, S., Robertson, C., Pierce, A., & Johnson, R. (1978). Cardiovascular effects of positive end-expiratory pressure in dogs. *J. Appl.Physiol., 44,* 743–750.

Chung, D. C. (1996). Ventricular interaction in a closed-loop model of the canine circulation. Unpublished master's thesis. Dept. of Electrical and Computer Engineering, Rice University.

Chung, D. C., Niranjan, S. C., Clark, J. W., Bidani, A., Johnston, W. E., Zwischenberger, J. B., & Traber, D.L. (1997). A dynamic model of ventricular interaction and pericardial influence. *Am. J. Physiol.*, 272 (*Heart Circ. Physiol.*, 41). H2942–H2962.

Coleridge, J. C. G., & Linden, R. J. (1959). The variations with respiration in effective left and right atrial pressures in the dog. *J. Physiol. London, 145,* 482–493.

Elad, D., & Kamm, R. (1989). Parametric evaluation of forced expiration using a numerical model. *Trans. ASME, 111,* 192–199.

Fewell, J., Abendschein, D., Carlson, J., Murray, J., & Rapaport, E. (1980). Continuous positive-pressure ventilation decreases right and left ventricular end-diastolic volumes in the dog. *Circ. Res., 46,* 125–132.

Folkow, B., & Neil, E. (1971). *Circulation.* New York: Oxford University Press.

Fredberg, J., & Stamenovic, D. (1989). On the imperfect elasticity of lung tissue. *J. Appl. Physiol., 67*(6), 2408–2419.

Golden, J., Clark, J. W., Jr., & Stevens, P. (1973). Mathematical modeling of pulmonary airway dynamics. *IEEE Trans. Biomed. Eng.,* BME-20(6), 397–404.

Greene, M. E., Clark, J. W., & Mohr, D. N. (March 1973). A mathematical model of left ventricular function. *Med. & Bio. Eng. & Comput.,* 126–134.

Greenwood, D. (1977). *Classical dynamics.* Englewood Cliffs, NJ: Prentice Hall.

Guccione, J. M., & McCulloch, A. D. (1991). Chapter 6. In L. Glass, P. Hunter, & A. McCulloch (Eds.), *Finite element modeling of ventricular mechanics in theory of the heart* (pp. 121–144). New York: Springer-Verlag.

Halperin, H. R., Tsitlik, J. E., Beyar, R., Chandra, N., & Guerci, A. D. (1987). Intrathoracic pressure fluctuations move blood during CPR: Comparison of hemodynamic data with predictions from a mathematical model. *Ann. Biomed. Eng., 15,* 385–403.

Hantos, Z., Daroczy, B., Csendes, T., Suki, B., & Nagy, S. (1990). Modeling of low-frequency pulmonary impedance in dogs. *J. Appl. Physiol., 68*(3), 849–860.

Hardy, H. H., Collins, R. E., & Calvert, R. E. (September 1982). A digital computer model of the human circulatory system. *Med. & Bio. Eng. & Comput., 20,* 550–564.

Hawkim, T., Michel, R., & Chang, H. (1982). Effect of the inflation on pulmonary vascular resistance by arterial and venous occlusion. *J. Appl. Physiol., 53*(5), 1110–1115.

Hildebrandt, J. (1970). Pressure-volume data of cat lung interpreted by a plastoelastic, linear viscoelastic model. *J. Appl. Physiol.*, *28*(3), 365–372.

Hughes, R. A., May, J., & Widdicombe, J. (1959). Stress relaxation in rabbits' lungs. *J. Physiol. London*, *146*, 85–97.

Jackson, A., & Milhorn, H. (1973). Digital computer simulation of respiratory mechanics. *Comput. Biomed. Res.*, *6*, 27–56.

Jonson, B., Beydon, L., Brauer, K., Mansson, C., Valind, S., Grytzell, H. (1993). Mechanics of respiratory system in healthy anesthetized humans with emphasis on viscoelastic properties. *J. Appl. Physiol.*, *75*(1), 132–140.

Khalil, H. K. (1992). *Nonlinear systems*. New York: Macmillan.

Kochi, T., Okubo, S., Zin, W., & Milic-Emili, J. (1988). Flow and volume dependence of pulmonary mechanics in anesthetized cats. *J. Appl. Physiol.*, *64*, 441–450.

Lau, V., Sugawa, K., & Suga, H. (1979). Instantaneous pressure-volume relationship of right atrium during isovolumic contraction in canine heart. *Am. J. Physiol.*, *236* (*Heart Circ. Physiol.*, 5). H672–H679.

Li, J. (1996). *Comparative cardiovascular dynamics of mammals*. New York: CRC Press.

Little, W. C., Badke, F. R., & O'Rourke, R. A. (1984). Effect of right ventricular pressure on the end-diastolic left ventricular pressure-volume relationship before and after chronic right ventricular pressure overload in dogs without pericardia. *Circ. Res.*, *54*, 719–730.

Liu, C., Niranjan, S. C., Clark, J. W., San, K., Zwischenberger, J. B., & Bidani, A. (1998). Airway mechanics, gas exchange, and blood flow in a nonlinear model of the normal human lung. *J. Appl. Physiol.* *84*(4), 1447–1469.

Lutchen, K. L., Suki, B., Hantos, Z., Daroczy, B., Petak, F., Kaczka, D., & Zhang, Q. (1994). Direct use of mechanical ventilation to measure respiratory mechanics associated with physiological breathing. *Eur. Respiratory J.*, *19*, 198–202.

Marquardt, D. W. (1963). An algorithm for least-squares estimation of nonlinear parameters. *J. Soc. Industr. Appl. Math.*, *11*, 431–441.

Melchior, F. M., Srinivasan, R. S., & Charles, J. B. (1992). Mathematical modeling of human cardiovascular system for simulation of orthostatic response. *Am. J. Physiol.*, *262* (*Heart Circ. Physiol.*, 31). H1920–H1933.

Milic-Emili, J., Mead, J., Turner, J. M., Glauser, E. M. (1964). Improved technique for estimating pleural pressure from esophageal balloons. *J. Appl. Physiol.*, *19*(2), 207–211.

Milnor, W. (1972). Pulmonary hemodynamics. In D. Bergel (Ed.). *Cardiovascular fluid dynamics* (Vol. 2). New York: Academic.

Mountcastle, V. B. (Ed.). (1968). *Medical physiology* (12 ed., vol. I). St. Louis, MO: Mosby.

Nunn, J. F. (1993). *Nunn's applied respiratory physiology.* Oxford: Butterworth-Heinemann.

Olansen, J. B. (1999). *Virtual bio-instrumentation: Integrating biomedical experimentation with systems-level modeling and analysis.* Ph.D. dissertation, Rice University.

Olansen, J. B., Clark, J. W., Khoury, D., Ghorbel, F., & Bidani, A. (2000). A closed-loop model of the canine cardiovascular system that includes ventricular interaction. *Comp. & Biomed. Res., 33,* 260–295.

Olansen, J. B., Ghorbel, F., Clark, J. W., Deyo, D., Zwischenberger, J. B., & Bidani, A. (1998). An automated LabVIEW-based data acquisition system for analysis of pulmonary function. *J. Clin. Eng., 23*(4), 279–287.

Olender, M., Clark, J. W., Jr., & Stevens, P. (1976). Analog computer simulation of maximum expiratory flow limitation. *IEEE Trans. Biomed. Eng.,* BME-23(6), 445–452.

Opdyke, D. F., & Brecher, G. A. (1950). Effect of normal and abnormal changes of intrathoracic pressure on effective right and left atrial pressures. *Am. J. Physiol., 160,* 556–566.

Palladino, J. L., & Noordergraaf, A. (1998). Muscle contraction mechanics from ultrastructural dynamics (Chap. 3). In G. M. Drzewiecki, & J. K-J. Li (Eds.), *Analysis and assessment of cardiovascular function* (pp. 33–57). New York: Springer-Verlag.

Pardaens, J., van de Woestijne, K., & Clement, J. (1972). A physical model of expiration. *J. Appl. Physiol., 33*(4), 479–490.

Paulsen, R. A., Clark, J. W., Murphy, P. H., & Burdine, J. A. (March 1982). Sensitivity analysis and improved identification of a systemic arterial model. *IEEE Trans. Biomed. Eng.,* Vol. BME-29, No. 3.

Powers, S., & Dutton, R. (1975). Correlation of positive end-expiratory pressure with cardiovascular performance. *J. Crit. Care Med., 3,* 64–68.

Qvist, J., Pontoppidan, H., Wilson, R., Lowenstein, E., & Laver, M. (1975). Hemodynamic response to mechanical ventilation with PEEP. *Anesthesiology, 42,* 45–55.

Rideout, V. C. (1991). *Mathematical and computer modeling of physiological systems.* Prentice Hall Biophysics and Bioengineering Series. Englewood Cliffs, NJ: Prentice Hall.

Robotham, J. L., Lixfield, W., Holland, L., Macgregor, D., Bryan, A., & Rabson, J. (1978). Effects of respiration on cardiac performance. *J. Appl. Physiol.: Respirat. Environ. Exercise Physiol., 44,* 703–709.

Ruskin, J., Bache, R. J., Rembert, J. C., & Greenfield, J. (1973). Pressure-flow studies in man: Effect of respiration on left ventricular stroke volume. *Circulation, 48,* 79–85.

Santamore, W. P., & Burkhoff, D. (1991). Hemodynamic consequences of ventricular interaction as assessed by model analysis. *Am. J. Physiol.,* 260 (*Heart Circ. Physiol.,* 29). H146–H157.

Santamore, W. P., Lynch, P. R., Meier, G., Heckman, J., & Bove, A. A. (1976). Myocardial interaction between the ventricles. *J. Appl. Physiol.,* 4(3), 362–368.

Sharp, J. T., Johnson, F. N., Goldberg, N., & van Lith, P. (1967). Hysteresis and stress adaptation in the human respiratory system. *J. Appl. Physiol.,* 23(4), 487–497.

Similowski, T., Levy, P., Corbeil, C., Albala, M., Pariente, R., Derenne, J., Bates, J., Jonson, B., & Milic-Emili, J. (1989). Viscoelastic behavior of lung and chest wall in dogs determined by flow interruption. *J. Appl. Physiol.,* 67(6), 2219–2229.

Slinker, B. K., Goto, Y., LeWinter, M. M. (1989). Systolic direct ventricular interaction affects left ventricular contraction and relaxation in the intact dog circulation. *Circ. Res., 65,* 307–315.

Stamenovic, D., Glass, G., Barnas, G., & Fredberg, J. (1990). Viscoplasticity of respiratory tissues. *J. Appl. Physiol.* 69(3), 973–988.

Suga, H. (1969). Time course of left ventricular pressure volume relationship under various end-diastolic volumes. *Japan. Heart J., 12,* 509–515.

Suga, H., Sagawa, K., & Shoukas, A. A. (1973). Load independence of instantaneous pressure volume ratio of the canine left ventricle and effects of epinephrine on the ratio. *Circ. Res., 32,* 314–322.

Suki, B., & Bates, J. (1991). A nonlinear viscoelastic model of lung tissue mechanics. *J. Appl. Physiol.,* 71(3), 826–833.

Suki, B., Barabasi, A-L., Lutchen, K. R. (1994). Lung tissue viscoelasticity: A mathematical framework and its molecular basis. *J. Appl. Physiol.,* 76(6), 2749–2759.

Summer, W., Bromberger-Barnea, B., Shoukas, A., Sagawa, K., & Permutt, S. (1976). The effects of respiration on left ventricular function. *Circulation, 54,* Suppl. II, 13.

Sun, Y., Bashera, M., Lucariello, R., & Chiaramida, S. (1997). A comprehensive model for right-left heart interaction under the influence of pericardium and baroreflex. *Am. J. Physiol.,* 272 (*Heart Circ. Physiol.,* 41). H1499–H1515.

Svantesson, C., John, J., Taskar, V., Evander, E., & Jonson, B. (1996). Respiratory mechanics of rabbits ventilated with different tidal volumes. *Respir. Physiol., 106,* 307–316.

Taylor, R. R., Covell, J. W., Sonneblick, E. H., & Ross, J., Jr. (1967). Dependence of ventricular distensibility on filling of the opposite ventricle. *Am. J. Physiol.*, *213*(3), 711–718.

Thiriet, M., Bonis, M., Adedjouma, A., Hatzfeld, C., & Yvon, J. (1987). Experimental and theoretical models of flow during forced expiration: Pressure and pressure history dependence of flow rate. *Med. Biol. Eng. Comput.*, *25*, 551–559.

Tomlinson, S., Tilley, D., & Burrows, C. (1994). Computer simulation of the human breathing process. *IEEE Eng in Med. Biol.*, 115–124.

Tsitlik, J. E., Halperin, H. R., Popel, A. S., Shoukas, A. A., Yin, F. C., & Westerhof, N. (1992). Modeling the circulation with three-terminal electrical networks containing special nonlinear capacitors. *Ann. Biomed. Eng.*, *20*, 595–616.

Ursino, M. (April 1999). A mathematical model of the carotid baroregulation in pulsating conditions. *IEEE Trans. Biomed. Eng.*, Vol. BME-46, No. 4.

Wada, S., Seguchi, Y., & Tanaka, M. (1991). Breathing-ventilation model, and simulation of high-frequency ventilation. *JSME Intern. J. Ser. 1*, *34*(1), 98–105.

Virtual Bio-Instrumentation, LabVIEW™, and the Internet

12

Access to Biomedical Virtual Instruments Anywhere
Why Use Networked Virtual Bio-Instrumentation?
Classifying an Internet-enabled Virtual Bio-Instrumentation System

Internet Technologies and Virtual Instrumentation
How Does Information Get Transferred from the Server to the Client?

Choosing a Technical Solution with LabVIEW
TCP/IP
UDP

The VI Server: Remotely Manipulating VIs

DataSocket

The LabVIEW Web Server

Controlling VIs Over the Web
CGI and HTML Forms
Java Applets

Access to Biomedical Virtual Instruments Anywhere

One of the main purposes of virtual bio-instrumentation, as we have been showing in this book, is to create more powerful, flexible, and cost-effective instrumentation systems built around a PC using software as the engine and interface. A virtual instrument can easily export and share its data and information with other software applications since they often reside on the same computer.

When National Instruments introduced LabVIEW™ over a decade ago, it launched a unique software tool for creating virtual instruments. The appeal of LabVIEW is largely tied to its graphical programming nature, since it is very easy to prototype and develop an application in a fraction of the time it would take to produce the same in a language like C++. There is an interesting parallel between the success of LabVIEW and the popularity of the Web: In both cases, it was not so much the underlying technology that was so innovative but rather the well-designed graphical interface that made it accessible. After all, almost everything you could do in LabVIEW could be done in C or assembly code years before LabVIEW was popular. The Internet's use goes back to the 1960s. Most of what it took to create these revolutionary tools, though, was the development of an intuitive, human-friendly interface. For LabVIEW, that meant programming by wiring graphical objects together, like building a breadboard circuit. For the Web, it meant a Web browser application that involves little more than pointing and clicking on images or words of interest that are hyperlinked to other places on the Web.

So you might say that using LabVIEW to create Internet-enabled applications brings some of the best user interface designs together. The possibilities are exciting for creating easy-to-use and intuitive networked applications that take virtual instrumentation to another level.

Why Use Networked Virtual Bio-Instrumentation?

Consider these scenarios:

A doctor gets a page while on call, alerting him to an onset of tachycardia from one of his patients hospitalized after heart surgery. Since he is currently in a clinic across town, he looks up the patient's live vital signs and history on a nearby terminal by logging in over a secure connection. He has the information to decide whether to hurry over to examine the patient further or to simply call the nurse with instructions. Fortunately, the LabVIEW program on his machine indicates the problem is a familiar pattern that he knows is typical of this patient's condition, and he can continue with his current work.

On one early morning at a university campus, a pre-med student gets ready to do her homework experiments as part of the bio-instrumentation lab course. She realizes the assignment is due this morning and the lab won't be open for another couple of hours. Besides, the dorm room is cozy. Drinking her coffee, she opens her Web browser and uses a Java applet interface to remotely connect to the lab equipment, which is running LabVIEW. Running the experiment in real-time from the Web browser, she inputs several parameters until she is satisfied with the tissue resistivity graphs. She saves the graphs and e-mails the results to her professor, turning in her assignment a few minutes before it was due.

A new high-power electron microscope has just been purchased by an internationally funded research agency for molecular biology research. Although the electron microscope is located in California, it is being made available to researchers in Russia. The Russian scientists do not need to travel to this facility, since they can control the settings, run experiments, and retrieve images remotely from specimens thanks to an Internet-enabled system. Because it is nighttime in the United States during business hours in Moscow, both American and Russian scientists have the microscope available to them over the Net almost every day.

Classifying an Internet-enabled Virtual Bio-Instrumentation System

The previous scenarios illustrate the possibilities of combining virtual bio-instrumentation and the Internet. What kinds of applications are possible

Table 12-1. *Functional categories of Internet-enabled instrumentation.*

Remote Monitoring	When a process can be observed from another location on the network. The observation is done with a client while the process runs on a server. In a pure remote monitoring scenario, the client cannot give any feedback or provide any inputs to the server process.
Remote Control	Usually includes the same capabilities as a remote monitoring system but also allows the remote user (the client) to send some data, messages, or inputs back to the server process that in turn affect the output.
Collaboration	In a collaboration scenario, multiple users from remote sites can use a client program to not only communicate and share information with the server process but also be aware of and share information with each other as part of the communication.
Distributed Computing	In its simplest form, it refers to a software process that runs on more than one computer or machine. Distributed computing is a way of sharing computing and hardware resources to accomplish a task that might be burdensome, inconvenient, or impossible to perform on one machine.

when you leverage network and Internet technology into your systems? What advantages are there to Internet-enabling your lab or patient monitoring processes? The answer usually falls into one or more of four categories, as outlined in Table 12-1.

Internet Technologies and Virtual Instrumentation

Before we get into some practical examples, let's look briefly at some of the core technologies that make it possible for Internet-enabling virtual bio-instrumentation systems. The very term *Internet* is a truncated form of *inter-connected network*. One basic definition of network is "two or more electronic devices linked in some way to permit the exchange of information." In information technology, a network is a series of points or nodes interconnected by communication paths. The Internet consists of thousands of interconnected networks. The Web is a client-server system—your Web browser is the client and any given Web site's server is the server.

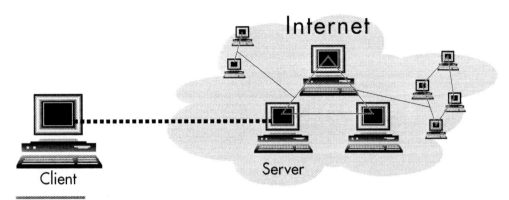

Figure 12-1.
The Web as server-client model.

In the client-server model, common in networked applications, one set of processes (called clients) requests services from another set of processes (servers). If you want to monitor a LabVIEW application from your laptop, you request that the information stored on the server be viewable from a client—in this case, your laptop (see Figure 12-1). As is the case in this scenario, a server usually waits for a client to initiate a connection. In many cases a server can service multiple clients at a time. You are already familiar with the way the Web functions as a client-server model. You likely are also familiar with e-mail and FTP (file transfer protocol) services that use the client-server model to transport information.

How Does Information Get Transferred from the Server to the Client?

The simple answer is that information is accessed, broken down into smaller pieces, sent electronically along various network pathways, reordered, reassembled, and voila! That, however, is accomplished through a slightly more complicated series of *communication protocols*, a set of rules or a language that allows two computers or devices to communicate with each other. Protocols exist at several *layers,* with high-level protocols (such as the Web's *HTTP* protocol) being built on lower-level protocols (*HTTP* relies on *TCP/IP*), which in turn may be relying on lower protocols (*TCP/IP* needs a physical layer protocol, such as *Ethernet*), in chained fashion. The most common communication protocols that you will encounter in building networked VIs are TCP/IP, UDP, HTTP, DSTP (DataSocket Transfer Protocol), and WAP.

Choosing a Technical Solution with LabVIEW

When designing an Internet-enabled instrumentation system, you will have a wide array of choices and tools at your disposal when you use LabVIEW as the primary data acquisition and analysis software. LabVIEW has a number of built-in capabilities for Internet connectivity, including

- **TCP/IP** and UDP functions;
- A built-in **Web server** that can create front panel images on the fly;
- LabVIEW's **VI Server**, a powerful framework for VIs or ActiveX applications to communicate over a network; and
- The **DataSocket** protocol for sharing live data over a LAN or the Internet.

With the use of add-on tools and other software development packages, LabVIEW can also be used with additional technologies such as

- **Java** applets to remotely control or monitor a VI,
- **ActiveX controls,**
- **CGI support in the G Web server, and**
- **E-mail, FTP, and telnet.**

TCP/IP

TCP stands for *Transmission Control Protocol;* IP stands for *Internet Protocol*. Together, they ensure that the right information gets to the right destination in the right configuration. TCP is the standard protocol for most Internet applications and is the backbone protocol of the Internet.

LabVIEW has full support for TCP/IP communication. You can find some good examples of TCP communication between VIs in the communication examples included with LabVIEW.

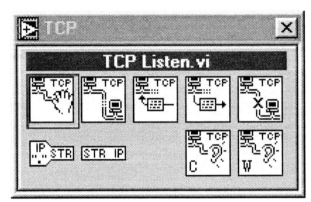

Figure 12-2.
TCP/IP palette in LabVIEW.

Note: Server and client VIs shown here can also be designed to communicate with clients or servers written in other languages, such as C, C++, or Perl. As long as both your LabVIEW VI and the external application agree on how messages will be passed back and forth, you can effectively communicate outside of LabVIEW. This is often a good technique to use even when both a VI and the external program reside on the same PC.

Since TCP/IP is a connection-based protocol, the server VI must be able to receive and process incoming requests for a connection, while the client VI opens a connection to the server. In a TCP/IP connection, one host must *listen* on a certain port (the listener), and the other host must *open a connection* to the IP address and port that the listener is active on. If a listener is not active before a connection is attempted, the connection will fail. No data can be read or written between the hosts until the connection has been established. Figure 12-2 shows the TCP/IP palette in LabVIEW.

Once a connection is made (a *socket connection*), either host can read and write to the other for as long as the connection is open. Either host can also terminate the connection. In a sense, the only thing that really distinguishes the server from the client is who plays what role in initiating the connection (the server listens, the client opens the connection), since there is no practical distinction or limitation between two connected hosts. Figure 12-3 illustrates the flow of communication between client and server.

TCP Block Diagram Functions

The following descriptions explain what each of the TCP VIs found in LabVIEW actually does.

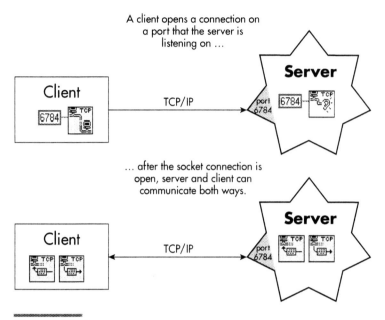

Figure 12-3.
TCP/IP VIs used for client-server communication.

The **TCP Listen** VI is usually the initialization function that is called by the server program. It sets up a listener on the specified *port* of your local machine. It will wait either forever (default) or the specified amount of time in *timeout ms*. This VI will not exit until either a connection or an error occurs. When a connection from a remote machine is made, the VI returns the IP address (*remote address*), the *remote port* the connection was made from, and a unique **connectionID** (a datatype LabVIEW uses to identify an active TCP/IP connection).

The **TCP Open Connection** is usually the initialization function that is called by the client program. It attempts to open a TCP connection with the specified remote IP *address* and *remote port*. The *timeout* specifies how long it should try to connect before giving up (a value of -1 will make it try forever). You can optionally specify a *local port* that the connection is originating from; otherwise LabVIEW chooses an unused one for you. This function returns the **connectionID** once the connection has been made.

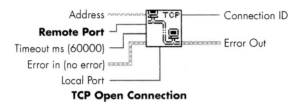

TCP Open Connection

The TCP Read function receives a maximum of **bytes to read** bytes from the specified TCP **connectionID**, returning the results as a binary string in **data out**. The *mode* is discussed later in this section.

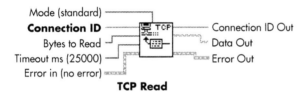

TCP Read

The TCP Write function writes the binary string **data in** to the specified TCP **connectionID.**

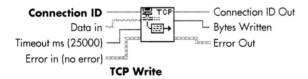

TCP Write

The TCP Close Connection function closes the connection specified by the **connectionID.** The **abort** input is ignored and doesn't do anything, according to LabVIEW documentation.

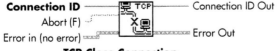

TCP Close Connection

Setting Up a TCP Listener

You can use either of two methods to wait for an incoming connection:

- Use the **TCP Listen** VI to create a listener and wait for an accepted TCP connection at a specified port. If the connection is successful, the VI returns a connection ID and the address and port of the remote TCP. The TCP Listen VI will wait indefinitely for a connection before exiting (see Figure 12-4).

- Use the **TCP Create Listener** function to create a listener, and then use the **Wait on Listener** function to listen for and accept new connections. **Wait on Listener** returns the same listener ID that was passed to the function as well as the connection ID for a connection. When you are finished waiting for new connections, you can use **TCP Close** to close a listener (see Figure 12-5). The **TCP Create Listener** function exits immediately and the **Wait on Listener** function waits indefinitely for a connection.

The advantage of using the second method is that you can cancel a listen operation by calling TCP Close. This technique is useful for applications in which you want to listen for a connection without using a timeout, but you want to cancel the listen when some other condition becomes TRUE (for example, when the user clicks a button).

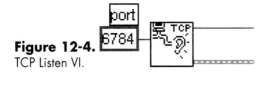

Figure 12-4.
TCP Listen VI.

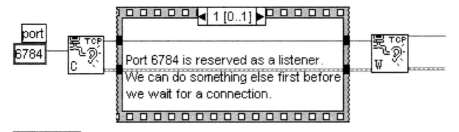

Figure 12-5.
TCP Create Listener with Wait on Listener function.

Caution: In LabVIEW, you cannot create more than one listener on the same port. If you try to call **TCP Listen** or **Create Listener** again on the same port before the connection is closed, you will get an error.

This doesn't mean you cannot have multiple connections to the same port. In fact, to handle multiple connections, you should not create a new listener for each separate client connection. Instead, for multiple connections, call **Create Listener** *once* and then call **Wait On Listener** repeatedly for each client.

Once a connection is established, data can be written or read from either connected party to the other. When a connection is closed, no more data can be read or written between the two hosts.

Something interesting you should notice about the TCP/IP read and write functions is that the only data type they accept is a *string*. So in LabVIEW you must convert everything you are sending across the network to a string type when using the TCP functions. This keeps TCP functions fairly efficient and able to work with other types of software, since they do not have to worry about converting or packaging the data. The downside is that your client and server need to know ahead of time how the data is going to be encoded, how many bytes you will be sending, and so on.

Usually you should use either the **TypeCast** function or the **Flatten To String** function from the **Advanced>>Data Manipulation** palette in LabVIEW to send your data over TCP/IP. If you are not familiar with these functions, consult the on-line documentation and examples, as you will need to use these functions with TCP applications. Figures 12-6 and 12-7 show the **Simple Data Server** and **Simple Data Client** VIs that you can find in the built-in LabVIEW examples.

UDP

Universal Datagram Protocol (UDP), like TCP, is a protocol built on top of IP. UDP uses the same Internet addressing scheme and port numbers to identify hosts. However, it is different from TCP in several respects:

- UDP is a *connectionless* protocol; that is, no connection needs to be established between two hosts prior to sending data; instead the data is *broadcast* to the destination(s).

- UDP does not guarantee delivery of the data packets.

- With UDP, data packets larger than a certain size (dependent on the network hardware) may arrive incomplete or not arrive at all.

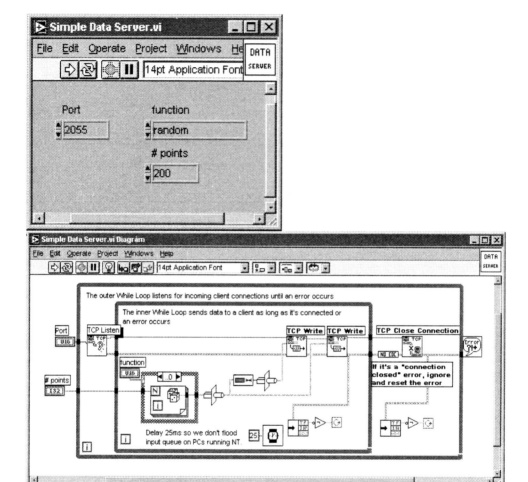

Figure 12-6.
Simple Data Server VI.

The basic tradeoff you get with UDP is significantly higher performance but less reliability than TCP. TCP first had to find the destination, negotiate and open a connection, start piping data, check to make sure the data arrived, and resend packets if necessary. UDP doesn't even check to see if the destination is reachable; it simply broadcasts the packets on the network and pushes them toward the destination. If the destination IP address is unreachable, or the port is not open as a listener, the packets are simply discarded.

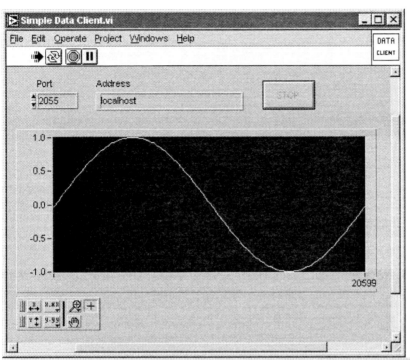

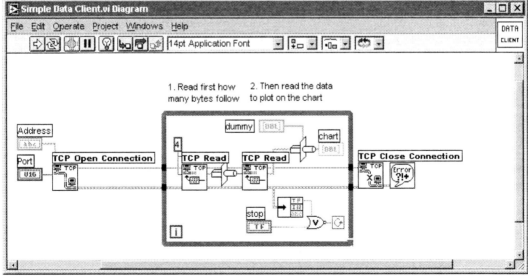

Figure 12-7.
Simple Data Client VI.

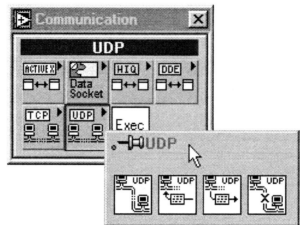

Figure 12-8.
The UDP palette.

One good analogy is to think of UDP as bulk mail or junk mail—it gets sent out in mass quantities at a cheap rate, but if it is undeliverable, the post office simply throws it away. In the same analogy, TCP/IP would be like Federal Express, more costly but with guaranteed tracking and delivery.

> *Hint:* As a simple rule of thumb, if you are designing an application that can do okay if it misses some data packets every now and then, and you want higher throughput performance than TCP, then you should consider using UDP.

Figure 12-8 shows the UDP VIs provided with LabVIEW.

The VI Server: Remotely Manipulating VIs

The VI Server gives you the capability to programmatically access features in LabVIEW either using the VI Server functions in a block diagram or through an ActiveX control. Don't let the name confuse you: VI Server is much more than just some type of networking server built into LabVIEW (although it is that as well). The VI Server functionality is really a way of introducing *object-oriented programming* into LabVIEW. For example, with VI Server, you can programmatically

- Load a VI into memory, run it, and then unload the VI without the need to have it statically linked as a sub-VI in your block diagram.

- Dynamically run a sub-VI that gets called at run time, by only knowing its name and connector pane structure (this is known as calling a VI by reference).

- Change properties of a particular VI, such as the size and position of the front panel window, and whether it is editable.

- Make LabVIEW windows move to the front of the screen.

- From the block diagram, call a sub-VI without waiting for it to finish executing (one of the few places you can get away with not obeying the normal dataflow paradigm!).

- Dynamically change the attributes (properties), such as color and size, of a front panel object.

So what do these features have to do with network applications? Not that much, if it weren't for one of the VI Server's most outstanding features: *network transparency*. Network transparency means you can do all of the mentioned manipulation of a VI or of LabVIEW itself on another machine across the network (including the Internet) in just the same way as if it were on your own machine. This means that, for example, you could have a data acquisition VI running at remote sites, while your local analysis VI gets information from the remote machines without having to write any special networking and without using the TCP/IP functions.

Figures 12-9 and 12-10 show you an example of how trivial it is to run a VI remotely (don't worry about what the functions are if you haven't seen them before; they're from the **Application Control** palette and we'll examine them in detail shortly).

The VI Server exposes its functionality in LabVIEW through block diagram functions like those in Figure 12-9. But it also allows its functionality to be accessed in Windows from external programs through an ActiveX automation client (e.g., a Visual Basic program or macro) and from a remote LabVIEW VI over TCP/IP. The following diagram illustrates this architecture. As you can see in Figure 12-10, the VI Server functionality is accessible by block diagram functions, by external ActiveX programs, or by a remote LabVIEW VI over TCP/IP. To enable the VI Server in LabVIEW, go to the **Tools** menu, choose **Options>>VI Server: Configuration** (see Figure 12-11). The functions for using the VI Server are in the **Application Control** palette as shown in Figure 12-12.

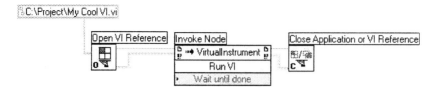

This tells the local machine to run My Cool VI.vi

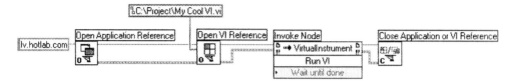

This tells the Lab VIEW across the Internet at the address lv.hotlab.com to run
My Cool VI.vi. The path to the VI refers to the remote drive.

Figure 12-9.
Network transparency feature in the VI Server.

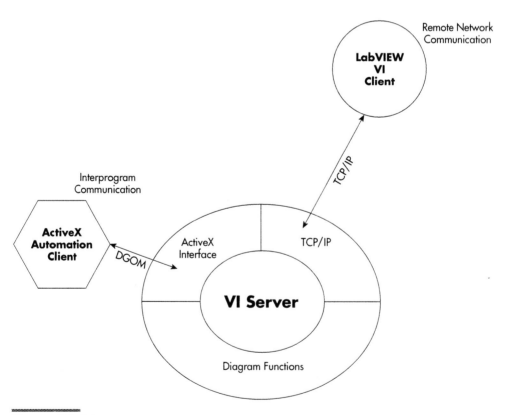

Figure 12-10.
Mechanisms for accessing the VI Server.

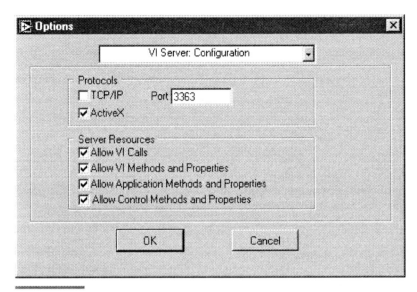

Figure 12-11.
VI Server configuration.

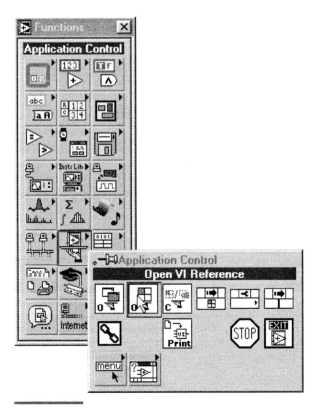

Figure 12-12.
VI Server functions.

DataSocket

So far we have looked at two different communication mechanisms in Lab-VIEW that rely on TCP/IP networks: TCP/IP VIs and the VI Server. Now we will examine a third mechanism: DataSocket.

DataSocket, according to National Instruments, is "a single, unified, end-user API based on URLs for connecting to measurement and automation data located anywhere, be it on a local computer or anywhere on the Internet. It is a protocol-independent, language-independent, and OS-independent API designed to simplify binary data publishing." Essentially, DataSocket is a technology that allows you to send and receive data over a network from a variety of software platforms (including LabVIEW) without worrying about the low-level implementation details. With DataSocket, for example, you can send formatted data back and forth between two LabVIEW machines on a network, or a LabVIEW and a LabWindows program, or even between several Web browser clients and LabVIEW.

Because one of the key components of DataSocket is the *DataSocket Server*, a small application that is external to the programming environment, your programs don't have to worry about how to manage TCP/IP connections, nor does performance in your specific application vary with the number of clients connected. In addition, DataSocket can handle several datatypes, including integers, floats, strings, and Booleans, as well as arrays of these. By letting DataSocket handle typecasting and conversions internally, you don't have to worry about sending header information or formatting your data in a special way to be transmitted over the network. Figure 12-13 shows the DataSocket architecture.

DataSocket has two main pieces that work together:

1. The *DataSocket Server,* and
2. The *DataSocket API* (Application Programming Interface) for clients.

The DataSocket Server is a standalone application (available for Windows only) that runs on a computer and will handle client connections. The client connections may write data to the server (known as DataSocket *publishers*) or read data (DataSocket *subscribers*) from any one of the publishers. The Data-Socket Server automatically handles the underlying network connections and data packet transmission, making it transparent to the clients. Figure 12-14 illustrates how these connections interact. We will examine the DataSocket Server in more detail shortly.

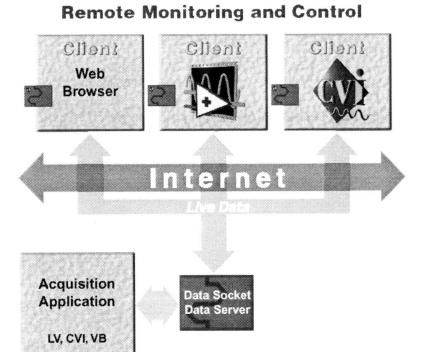

Figure 12-13.
DataSocket architecture.

The DataSocket clients implement the DataSocket API to talk to the server. This API is available in the following implementations:

- LabVIEW VIs
- LabWindows/CVI C library
- ActiveX control
- JavaBean

The DataSocket Server is automatically installed with LabVIEW for Windows. Because the DataSocket Server is implemented with an ActiveX interface, the DataSocket Server is only available on the Windows platform (so, while unfortunately MacOS and Unix users can't currently run a DataSocket *Server* on their machines, it is possible to create DataSocket *clients* on other platforms).

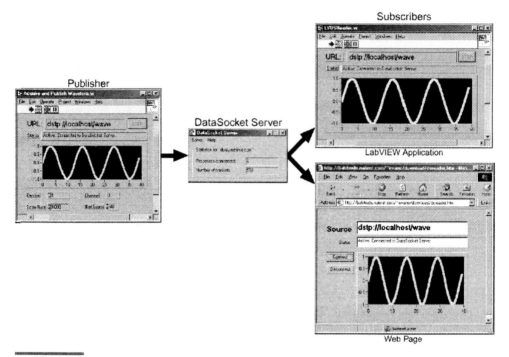

Figure 12-14.
DataSocket being used to pass data to a VI and a Web browser ActiveX control.

Figure 12-15.
The DataSocket Server.

To start the DataSocket Server, simply go to **Start>>Program Files>>National Instruments DataSocket>>DataSocket Server.** You will see a screen like that in Figure 12-15.

DataSocket is very easy to use; you can use DataSocket connections directly from the front panel without any block diagram programming! Each front panel control or indicator can publish or subscribe to data through its own DataSocket connection. Front panel DataSocket connections publish only the data, not a graphic of the front panel control, so the VIs that subscribe through a DataSocket connection can perform their own operations on the data.

You can use front panel DataSocket connections to publish or subscribe to live data in a front panel object. When you share the data of a front panel object with other users, you publish data. When users retrieve the published data and view it on their front panel, users subscribe to the data. To set a front panel's DataSocket connection, from a pop-up menu, select **Data Operation>>DataSocket Connection . . . .** You will get a dialog box like that shown in Figure 12-16. You can then choose to have this item's data be published, or subscribed to, or both. You will always need to specify a valid DataSocket Server (which could be running on your own machine or some other one on the network).

You can also set up DataSocket reads and writes (subscribing and publishing) programmatically in the block diagram. The DataSocket function palette is part of LabVIEW; you can find it in **Functions>>Communication>> DataSocket** (see Figure 12-17). The DataSocket palette contains the **Data-Socket Read**, **DataSocket Write**, and the **Variant** sub-palette of functions.

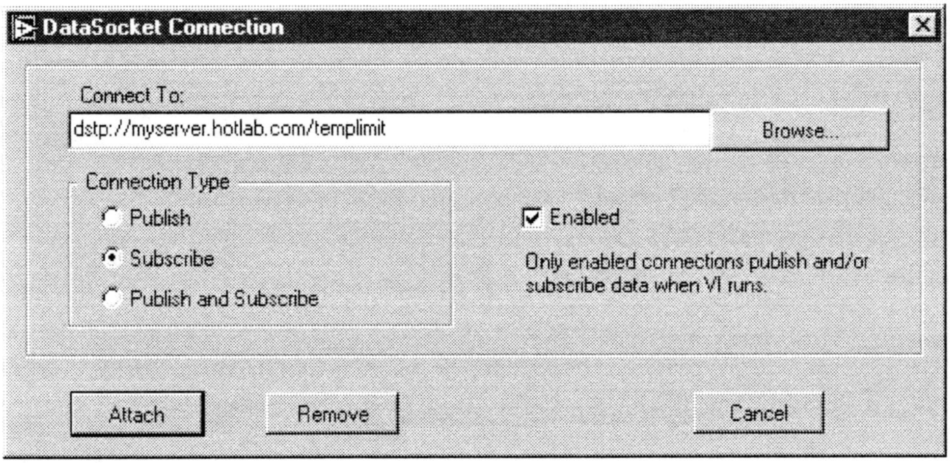

Figure 12-16.
DataSocket connection to a VI's front panel object.

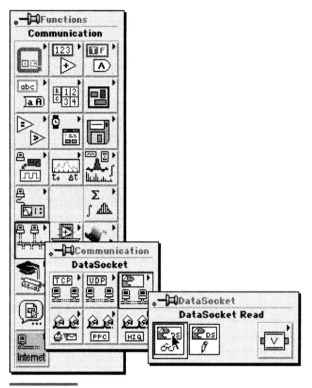

Figure 12-17.
The DataSocket palette.

The LabVIEW Web Server

So far we've seen the mechanisms available in LabVIEW for communication between VIs across a network. But what if you want your VIs to communicate and publish their data to the Web, not just a VI?

One of the easiest ways of making a VI's front panel viewable over a Web browser is by using the **LabVIEW Web Server.** With the LabVIEW Web Server, you can *dynamically* create Web pages with images of a VI's front panel on the fly, without the need for any special coding in your block diagram. You can turn on the LabVIEW Web Server at **Tools>>Options>>Web Server: Configuration** (see Figure 12-18).

With the LabVIEW Web Server, there is nothing to program on the block diagram. Unlike the VI Server methods, which programmatically create and

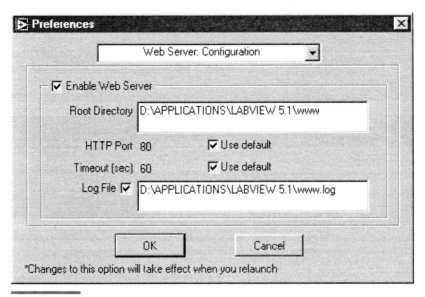

Figure 12-18.
LabVIEW Web Server configuration.

save a static HTML document into your server's file system, the LabVIEW Web Server dynamically creates images of a front panel in response to an HTTP request. The VI to be displayed in a Web browser must be loaded into memory in LabVIEW to be served by the LabVIEW Web Server. The Lab-VIEW Web Server can provide either a static (snapshot) image of a VI's front panel, or an image of the VI's front panel that automatically gets refreshed (monitor). To see how easy it is to publish VIs to the Web, try the following steps:

1. Make sure you've turned on the built-in Web server, as described earlier.
2. Open the VI Temperature System Demo.vi (from the LabVIEW examples, in `/examples/apps/tempsys.llb`).
3. Open your Web browser.
4. To see a static image of the VI, type this URL in your Web browser (on the same machine): `http://127.0.0.1/.snap?Temperature+System+Demo.vi` (You need to type a + to replace the spaces in the VI name, since spaces are not allowed in a URL; see Figure 12-19.)
5. To see a self-refreshing image, type this URL: `http://127.0.0.1/.monitor?Temperature+System+Demo.vi`

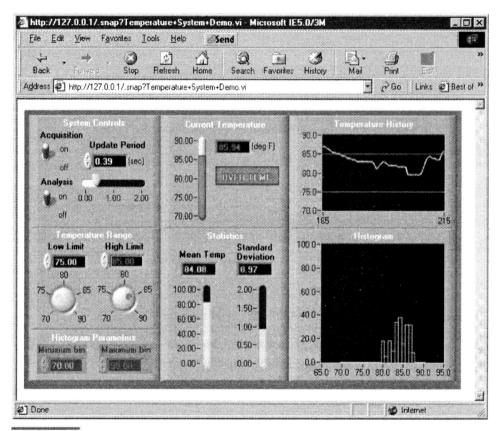

Figure 12-19.
Web browser viewing VI image.

Note that these dynamic documents returned by the LabVIEW Web Server are simply front panel images, they are not HTML documents. However, it is very straightforward to create your own HTML documents that embed dynamic images of a VI's front panel; you do this by simply specifying the URL as the image source in an <IMG> tag.

Caution: The LabVIEW Web Server only lets you *monitor* a LabVIEW VI, you cannot *control* the VI .

In addition to the LabVIEW Web Server, you can enhance some of LabVIEW's Internet capabilities with the **Internet Developer's Toolkit for G** from National Instruments. Table 12-2 summarizes which application-level protocols are bundled with the corresponding software.

Table 12-2. *Protocols bundled with LabVIEW 6.i or Internet Toolkit.*

LabVIEW 6.i	Internet Toolkit
	FTP
HTTP (Web) Server	**SMTP** (e-mail)
DSTP DataSocket	**Telnet**
	HTTP with added capabilities, such as CGI and password-protected directories

Controlling VIs Over the Web

As you now know, displaying data on the Web from LabVIEW is trivial thanks to the built-in Web server. You can even display animated front panels on Netscape browsers using the special .monitor? URL. However, if you need to *control* or otherwise have users interact with your LabVIEW system over the Web, then you must choose between CGI, Java applets, and ActiveX for your solution. Table 12-3 contrasts the differences among these three.

The following sections describe CGI, Java, and ActiveX in more detail.

CGI and HTML Forms

CGI stands for *Common Gateway Interface*. Essentially CGI refers to the communication protocol by which a Web server communicates with other applications and databases. With CGI, you can both *get* user data from the Web browser (for example, from a form a user fills out), as well as *create* HTML data on the fly (for example, an order confirmation page).

CGI just defines an interface; it isn't a programming language or an application. This sometimes confuses people because the term *CGI script* or *CGI application* is used to describe the outside application that uses CGI to communicate with a Web browser. The diagram in Figure 12-20 illustrates this relationship.

CGI scripts (or applications) can be written in almost any language. The most popular language for writing CGI scripts is Perl, a powerful and elegant scripting language. You can also write CGI scripts in C, C++, FORTRAN, Pascal, AppleScript, and, of course, LabVIEW. The **Internet Toolkit** for LabVIEW provides CGI VIs you can use.

Table 12-3.

	CGI	Java Applets	ActiveX
Ability to interact with controls (knobs, switches)	Possible with imagemaps, but very limited; requires reloading of page	Yes	Yes
Ability for user to type in text or form data	Very easy with HTML forms; this is mainly what CGI is designed for	Yes, but more complex than CGI	Yes, but more complex than CGI
Ability to display live, animated data (e.g., strip charts, gauges).	No	Yes	Yes
Can use DataSocket	Not directly	Yes	Yes
Considerations for browser downloading program	Nothing special to download; CGI is server-based	Support from both major browsers. Browser automatically downloads the applet and begins running it. Applets are typically an order of magnitude smaller in code than ActiveX controls (30 to 500 K).	Only works with Internet Explorer browser. ActiveX controls are usually large (over 1 MB), must install themselves in the Registry, and usually require a reboot of the computer the first time the visitor visits the page.
Multiplatform support	Yes. All browsers work with CGI.	Yes. Most browsers work with Java.	No. Only Internet Explorer on Windows can use ActiveX controls.

Table 12-3 (continued)

	CGI	Java Applets	ActiveX
Tools recommended for development	Internet Toolkit for G (National Instruments)	Virtual Instrumentation Beans (ErgoTech) plus a Java IDE such as IBM VisualAge for Java	ComponentWorks (National Instruments) plus Visual Basic
Security considerations	Possible security threats to server if CGI program is not designed right; none to client	Limited security concerns to client or server, since sandboxing limits what Java applets can do	Possible security threats to client from unstable or malicious ActiveX controls
Ability to sign code	Not applicable	Yes	Yes
Learning curve for development	Medium	High	Medium with Visual Basic

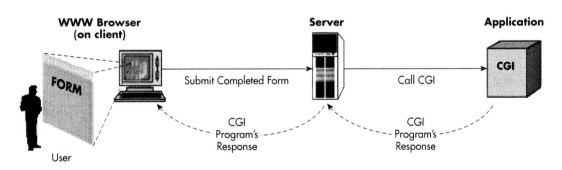

Figure 12-20.
CGI data flow.

Information can be passed to a CGI application through the use of HTML forms, imagemaps (images with defined clickable areas), as well as Java applets. We'll talk about forms and imagemaps as well.

CGI applications are numerous; you have probably used a CGI application many times without knowing it. Most forms you fill out on the Web are passed to a CGI application. Examples include search engines, information-request forms, many e-commerce sites, interactive sites, and so on. Using Lab-VIEW with CGI opens the door to creating an interactive Web site that can communicate with your LabVIEW application. For example, you could use CGI with LabVIEW to

- Create an interactive Web-based laboratory, by allowing users to input data on a Web form, pass that data to a LabVIEW VI for executing, and return the results back to the Web browser.
- Have LabVIEW create dynamic Web pages on the fly that contain more real-time data than just panel images; you could create tables, formatted text, images, counters, and so forth.
- Interact with a database and LabVIEW over the Web; for example, using CGI to create instant composite Web reports from database data and LabVIEW results.
- Add security: log-on features for a protected Web site.
- Use cookies to track visitors and store state information.
- Simulate front panel button clicks by combining LabVIEW's ability to generate front panel images with imagemaps.

CGI Example

CGI and CGI applications don't have HTML tags, because they're not an HTML element. Usually the CGI application is named in a form tag, telling the browser to call a CGI application to process the form input values; for example,

```
<FORM ACTION="/cgi-bin/ProcessMyForm.vi"
NAME="MyForm" METHOD="POST">>
```

The browser in Figure 12-21 shows an example of a VI controlled through an imagemap and an HTML form that interfaces with CGI VIs on the Internet Toolkit's Web server.

Tools and Resources for CGI

The tools you use to write your CGI application will depend first on what language you will write it in, and second, what type of Web server you will be using. If you decide to write CGI applications in LabVIEW, we refer you to *In-*

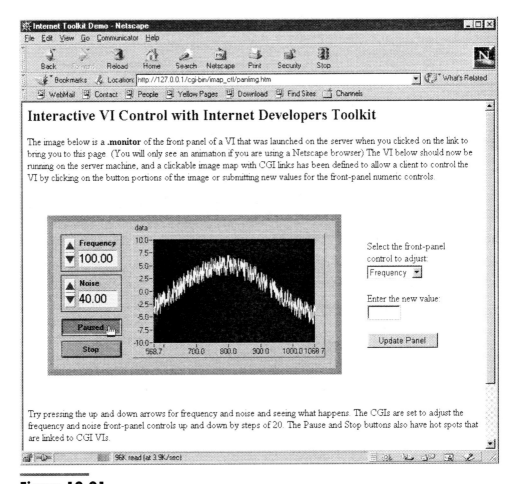

Figure 12-21.
Common gateway interface to a VI.

ternet Applications in LabVIEW by Jeffrey Travis. Here are some other recommended tools:

- **General information on CGI:** You can get information on the CGI standard from the folks who developed it at the National Center for Supercomputing Applications, hoohoo.ncsa.uiuc.edu /cgi.

- **LabVIEW and CGI:** Use the *Internet Developers Toolkit for G* from National Instruments. You can get more information at www.ni.com/ labview/internet.

- **Perl and CGI:** To some people, Perl and CGI are inseparable. Perl is cross-platform, it's open source, it's fast, it's *free*. The Perl motto is that

"Perl makes simple things easy and hard things possible." It's well suited especially for parsing text and handling large scripts. Find out more or download the Perl interpreter for your OS at www.perl.org.

Java Applets

Java is an object-oriented programming language developed by Sun Micro-systems. What makes Java unique, besides the massive amount of publicity that has made it a household name, is its ability to create programs that can run on a variety of platforms, as a standalone executable, on an embedded de-vice, or inside a Web browser. The Java motto is "write once, run anywhere." This concept is not new to us LabVIEW developers! National Instruments has also embraced platform portability, allowing us to seamlessly write LabVIEW VIs once and (with some exceptions) run them on a Windows, Mac, or Unix machine.

The key to understanding how Java works in practice is to realize that a Java application must always run inside a *Java Virtual Machine* (VM; analogous in some ways to the LabVIEW Run-Time engine). A Java VM is a shell that pro-cesses the Java bytecode and transforms it into platform-specific machine code at run time. So a variety of Java VMs exist for Windows, Mac, and Unix. Java VMs don't compile the code (the code is already compiled); they just have the right hooks into the specific OS to run the compiled Java application.

An *applet* is special kind of Java application. Java applets are arguably the most widely used type of Java program. A Java applet normally runs inside a Web browser, such as Netscape Navigator or Internet Explorer, instead of running on the desktop. Both these browsers incorporate their own Java VM embedded into the browser. In addition, Java applets can easily be distributed over a network by simply connecting to a Web site that contains an applet reference in the HTML code. And because Java applets are (in theory) inde-pendent of the browser or platform, any browser on any platform that has a compatible Java VM can run the applets. Java applets have many security restrictions that do not apply to full Java applications. For example, normally applets cannot read or write to the user's hard drive, cannot open a network connection to an IP address other than the one the applet originated from, and so on. These security restrictions are known as *sandboxing*.

One other important Java technology you should know about is *JavaBeans*. The JavaBeans API defines a software component model, in the same way that ActiveX is a component object model. Essentially, when someone compiles her Java code into a JavaBean, it becomes a black-box component that you can use in your own applications without having to know how the inside works. For example, if someone gives you a JavaBean that sends e-mail, you could just plug it into your Java app without recoding or compiling any of it. National Instruments provides a DataSocket interface as a JavaBean that can be used to implement that DataSocket API.

Java can be used to create almost any kind of application, just as you could with C++, but Java applications have the additional benefit of being optimized for network distribution and incorporate a high level of security. Any Java application can be turned into a Java applet, which is easily embedded into a Web page by simply providing the right HTML tag. When you visit a Web page that specifies a Java applet, the Java code is downloaded to your computer (typically a small file under 150 K) and begins executing immediately. Java applets can be used in many applications, including

- Real-time charts and other graphical displays for data like stock quotes, weather, lab readings, news tickers, and so on;
- Virtual conference and chat rooms;
- Interactive forms for e-commerce and on-line banking;
- 3D visualization;
- Database access and reports;
- Remote control of instrumentation; and
- Remote management of process control and factory automation.

Java offers some of the most exciting possibilities for LabVIEW for remote monitoring and control over the Web. Java 2, the current version, includes a great deal of flexibility in building GUIs just as elaborate as those you can build in LabVIEW. In theory you could recreate the front panel of any VI and make it available remotely to anyone with a Web browser, if you can hook it into your LabVIEW application.

Java, like LabVIEW, is multiplatform, and Java applications will run on Windows 95/98/2000/NT, MacOS, Linux, HP-UX, or Solaris. You can even run Java applications on many embedded and real-time OS like VxWorks, QNX, and so on, making it possibly the most portable application language. Unlike ActiveX, which essentially limits remote clients to Windows and Internet Explorer, you open your system to a much broader user base.

Java applets are also supposed to run on any modern Web browser (Netscape and IE 4.0 version or higher) on any platform. Having said that, reality is never as smooth as theory. Although Java applets in theory are write once, run anywhere, the fact is they depend on the Java VM being developed correctly for a particular platform or browser. So while the specs exist to make Java truly platform-neutral, not all vendors implement or support the Java VM in the same way. At press time, for example, Netscape still does not support much of the core Java 1.1 language on the Macintosh. Microsoft placed some of its own Windows-only features for Java on its Internet Explorer browser. This doesn't mean Java applets won't work for all these browsers, it just means in practice you will need to test the target platforms.

One of the most compelling reasons to use LabVIEW and Java is the availability of the DataSocket JavaBean, which makes it possible to use DataSocket and a Java applet or application as a remote client.

Java Example

Figures 12-22 and 12-23 show a custom Java applet that remotely controls a LabVIEW-based BioBench™ virtual instrument.* In this case, not only is the data displayed in real time, but the remote user can interact with the panel, by selecting the waveforms to view, logging data, and so forth, all from the Web browser.

Tools and Resources for Java

- **Remote Web control: LabVNC** is a useful tool for remote Web control of LabVIEW front panels. LabVNC is free and open source, allowing you to use Java applets to remotely interact with your VIs. It is included on the accompanying CD-ROM or you can download it at http://jeffreytravis.com/lost.

- **Java language:** There are myriad books and resources on Java. One good place to learn about Java is Sun's Web site at http://java.sun.com. One of the best books on Java we have encountered is *Just Java* (4th ed.), by Peter van der Linden (Prentice Hall).

- **Java development:** You can download for free the Java Software Development Kit (SDK) from Sun at http://java.sun.com, which is available

---

Note: This Web-enabled BioBench demonstration was developed by Jeffrey Travis Studios in cooperation with Premise Development Corp. For more information, please go to http://jeffreytravis.com.

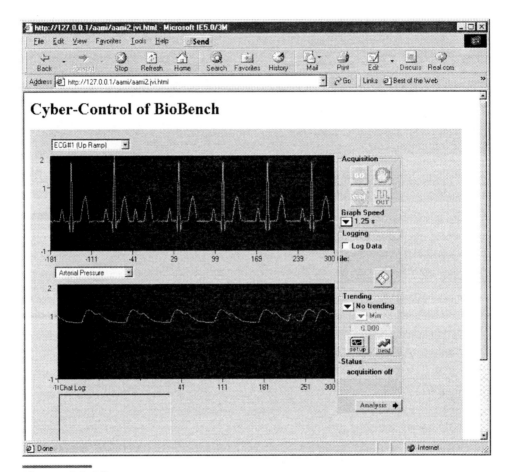

Figure 12-22.
Java applet running a BioBench front panel.

for Windows and Unix flavors. Mac users can get a MacOS SDK for Java at www.apple.com/macos/java. Commercial development environments for Java are available from Metrowerks (www.metrowerks .com), Inprise (formerly Borland; www.inprise.com), Symantec (www .symantec.com), and Microsoft (www.Microsoft.com).

• **Java and LabVIEW:** National Instruments has a JavaBean for DataSocket (www.ni.com/datasocket).

• **Virtual instrumentation Java components:** Ergotech (www.ergotech .com) makes a product called Virtual Instrumentation Beans (VIB),

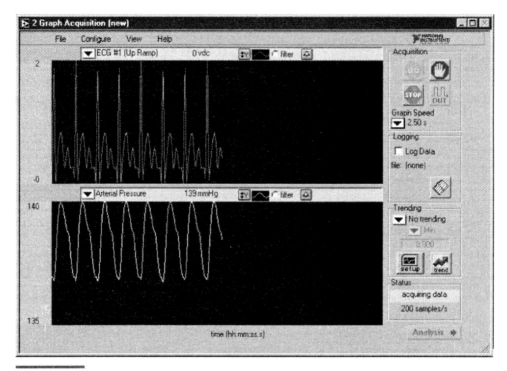

Figure 12-23.
BioBench VI being controlled remotely.

which you can use in conjunction with the DataSocket Bean to build Java applets that communicate with LabVIEW.

ActiveX Controls

ActiveX is a Microsoft technology that defines a component-based architecture for building applications that can communicate with each other. Do you remember OLE (Object Linking and Embedding)? ActiveX is the latest version of OLE, with a snazzier name for the Internet. In fact, when working with ActiveX, you may still see references to OLE components or OLE variants, which are part of ActiveX and perhaps haven't been rechristened yet. ActiveX is mostly a Windows-only technology and is part of Microsoft's DNA (Distributed Network Applications) architecture.

ActiveX components, or controls as they are sometimes called, are pieces of reusable code that can be embedded into another ActiveX-capable application, or *ActiveX container.* Any ActiveX component has to be inside an ActiveX container for it to run. ActiveX components can be combined as building blocks to build more complex applications. In addition, one big advantage of using ActiveX controls is that you don't need to see the source code or make modifications to use them in your application; they function like a black box, allowing you to execute their methods and set their properties. ActiveX components can be developed in different languages, such as C++ or Visual Basic.

Strictly speaking, ActiveX's counterpart is a JavaBean, in the sense that they are reusable software components for any generic application. However, when we talk about Web-enabled applications, it's more common to compare an ActiveX control (which can be embedded in a Web browser) to a Java applet (also embedded in a Web browser, but may or may not rely on JavaBean technology).

ActiveX Example

An example of an ActiveX control is a **Date&Time** picker control (see Figure 12-24), which is one of many components you can install on Windows (ActiveX components, once installed, permanently stay in your Windows directory and registry). An example of an ActiveX container is Microsoft's Internet Explorer, which is why it's very simple to embed an ActiveX control in a Web page. If you've worked with ActiveX and LabVIEW, you know that LabVIEW (for Windows) also provides you with ActiveX containers that you can place on your front panel. You could set up an application on the server that then reads the date provided by the Date&Time selector from the remote user.

This ActiveX control isn't just available for Internet Explorer; it can be embedded into any ActiveX container. For example, we could just as easily put it into LabVIEW's front panel as shown in Figure 12-25. Internet Explorer is not only an ActiveX container but also an ActiveX component, meaning you can embed the Web browser into other applications, including LabVIEW.

One limitation of ActiveX is that, being a Microsoft technology, it (not surprisingly) tends to favor only Microsoft products, like Internet Explorer. If your remote users use Netscape as their browser, they will have to download a special plug-in before they can use ActiveX controls in their Web browser.

Figure 12-24.
An ActiveX control embedded into the Web page (Internet Explorer browser).

Generally, ActiveX controls are better suited for intranets and are not as common on the Internet, since they rely on a narrower browser specification.

ActiveX has many powerful possibilities with LabVIEW (assuming you don't need multiplatform support). LabVIEW supports communication with ActiveX components through the functions on the **Communication>> ActiveX** palette.

But the most obvious application for remote monitoring and control on the Web using ActiveX controls is DataSocket. With Visual Basic, you can easily

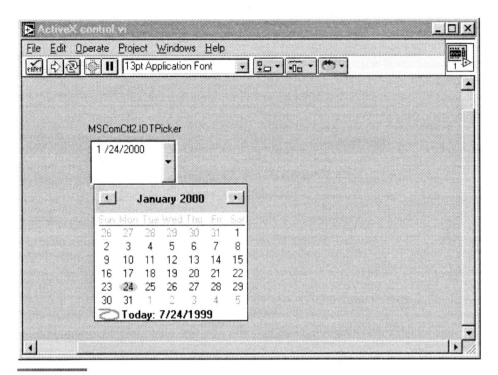

Figure 12-25.
The same ActiveX control inside a VI front panel container.

create an ActiveX control that uses National Instruments ComponentWorks GUI objects and the DataSocket components to handle communication. By building an ActiveX control that uses the DataSocket protocol, you greatly simplify the LabVIEW programming needed. The DataSocket Server can treat an ActiveX control on the Web just like any other DataSocket client, meaning that nothing changes on the LabVIEW side.

Tools and Resources for ActiveX

To build ActiveX components, you can use Microsoft's Visual Studio environment—more specifically, you can use a development kit like Microsoft Visual C++ or Visual Basic. Visual Basic is often a preferred tool because it's easier to use and still remains quite powerful. There are a number of third-party add-ons, tools, and utilities available on the Web as well.

- To learn more about ActiveX in general, go to www.microsoft.com/activex.

- You can develop ActiveX components for Web control using **Visual Basic Control Creation Edition**, which is a free download at http://msdn.microsoft.com/vbasic/downloads/cce.

- In order to use virtual instrument controls (knobs, charts, etc.), you'll need *Measurement Studio* from National Instruments. Measurement Studio is a set of ActiveX controls for instrumentation that you can use with environments like Visual Basic. You can get more information at http://ni.com/cworks.

XML

First, some background: the mother lode of markup languages is called SGML (Standard Generalized Markup Language), which defines how to create markup languages. For example, HTML is a subset of SGML (although it doesn't follow all its rules). XML is a simplified daughter of SGML.

XML (eXtensible Markup Language) is a meta-language that lets you create, describe, format, that is, *extend,* your own markup tags. In HTML, by contrast, every tag is static and part of the HTML specification; you cannot change the effect of the <BODY> or <TABLE> tags. With XML, you can create your own tags, configure what they mean, and design what effect they have; for example, <SuperHeading>, <CrazyFont>, <StockPrice>.

Each of your XML tags is defined in your own *Document Type Definition* (DTD) and *stylesheets* (such as CSS, or the more generic XSL, eXtensible StyleSheet Language).

Applications that can display XML content are not limited to modern Web browsers; other custom applications can interpret and parse XML as well, making it more powerful and suitable. XML content usually needs three files:

- XML document, which contains the data with your custom tags.

- Stylesheet, which defines how document elements should be formatted when displayed.

- Document Type Definition, which defines the rules for your XML elements—data are defined and logically related.

Because XML offers so many advantages over HTML, there is a push in the industry to replace or redefine HTML with XML. However, XML is still in a state of change and is not yet widely adopted in most Net applications. Its future solidification remains to be seen.

Note: Earlier versions of Web browsers do not handle XML. Internet Explorer 5.0 and Netscape 6 have good support for XML.

XML Example

Here is a simple example, contrasting HTML with XML. This is a simplified Web page that displays instrumentation data:

```
<h1>Today's Temperature Logger Results from Chamber A</h1>
<p>
Date / Time / Avg. Temp / Max Temp / Min Temp
<p>
12-Feb-2000 08:24, 23.4, 27.5, 21.9 <br>
13-Feb-2000 12:15, 24.1, 29.0, 22.2 <br>
<p>
```

The same information could be marked up with XML as follows:

```
<h1 chamber>Today's Temperature Logger Results from Chamber A
</h1 chamber>
<p>
Date / Time / Avg. Temp / Max Temp / Min Temp
<p>
<datetime>12-Feb-2000 08:24</datetime>,
<temp>
<max>23.4,</max><avg>27.5</avg>,<min>21.9</min>
</temp> <br>
<datetime>13-Feb-2000 12:15</datetime>,
<temp>
<max>24.1</max><avg>29.0</avg>,<min>22.2</min>
</temp> <br>
<p>
```

In our stylesheet and DTD, we would define the rules for <temp>, <max>, and <datetime> tags. Notice how the <temp> tags use nested tags that are attributes. By using XML, all of a sudden the data is much more meaningful to a search engine or an analysis program. For example, it would be fairly easy to write a script that answers a question like "What day in the month of February did the temperature reach its highest point?" by parsing the XML tags. This would be much more tedious and nonportable if we only used HTML.

XML could offer some exciting possibilities for LabVIEW-generated documents. Currently, LabVIEW can generate HTML documents that are easily readable on a Web browser, but suppose it could generate XML with meaningful tags that could be interpreted by other applications? If you have other XML-enabled applications, you could use LabVIEW and the Internet Toolkit's HTML VIs to generate XML documents.

With XML, your LabVIEW reports could describe and define what kind of data is in the document, not just simply display it. If we had to guess, we'd bet someone will make a toolkit out of this, or National Instruments will incorporate it as a future feature in LabVIEW.

Tools and Resources for XML

You can get the most current XML info at www.w3.org/xml. Another promising development is XHTML, which is a reworked HTML 4.0 that follows XML standards. More information on XHTML can be found at www.w3 .org/TR/xhtml1/.

The Wireless Web: WAP and WML

The recent explosive growth of smart cellular phones, text-enabled pagers, and handheld wireless devices like the Palm and Handspring have encouraged the development of standards for accessing the Web in new ways. Because most of these devices have a very small screen and limited bandwidth compared to a PC, the traditional protocols like HTTP and HTML for the Web do not work too well. The *Wireless Access Protocol* (WAP; see Figure 12-26) defines things like how information from files and servers can be made accessible to wireless devices, much in the same way that HTTP defines how Web servers and browsers communicate.

For most wireless devices, documents are not HTML documents, but *Wireless Markup Language* (WML) documents. WML is an XML-based specification that defines tags for how data and events will be displayed and handled in a wireless device. While LabVIEW does not currently have any support for WML or WAP, it is possible to still connect VIs to a WAP server by writing custom software with commercial toolkits. For more information on WML and WAP, go to www.allnetdevices.com/. For WAP development toolkits and phone emulators, go to http://updev.phone.com/.

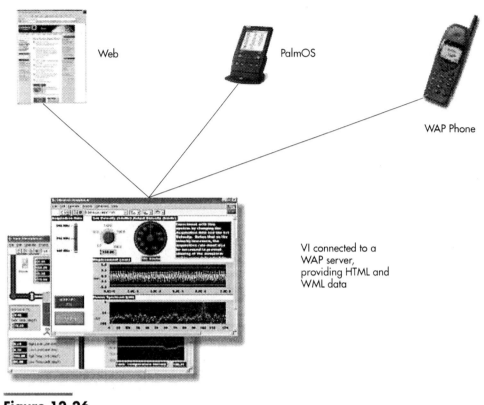

Web   PalmOS

WAP Phone

VI connected to a
WAP server,
providing HTML and
WML data

Figure 12-26.
Wireless Access Protocol.

Designing VBI for Remote Network Access

A very common question people have when designing an Internet-enabled system is "which technologies should I use for my application?" The flowchart in Figure 12-27 shows you a general decision process for deciding what technologies to use. The decision process assumes that you will be using LabVIEW as a data server in some capacity and doesn't take into account interfacing with other software systems (besides the Web).

Web versus VI Clients

For a distributed computing system that uses LabVIEW, you will almost invariably use all VI-based components. For a client-server system, you can see

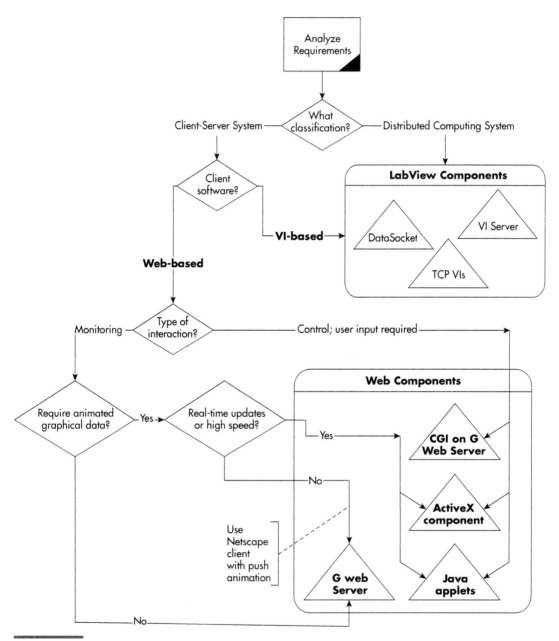

Figure 12-27.
Decision tree for Internet-based LabVIEW development.

Table 12-4. *The pros and cons of Web-based versus VI-based clients.*

	Web-based Clients	VI-based Clients
Pros	• No special software to install; remote users only need a browser • Lower maintenance and upgrade costs, since all code resides on server • Easier for clients to use • Security: no source code ever is on the client machines • Wide platform support	• Remote control and interactivity much easier to program, since it's all in LabVIEW • More flexibility in choice of user interface, behavior • Security: only users with access to the client VIs can use the system
Cons	• Remote control is far more complex to implement; requires CGI, Java, or ActiveX programming • Security: anyone with a browser could potentially access the system	• Requires that remote user have the current version of LabVIEW installed, or distribute large executables • Security: remote users may have access to source code • Costlier and more complex maintenance: changes may require all users to download a new client program • If executables are used instead of VIs, multiplatform support requires providing separate executables

that the major decision is choosing a Web-based solution versus a VI-based solution. Table 12-4 compares these two choices.

DataSocket versus VI Server

Whether you're implementing a distributed system or a VI-based client-server system, you will end up using DataSocket, VI Server, or TCP VIs, or a combination of these. Using TCP VIs is really necessary only if you're designing your own communication protocol, or if you need to interface to other code. Sometimes people are confused about when to use VI Server and when to use DataSocket. The truth is that they're really intended for different purposes. Here is the general rule:

- Use VI Server when you need to share *processes* and *interfaces*; for example, if you want to be able to call a VI's functionality remotely, or display its front panel.
- Use DataSocket when you want to share *live data*, independently of how you're displaying, storing, or obtaining it. Also you should use DataSocket if you want data to be available to different type of clients (VIs and the Web).

Enterprise Connectivity: The Big Picture

For the most part in this book, we've taken an approach for simple do-it-yourself networked systems that use LabVIEW as a server and a VI or Web browser as a remote client. Sometimes, though, especially for applications that require storing and processing lots of data, you need to design systems from the ground up with experts who understand the different tiers of an enterprise-level network. This often includes working with SQL databases such as Oracle or Sybase, tying together the LabVIEW labs to a common network, and implementing remote client access. Figure 12-28 illustrates this architecture.

For example, suppose you had several test labs using LabVIEW to do functional testing on widget products. Adding Internet capability would allow you to monitor the current test status of any of the labs using a Web browser. However, by tying a database into the network, you could not only obtain historical test information from the same Web browser, but you could examine all kinds of interesting relationships. For example, you could compare average pass/fail times between different labs, or run a query comparing historical results with current real-time tests.

So, of particular importance in the enterprise system is the back-end database. In many cases, you will want to store your data acquisition test results in a database rather than on your local drive. Databases allow far more efficient storage and retrieval of large amounts of data, particularly for situations of concurrent publishers and subscribers. In addition, databases allow programmers to create all types of queries that can examine relationships between elements in the database.

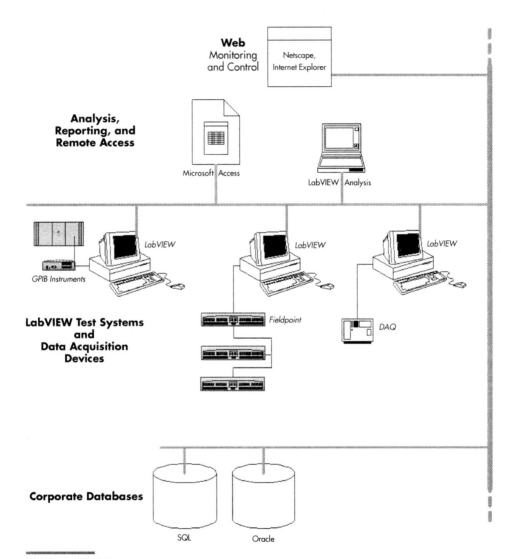

Web
Monitoring
and Control

Netscape,
Internet Explorer

**Analysis,
Reporting, and
Remote Access**

Microsoft Access

LabVIEW Analysis

LabVIEW

LabVIEW

LabVIEW

GPIB Instruments

**LabVIEW Test Systems
and
Data Acquisition
Devices**

Fieldpoint

DAQ

Corporate Databases

SQL   Oracle

Figure 12-28.
Enterprise connectivity.

Although there are a variety of database vendors and brands, from the simple Microsoft Access to high-end Oracle systems, most of them can work with a common language known as *SQL*. SQL stands for Structured Query Language and is principally used to write queries that are sent to the database in order to retrieve the desired data. It's possible to use SQL statements with a variety of different databases, partly thanks to *ODBC* (Open DataBase Connectivity), a glue layer that abstracts SQL statements from an application and lets them work with a variety of database drivers.

Another way databases are used is to allow Web access to the data remotely. By writing Web server scripts (using CGI or ASP, for example), a remote user with a Web browser can perform queries and pull up data or charts that are displayed right on the browser. If your LabVIEW-based application handles a large amount of data and you'd like to have Web access to not only the real-time processes but also historical data, consider tying a database to the LabVIEW system (as a publisher of data) and a Web server script (as the subscriber of data). Your Web scripts could even be written in LabVIEW as CGI VIs!

National Instruments sells an SQL Toolkit for LabVIEW that has support for over 70 databases, allowing LabVIEW to read and write to almost any SQL database. You can also obtain the Enterprise Connectivity Toolkit, which bundles the SQL Toolkit, the Internet Toolkit, and the SPC (Statistical Process Control) Toolkit for LabVIEW in one package. For more information about these toolkits, visit the National Instruments Web site (www.ni.com).

Future Potential

13

From the inception of this book, it has been our sincere wish that you, the reader, would gain insight and appreciation for the many ways in which virtual bio-instrumentation can be applied to the biomedical and healthcare industry. We have striven to provide you with real-world and practical applications of virtual instrumentation in biomedicine, healthcare, and related industries. This book intentionally covers a broad range of topics to illustrate how computer-based instrumentation, analysis, and presentation empowers engineers, researchers, medical device manufacturers, clinicians, and administrators to create user-defined tools and solutions to acquire, manage, analyze, and archive information.

Although the future is very difficult to predict, there are several technologies and trends that we believe will continue to impact biomedical research and the delivery of healthcare. VBI can be a significant part of these developing fields if people like you apply it in earnest. The following top ten list highlights some of these emerging trends and technologies that we predict will make significant strides in the near future.

1. *Information management and advanced user interface capabilities will be used to gather data to help physicians care for their patients* and provide information in easy-to-access formats. One of the biggest challenges facing

clinicians today is managing the flow of information across the entire enterprise. Specifically, there is an abundance of clinical information such as medical images, lab results, and vital signs data that need to be brought together to help doctors and nurses make the most effective treatment decisions for their patients. These advances will also help patients take a more active role in their disease management—in the same way that diabetics take an active part in monitoring the status of their disease. It is interesting to note that an estimated 150 million Americans now have access to the Internet from their homes, a 32% increase over the same time last year. Over half the people on-line are searching for healthcare information and assistance with care. The fastest growing Internet user segment is the 55 and older age group.

2. *Telemedicine and communications technologies* will eliminate time and space and seamlessly connect patients with care providers through advanced technologies such as wireless telemetry systems and the Internet. Telemedicine holds much promise to help improve patient care, particularly in crossing barriers faced by rural and other less accessible patient populations.

3. *Sensors and smart devices* will be so sophisticated that they will recognize and adapt automatically to the changing needs of the patient. As virtual instruments can now be embedded into microprocessors and run under real-time operating systems, we believe that VBI technology will play a significant role in the area of integrated sensors and smart devices.

4. *Device miniaturization breakthroughs* will continue to dramatically reduce the size of implantable devices—yet increase their efficacy, longevity, and information communication capabilities. Over the past few years, we have witnessed dramatic impact and benefits of minimally invasive and robotic surgery.

5. *Imaging, visualization, and navigation advances* integrated into products will help physicians accurately assess the location and characteristics of the tissue or organ of interest, decide on the course of action, and precisely deliver a desired therapy to that point. Enhanced diagnostic capabilities, three-dimensional modeling, digital image storage, and teleradiology and Picture Archiving and Communication Systems (PACS) will continue to revolutionize diagnostic and therapeutic procedures.

6. *New implanted diagnostic systems* will fully integrate sensors, electronics, and telemetry to enhance a physician's ability to determine patient status and conveniently access meaningful diagnostic information—continuously or at whatever monitoring intervals they select.

7. *Unique technology convergences* will integrate the capabilities of two or more therapeutic or monitoring devices to help physicians more comprehensively and effectively treat complex chronic diseases.

8. *New developments in electrostimulation therapy* will build on today's knowledge and understanding of the use of electrical stimulation in physiological systems to further enhance and discover yet unknown therapeutic solutions for patients with chronic disease.

9. *New ideas in tissue engineering and use of advanced materials* will allow researchers to develop therapies that combine traditional engineering and living cells, promote the in-growth of cells in tissue scaffolds, grow new structures, or even replace an implantable therapy with restorative living tissue. The potential benefits of these technologies in the areas of organ transplants and replacement body parts are significant but must be balanced with a whole host of ethical issues.

10. *Bold forward strides in biological and gene therapies* will enable scientists and researchers to appropriately integrate molecular biology-based and gene therapies into evolving medical technologies. The growing field of genomics is the result of 15 years of global research and an investment exceeding $3 billion. To date, over 100,000 genes and 3 billion gene sequences have been identified. Today, scientists have a working map of the human genome. This map is being used to target 3,000–4,000 genetic diseases. The need to acquire, store, and analyze this growing body of knowledge is tremendous and as such will continue to rely heavily on computer-based technologies.

We strongly believe that virtual bio-instrumentation and other computer-based technologies will continue to play an important role in the design, validation, deployment, and utilization of these emerging technologies. Hopefully, this book has provided some quality insight and reference material that will help our readers position themselves at the forefront of these escalating endeavors.

Independent Solution Articles

Telerobotic Camera System for Operating Rooms

by Luc Dubé and Glen Levesque, Coordinators, Biomedical Services, CRCHUM

The Challenge: Creating a remotely controlled video camera to monitor operating room procedures when the positions of the patient, surgical lamps, and surgeon change during the procedures.

The Solution: Using National Instruments LabVIEW 6.0*i* and DAQCard-1200 to design a simple, yet flexible, operating room robotic camera system and graphical user interface (GUI) for remote system control.

Introduction

The ongoing advances and widespread availability of telecommunications and information systems have made interactive "telemedicine" increasingly attractive and accessible. Instead of using surgical room amphitheatres or live video produced by a video crew, surgeons often employ fixed cameras installed throughout an operating room to record surgical procedures. However, this solution is often impractical because the positions of the patient, surgical lamps, and surgeons change depending on the type of surgical procedure performed. The ideal operating room camera system must have a certain number of degrees-of-freedom that users can control from remote locations when recording various surgical procedures.

Design

To move a video camera over the operating field, the robotic arm of this camera system consists of a two degrees-of-freedom rotative, prismatic manipulator coupled with a two degrees-of-freedom "pan-tilt-zoom" Panasonic wv-cs 404 camera. Users can either mount the robotic camera system underneath and on the same rotational axis as the surgical lamps, or attach it directly to the operating room's ceiling. With this strategic configuration, users can easily position the camera on a variable circular plane over the operating field. To simplify system construction and minimize the total central axis, the system height elongation axis is slightly longer than the rotational axis, giving the unit the freedom to extend the camera towards and away from the main arm for proper camera position above the patient.

Users can control the robotic camera system manually or through a GUI. In manual mode, a user uses two, two-axis joysticks to control the length and rotational position of the arm, as well as the pan and tilt of the camera. He or she manually controls the zoom, focus, and other camera parameters from the camera control unit. However, while users can obtain all possible system configurations using the manual control mode, it is somewhat difficult to coordinate the allowable degrees-of-freedom to obtain the desired point of view.

To automate the process, we developed a PC-based system controller and GUI to easily control all system motions using National Instruments LabVIEW 6.0*i*. LabVIEW 6.0*i* helped us quickly assemble and program a functional prototype to control the system and acquire

Figure 13-1.
Using National Instruments LabVIEW 6.0*i* and DAQCard-1200, the TELEROBOT is controlled remotely and acquires images of operating room procedures.

data. Also, with the flexible and easy-to-learn graphical programming interface of LabVIEW 6.0*i*, we could quickly and easily modify the application, decreasing time spent on the experimental process.

The GUI mimics the actual position of the camera arm virtually by reproducing all system components and parameters on the control monitor. We also reproduced top and side views of the camera manipulator and operating table. We then incorporated the direct and inverse kinematics of the manipulator into the interface to operate the system in Cartesian coordinates, a common system used to quantify the distance of points from lines or planes. Users can also program different points of view, which the system can later recall to reposition the camera at the stored locations by using a simple recall position function.

Implementation

To facilitate evaluation of the prototype system, we installed it in an operating theatre equipped with an observation room. The camera manipulator was mounted directly to the ceiling of the operating room, eliminating the need to modify the current surgical lamp in-

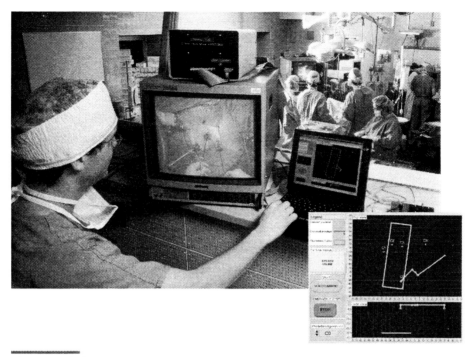

Figure 13-2.
Surgeons using a laptop computer and National Instruments products can control the camera and view surgeries remotely.

stallation. We then positioned the control center and system hardware in the adjacent observation room. The camera control unit manipulated the system cameras and transmitted captured images to two monitors. The first monitor displayed the surgical procedure captured by the robotic camera system. The second monitor displayed images obtained from fixed operating room cameras, giving visual feedback on the robotic camera system position and a full view of operating room activities.

Results

The prototype system initially filmed a minimally invasive surgery—a laparoscopic kidney removal in which the healthy donor's kidney was prepared and removed using conventional laparoscopic techniques. While the standard laparoscopic camera captured the inner-abdominal procedure, the telerobotic camera system captured an outside view of the surgical site insertions and focused on final kidney preparation and removal. Although preliminary, the images captured during this procedure required only minimal system adjustments. More importantly, the surgeons were very satisfied with the filming results.

This robotic camera system, named TELEROBOT, is an ongoing research project that will develop into other solutions. With LabVIEW 6.0*i* and DAQCard-1200, we have a flexible and versatile platform that meets ongoing project demands. LabVIEW 6.0*i* is lowering our costs by reducing development time and eliminating our need for other software platforms.

For more information contact

Luc Dubé, Coordinator, Hôpital Notre-Dame du CHUM,
1560 Sherbrooke Est., Montréal, Qc, H2L 4MI
tel (514) 281-6000 x7576, e-mail dube.luc@sympatico.ca

Glen Levesque, Coordinator, Biomedical Services, Hotel-Dieu du CHUM,
tel (514) 843-2611 x5042, e-mail glenl@colba.net

Appendix
Typical Math Model
Parameter Values

Using the model equations described in Chapter 11, the mathematical model of the closed-loop cardiovascular and pulmonary systems can be simulated using the parameters listed in the following tables.

Canine Model Parameters

Cardiovascular Model

Table A-1. *Parameters used to define the Gaussian characterization of ventricular and atrial elastance in the canine cardiovascular model.*

Parameter	Units	$i = 1$	$i = 2$	$i = 3$	$i = 4$	Description
$A_{v,i}$	N/A	0.430	0.360	0.500	0.550	Gaussian curve magnitude
$B_{v,i}$	sec	0.045	0.035	0.037	0.036	Gaussian curve width
$C_{v,i}$	sec	0.175	0.230	0.275	0.300	Gaussian curve delay
A_a	N/A	0.900	N/A	N/A	N/A	Gaussian curve magnitude
B_a	sec	0.018	N/A	N/A	N/A	Gaussian curve width
C_a	sec	0.065	N/A	N/A	N/A	Gaussian curve delay

Table A-2. *Parameters used to characterize the pressure-volume relationship of the atria and pericardium in the canine cardiovascular model.*

Parameter	Units	LAF	RAF	PCD	Description
E_{ES}	mmHg/ml	1.10	1.00	N/A	Systolic maximal elastance
V_D	ml	5.00	5.00	N/A	Systolic volume intercept
P_O	mmHg	0.50	0.35	1.00	Diastolic scale parameter
λ	ml^{-1}	0.05	0.05	0.03	Diastolic exponential parameter
V_O	ml	5.00	5.00	45.00	Diastolic volume offset

Table A-3. *Fixed parameters used to characterize the resistive and compliant elements of the canine cardiovascular model.*

Parameter	Description	Value	Units
R_M	Resistance: mitral valve	0.0225	mmHg*sec/ml
R_{AOP}	Resistance: proximal aorta	0.0544	mmHg*sec/ml
R_{COR}	Resistance: coronary circulation	89.9000	mmHg*sec/ml
R_{CRB}	Resistance: cerebral circulation	29.9700	mmHg*sec/ml
R_{AOD}	Resistance: distal aorta	0.0093	mmHg*sec/ml
R_{SA}	Resistance: large systemic arteries	0.0577	mmHg*sec/ml
R_{SV}	Resistance: systemic veins	0.4200	mmHg*sec/ml
R_{VC}	Resistance: vena cava	0.0050	mmHg*sec/ml
R_{RA}	Resistance: right atrium	0.0800	mmHg*sec/ml
R_{TC}	Resistance: tricuspid valve	0.0350	mmHg*sec/ml
R_{PAP}	Resistance: proximal pulmonary artery	0.0110	mmHg*sec/ml
R_{PA}	Resistance: pulmonary arterioles	0.0430	mmHg*sec/ml
R_{PS}	Resistance: pulmonary shunt	11.3000	mmHg*sec/ml
R_{PC}	Resistance: pulmonary capillaries	0.0900	mmHg*sec/ml
R_{PV}	Resistance: pulmonary veins	0.0600	mmHg*sec/ml
R_{LA}	Resistance: left atrium	0.0800	mmHg*sec/ml
R_{TAOD}	Viscoelastic element: distal aorta	0.0285	mmHg*sec/ml
C_{SC}	Compliance: systemic capillaries	0.0577	ml/mmHg
C_{SV}	Compliance: systemic veins	0.4200	ml/mmHg
C_{VC}	Compliance: vena cava	0.0050	ml/mmHg
C_{PC}	Compliance: pulmonary capillaries	0.0577	ml/mmHg
C_{PV}	Compliance: pulmonary veins	0.4200	ml/mmHg

Human Model Parameters

Cardiovascular Model

Table A-4. *Parameters used to define the Gaussian characterization of ventricular and atrial elastance in the human cardiovascular model.*

Parameter	Units	i = 1	i = 2	i = 3	i = 4	Description
$A_{v,i}$	N/A	0.370	0.360	0.500	0.560	Gaussian curve magnitude
$B_{v,i}$	sec	0.045	0.035	0.037	0.036	Gaussian curve width
$C_{v,i}$	sec	0.175	0.230	0.275	0.300	Gaussian curve delay
A_a	N/A	0.900	N/A	N/A	N/A	Gaussian curve magnitude
B_a	sec	0.018	N/A	N/A	N/A	Gaussian curve width
C_a	sec	0.065	N/A	N/A	N/A	Gaussian curve delay

Table A-5. *Parameters used to characterize the pressure-volume relationship of the atria and pericardium in the human cardiovascular model.*

Parameter	Units	LAF	RAF	PCD	Description
E_{ES}	mmHg/ml	0.400	0.40	N/A	Systolic maximal elastance
V_D	ml	35.000	35.00	N/A	Systolic volume intercept
P_O	mmHg	0.500	0.22	1.000	Diastolic scale parameter
λ	ml^{-1}	0.038	0.06	0.009	Diastolic exponential parameter
V_O	ml	35.000	35.00	135.000	Diastolic volume offset

Table A-6. *Fixed parameters used to characterize the resistive and compliant elements of the human cardiovascular model.*

Parameter	Description	Value	Units
R_M	Resistance: mitral valve	0.0095	mmHg*sec/ml
R_{AOP}	Resistance: proximal aorta	0.0084	mmHg*sec/ml
R_{COR}	Resistance: coronary circulation	35.9000	mmHg*sec/ml
R_{CRB}	Resistance: cerebral circulation	12.0000	mmHg*sec/ml
R_{AOD}	Resistance: distal aorta	0.0090	mmHg*sec/ml
R_{SA}	Resistance: large systemic arteries	0.0230	mmHg*sec/ml
R_{SV}	Resistance: systemic veins	0.1300	mmHg*sec/ml
R_{VC}	Resistance: vena cava	0.0020	mmHg*sec/ml
R_{RA}	Resistance: right atrium	0.0320	mmHg*sec/ml
R_{TC}	Resistance: tricuspid valve	0.0150	mmHg*sec/ml
R_{PAP}	Resistance: proximal pulmonary artery	0.0040	mmHg*sec/ml
R_{PA}	Resistance: pulmonary arterioles	0.0172	mmHg*sec/ml
R_{PS}	Resistance: pulmonary shunt	4.5200	mmHg*sec/ml
$R_{PC,0}$	Resistance: pulmonary capillaries	6.7000	mmHg*sec/ml
R_{PV}	Resistance: pulmonary veins	0.0140	mmHg*sec/ml
R_{LA}	Resistance: left atrium	0.0420	mmHg*sec/ml
R_{TAOD}	Viscoelastic element: distal aorta	0.0125	mmHg*sec/ml
C_{SC}	Compliance: systemic capillaries	0.4000	ml/mmHg
C_{SV}	Compliance: systemic veins	15.000	ml/mmHg
C_{VC}	Compliance: vena cava	3.5000	ml/mmHg
C_{PC}	Compliance: pulmonary capillaries	1.2000	ml/mmHg
C_{PV}	Compliance: pulmonary veins	0.4000	ml/mmHg

Pulmonary Model

Table A-7. *Adjustable parameters used to characterize the resistive and compliant elements of the full pulmonary model.*

Parameter	Description	Value	Units
$P_{TM,max}$	Max. transmural pressure	33.1060	mmHg*sec/ml
$V_{C,max}$	Max. volume: collapsible segment	0.1863	mmHg*sec/ml
$V_{C,term}$	Collapsible seg. P decay parameter	0.0451	mmHg*sec/ml
R_{VE}	Viscoelastic Resistance: lung tissue	0.7841	mmHg*sec/ml
C_{VE}	Viscoelastic Compliance: lung tissue	0.0150	mmHg*sec/ml
R_{RSA}	Resistance: large systemic arteries	−9.9650	mmHg*sec/ml
$R_{SM,i}$	Small a/w resistance param.: inspiration	1.8498	mmHg*sec/ml
$R_{SM,e}$	Small a/w resistance param.: expiration	49.5990	mmHg*sec/ml
$R_{SC,min}$	Small a/w resistance parameter	0.2000	mmHg*sec/ml
$R_{SC,max}$	Small a/w resistance parameter	1.0845	mmHg*sec/ml
V_{crit}	Small a/w resistance parameter	0.8000	ml/mmHg
V_{star}	Small a/w resistance parameter	3.9202	mmHg*sec/ml
P_{PLA}	Lung elastic recoil parameter	0.1753	mmHg*sec/ml
$P_{L,off}$	Lung elastic recoil parameter	−0.5320	mmHg*sec/ml
P_{LS}	Lung elastic recoil parameter	0.9908	mmHg*sec/ml
K_1	Resistance parameter	0.3802	mmHg*sec/ml
K_2	Resistance parameter	0.4816	mmHg*sec/ml
K_3	Resistance parameter	0.3638	mmHg*sec/ml
TLC	Total lung capacity (subject dependent)	5.1900	L
RV	Residual volume (subject dependent)	1.3600	L

Index

About the Authors

Jon B. Olansen received his B.S. degree in aerospace engineering and M.S. degree in mechanical engineering from the University of Notre Dame in 1987 and 1989, respectively. He spent several years serving as a mechanical systems flight controller in NASA's Mission Control Center, supporting 32 Space Shuttle missions. He has received a number of commendations for his service, including NASA's Public Service Medal, Space Flight Safety Award, and the Astronauts "Silver Snoopy" Personal Achievement Award.

Jon earned his Ph.D. in mechanical engineering from Rice University in 2000, after conducting extensive research focused on automated biomedical experimentation, simulation and analysis in the areas of electrophysiology and cardiovascular hemodynamics as well as cardiopulmonary dynamics in normal and ventilated patients. He was the inaugural recipient of the National Instruments Fellowship while at Rice, using LabVIEW and National Instruments hardware to establish and teach graduate level biomedical instrumentation courses.

Jon founded and currently serves as president of VBI Development Company, a firm specializing in computer-based biomedical and clinical research applications. He also has returned to the Johnson Space Center as a liaison for NASA's Astronaut Office. In this capacity, Jon provides operational assessments of all human-based biomedical research hardware and experiments being developed for the International Space Station.

Eric Rosow has over 16 years of experience in biomedical engineering and life science applications of virtual instrumentation. Eric is director of biomedical engineering at Hartford Hospital, a role through which he introduced virtual instrumentation into the hospital environment. He holds six patents and has published more than 45 academic papers. He also is a co-founder of Premise Development Corporation, a software company that creates award-winning software-based tools to help small-to-large organizations remain efficient, competitive, and profitable. During his career, Eric has co-developed numerous virtual instrument solutions (including National Instrument's *BioBench*™ for leading healthcare institutions throughout the world.

Eric earned a B.S. degree in mechanical engineering at Trinity College (Hartford, CT), where he was awarded the American Society of Engineers' Prize as well as Faculty Honors. He received his M.S. degree from the Hartford Graduate Center/Rensselaer Polytechnic Institute, where he was recognized with the Outstanding Student Award for the School of Engineering and Science. Among his professional affiliations are memberships in the Institute of Electrical and Electronics Engineers, the Engineering in Medicine and Biology Society, the Data Warehouse Institute, the Health Information Management Systems Society (HIMSS), the Association for the Advancement of Medical Instrumentation (AAMI), the American College of Clinical Engineering (ACCE), and the American College of Healthcare Executives (ACHE).

He has been recognized by several organizations for his pioneering work in virtual instrumentation, including AAMI's Biomedical Instrumentation and Technology's Outstanding Management and Technology Paper, National Instrument's "Best in Category" and "Best Overall" Application of Virtual Instrumentation, and the American College of Clinical Engineering's Professional Achievement Award. Eric also has served as a member of the U.S. Olympic Committee's Sports Equipment and Technology Committee (SETC) and is a former world-class rower with several national titles, a Pan American Games gold medal, and a fourth place at the World Rowing Championships to his credit.

Timely, Technical Information for LabVIEW™ Users!

Subscribe to LabVIEW Technical Resource!
Receive a new issue each quarter full of
technical solutions to get the most
out of LabVIEW. Each issue
includes a Resource CD packed
with VIs, Utilities and Source
Code. Tools, Tips, Techniques,
Tutorials, and more.

Written *by* LabVIEW programmers
for LabVIEW programmers.

To order your subscription or Library of Back Issues on CD-ROM,
please fill out the attached order form.

tel: 214.706.0587 • fax: 214.706.0506 • email: ltr@ltrpub.com

*"The examples on the Resource CD contain real solutions
to real-world problems... no LabVIEW programmer
should be without LTR!"*

Jeffrey Travis – Author of "Internet Applications in LabVIEW" and co-author of "LabVIEW For Everyone"

*"Please renew my subscription, which paid
for itself many times last year!"*

David Moschella – Pronto Product Development Corporation

*"Saved me hours if not days of
development time!"*

Brad Hedstrom – Applied Biosystems

visit our website

WWW.LTRPUB.COM

LABVIEW™ TECHNICAL RESOURCE

THE ONLY LABVIEW SUBSCRIPTION WITH VI SOFTWARE INCLUDED

ORDER FORM

WHAT IS LTR?

LabVIEW Technical Resource (LTR) is a quarterly journal for LabVIEW users and developers available by subscription from LTR Publishing, Inc. Each LTR issue presents powerful LabVIEW tips and techniques and includes a Resource CD packed with VI source code, utilities, and documentation. Technical articles on LabVIEW programming methodology, in-depth tutorials, and time-saving tips and techniques address everyday programming issues in LabVIEW.

In its eighth year of publication, LTR has subscribers in over 50 countries and is well-known as a leading independent source of LabVIEW-specific information.

Purchase the LabVIEW Technical Resource CD Library of Back Issues, Version 3.0 and browse this searchable CD-ROM for easy access to over 250 articles and VIs from LTR Volumes 1-8.

To subscribe to the LabVIEW Technical Resource or to order the CD-ROM Library of Back Issues, fax this form to LTR Publishing at **(214) 706-0506.**

Visit the LTR web page at **www.ltrpub.com** to download a free sample issue.

TEL: 214-706-0587   FAX: 214-706-0506

CONTACT INFORMATION

Name _____   Company _____

Address_____

City _____   State _____

Country _____   Zip/Post Code _____

Tel (required) _____   FAX _____   E-mail _____

ORDER INFORMATION

QTY	MAC/PC	PRODUCT	U.S.	INTL.	EXTENDED PRICE
		1 year subscription (4 issues / 4 Resource CDs)	$95	$120	
		2 year subscription (8 issues / 8 Resource CDs)	$175	$215	
		CD-ROM library of back issues (28 issues / over 250 VIs)	$350	$375	
		Back issues – [Article Index available at **www.ltrpub.com**]	$25	$30	
		Server Version CD-ROM library of back issues Version 3.0 (5 user license)*	$495	$530	
		10 user license Add-On pack (for Server Version)*	$295	$325	

*contact LTR for additional licensing information.

SUBTOTAL	
TX TAX @ 8.25%	
TOTAL	

PAYMENT INFORMATION

✔	PAYMENT METHOD
	Check enclosed (U.S. BANK ONLY* – Make check payable to LTR Publishing) (Texas residents please add 8.25% sales tax)
	Bill company(U.S. Only) / (fax of P.O. required) ▶ PO#
	Visa / MC / AMEX Card Number ▶ Exp.
	Signature
	* Wire information available for international orders

Fill out the form above and Fax it to: 214-706-0506 with your credit card information and signature,
OR fill out the form above and send order form with U.S. check to:

**LTR Publishing, Inc., 860 Avenue F, Suite 100 Plano, Texas 75074.
Tel: 214.706.0587 • Fax: 214.706.0506 • email: ltr@ltrpub.com**

You may also include your own Federal Express or Airborne #. If you are ordering
a product for delivery within Texas, please include Texas Sales Tax at 8.25%

WWW.LTRPUB.COM

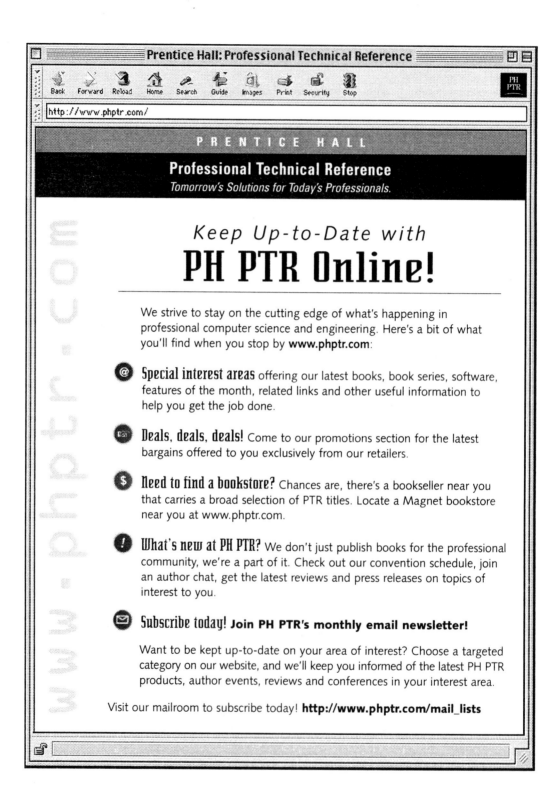